Land Use/Land Cover and Natural Hazards: Interactions, Changes, and Impacts

Land Use/Land Cover and Natural Hazards: Interactions, Changes, and Impacts

Editors

Matej Vojtek
Andrea Petroselli
Raffaele Pelorosso

MDPI • Basel • Beijing • Wuhan • Barcelona • Belgrade • Manchester • Tokyo • Cluj • Tianjin

MDPI

Editors

Matej Vojtek
Constantine the Philosopher
University in Nitra
Nitra
Slovakia

Andrea Petroselli
Tuscia University
Viterbo
Italy

Raffaele Pelorosso
Tuscia University
Viterbo
Italy

Editorial Office
MDPI
St. Alban-Anlage 66
4052 Basel, Switzerland

This is a reprint of articles from the Special Issue published online in the open access journal *Land* (ISSN 2073-445X) (available at: https://www.mdpi.com/journal/land/special_issues/lulc_natural_hazards).

For citation purposes, cite each article independently as indicated on the article page online and as indicated below:

LastName, A.A.; LastName, B.B.; LastName, C.C. Article Title. *Journal Name* **Year**, *Volume Number*, Page Range.

ISBN 978-3-0365-6738-9 (Hbk)
ISBN 978-3-0365-6739-6 (PDF)

Contents

About the Editors

Matej Vojtek

Matej Vojtek works as an Associate Professor in the Department of Geography, Geoinformatics and Regional Development at Constantine the Philosopher University in Nitra (Slovakia) and as a research scientist in the Institute of Geography at Slovak Academy of Sciences, Bratislava (Slovakia). His research interests include flood potential, flood hazard, and flood risk mapping and assessment at various spatial scales (local, regional, national), surface runoff assessment, and landslide hazard mapping and assessment using geospatial technologies (geographic information systems, remote sensing, global navigation satellite systems). He has participated in several national and international research projects as a leader or a project member. He is the author of more than 80 publications, including original research articles in impacted journals, e.g., *Geoscience Frontiers, Journal of Hydrology: Regional Studies, Journal of Environmental Management* or *Hydrological Sciences Journal*, and educational works, such as university textbooks.

Andrea Petroselli

Andrea Petroselli is an Associate Professor at Tuscia University (Italy), Department of Economics, Engineering, Society and Business Organization. He is an expert in modeling and monitoring hydrological processes. Recent research topics range from infiltration modeling to rainfall–runoff modeling. He is a member of GISTAR, the GIS Terrain Analysis Research Group (www.gistar.org), a web portal for researchers and professionals involved in the investigation, development, and application of GIS-based terrain analysis tools for hydrologic and geomorphic models, and a member of MechHydroLab, Mechanical Engineering for Hydrology and Water Science (www.mechydrolab.org/), a multidisciplinary laboratory composed of mechanical engineers, hydrologists, and water scientists with the goal of combining mechanical engineering technologies and hydrological sciences toward the development of novel experimental systems for advanced environmental monitoring.

Raffaele Pelorosso

Raffaele Pelorosso is qualified as an Associate Professor by National Scientific Qualification in urban and landscape planning and design (Disciplinary Sector 08/F1) and agricultural, forest and biosystems engineering (Disciplinary Sector 07/C1). He has held several lectures in ecology, cartography, environmental and urban planning at Tuscia University. His main research interests include environmental modelling, performance-based urban planning, urban stormwater and climate regulation by means of green and grey infrastructure, low-entropy systems, climate adaptation, nature-based solutions and ecosystem services, landscape perception and participatory planning, land use/land cover changes, rural and peri-urban landscapes, environmental impact assessment, and landscape connectivity. He is the author of more than 100 scientific works and he has peer-reviewed for many international journals, such as *Land Use Policy, Landscape and Urban Planning, Frontiers in Built Environment, European Planning Studies, Habitat International, Journal of Cleaner Production.*

land

MDPI

Article

Land Take and Landslide Hazard: Spatial Assessment and Policy Implications from a Study Concerning Sardinia

Federica Isola, Sabrina Lai, Federica Leone * and Corrado Zoppi

Department of Civil and Environmental Engineering, and Architecture-University of Cagliari, 09123 Cagliari, Italy
* Correspondence: federicaleone@unica.it

Abstract: Land take and soil sealing imply land cover transitions that may possibly result in decreased capacity to resist landslides; hence, this study focuses on the relations between land-taking processes and landslide hazard by addressing the following research question: "To what extent do land-taking processes increase landslide hazard?" The impact of land take is assessed through a regression model which relates the level of landslide hazard to a set of land cover variables which include artificialized land; that is, land taken up through urbanization processes, and a set of covariates that represent land cover types grouped in accordance with the LEAC (land and ecosystem accounting) classification. This methodological approach is implemented into the spatial context of Sardinia, an insular Italian region, and shows that not only the amount of taken up artificialized land, but also other types of land covers, are likely to increase the magnitude of landslide hazard. A set of implications concerning planning policies related to land cover and land cover transitions are discussed in the concluding section, where policy recommendations are identified in order to mitigate the impacts of land cover transitions on landslide hazards.

Keywords: land take; landslide hazard; land cover change

Citation: Isola, F.; Lai, S.; Leone, F.; Zoppi, C. Land Take and Landslide Hazard: Spatial Assessment and Policy Implications from a Study Concerning Sardinia. *Land* **2023**, *12*, 359. https://doi.org/10.3390/land12020359

Academic Editors: Matej Vojtek, Andrea Petroselli and Raffaele Pelorosso

Received: 30 December 2022
Revised: 18 January 2023
Accepted: 24 January 2023
Published: 28 January 2023

1. Introduction

According to the European Environment Agency, "Land take is the process in which urban areas and sealed surfaces occupy agricultural, forest or other semi-natural and natural areas" [1] (p.117). "The most intense form of land take is soil sealing, which is an essentially irreversible process that leads to the destruction or covering of soils by buildings and other construction, and layers of completely or partly impermeable artificial material (asphalt, concrete, etc.). Soil sealing accompanies land take, but areas subject to land take are usually not entirely sealed" [2]. At the end of 2021, the European Commission approved the new European Union (EU) Soil Strategy for 2030, which highlights how healthy soils are essential for achieving the objectives of the European Green Deal concerning climate and biodiversity. The strategy defines a general framework and concrete measures to protect and restore soils to ensure their sustainable use [3]. One of the long-term objectives to be achieved by 2050 is to reach "no net land take" [3] (p. 3). The general framework of the strategy termed "land take hierarchy" concerns four consequential actions: avoid, reuse, minimize, and compensate. The first action, i.e., "avoid", aims at preventing further land take as much as possible. If land take cannot be prevented, then the second action, i.e., "reuse", should be implemented, with a view to reusing land that has already been urbanized or sealed; for example, through soil remediation or densification. If land take cannot be prevented and land cannot be reused, the third action should be looked at, to minimize the effects of land take by impermeabilizing land that is already in unfavorable conditions. If all the precedent actions cannot be taken, the fourth action provides for applying compensation and mitigation measures.

In relation to the EU-28 (which means the 27 EU Member States plus the United Kingdom), although land take decreased from 2000 to 2018, in the period 2012–2018 it reached the amount of 539 km^2 per year; 78% of land take takes place in agricultural areas, such as arable lands and permanent crops (50.5%), pasture and mosaic farmland (27.2%), forests and transitional woodlands (14.3%), and grasslands (6%) [4]. The main causes of land take are to be attributed to expansion of industrial and commercial areas, and to enlargement of residential zones and construction sites [2]. In Italy, in the 2020–2021 period, land take took a value of 69.1 km^2, corresponding to an average value of 19 hectares per day. Only a tiny fraction of this growth in artificial surfaces was compensated by the restoration of natural areas, which equaled 5.8 km^2, due to change from consumed soil to unconsumed soil through to the recovery of building sites, areas, and surfaces in cases where "reversible" land take [5] took place.

Moreover, land-taking processes entail several problems, such as the loss of multifunctional and fertile soils, biodiversity degradation and loss of ecosystem services (ESs) [6–8]. ESs are defined as benefits provided by ecosystems to human beings [9]. Since 1970, when the term "ecosystem service" was coined [10], this theme has increasingly been studied and analyzed in conceptual terms [11,12], in terms of classification [9,13], and in relation to their assessment and mapping [14,15]. Moreover, ESs provide protection against hydrogeological hazards [16]. According to Notaro and Paletto [17], this protection can be direct or indirect, where the former concerns the defense against natural phenomena such as floods and landslides.

A landslide is defined as the movement of materials such as rock, soil, debris, and artificial fill downward and outward along a slope [18] when the forces of gravity exceed the slope resistance [19]. Moreover, although landslides occur primarily in mountainous areas, this phenomenon may happen in low-relief zones [20]. The United States Geological Survey (USGS) [20] identifies three types of landslide causes: geological, morphological, and human. Indeed, the drivers of landslides may be either natural, caused by the intrinsic properties of rocks and soils or by physical processes, such as heavy rains and seismic activity, or artificial when human activities bring about changes in slope stability, as happens through deforestation or excessive soil sealing [21].

From this perspective, landslides are strongly connected with land use and land cover dynamics [22] and, particularly, with human-driven processes [23]. Human-induced activities, such as land use/land cover changes, may alter vegetation structure and modify soil characteristics and hydrogeological processes [24,25]. Indeed, despite the slowness that characterized geological and geomorphological changes, land use/land cover changes can occur in a short period of time due to their high dynamicity [26,27]. Land use/land cover changes can influence landside events in terms of frequency and spatial configuration due to their potential negative impacts on hydrological and mechanical processes involving soils [28,29].

The relation between land use/land cover changes and landslides has been studied by various authors [23,30,31]. Hao et al. [23] investigated the extent to which the landslide disaster occurred in 2018 in Kerala, India, was influenced by land use/land cover changes through a comparison between land use/land cover changes before (2010) and after (2018) the disaster. Pisano et al. [30] investigated how land cover changes influenced landslide susceptibility in the past and how they might influence future events, by carrying out a landslide susceptibility analysis implemented through a spatial multi-criteria evaluation in relation to three past land cover maps (1954, 1981, and 2007) and three future scenarios (one in 2030 and two in 2050) in the Rivo Basin, Italy. Muñoz-Torrero Manchado et al. [31] studied the influence of deforestation and related agricultural activities on landslide susceptibility using remote-sensing techniques and free satellite data in Nepal.

However, although several authors have studied the relation between land use/land cover changes and landslide hazard, the relation between land-taking processes and landslide hazard is still under-researched. Therefore, this study aims at analyzing the relations between land-taking processes and landslide hazard in order to understand to what extent land-taking processes increase landslide hazard through a regression model that relates the level of landslide hazard to a set of land cover variables that includes artificialized land; that is, land taken up through urbanization processes, and a set of covariates that represent the land cover types associated with the LEAC (land and ecosystem accounting) classification. The methodological approach is implemented in the Sardinia Region, Italy.

The study is structured into six sections as follows. The second section describes the study area, the methodological approach used, and the input data for the regression model (landslide hazard, LEAC land cover groups, geological characteristics and elevation). The results are presented in the third section and discussed in the fourth section. The fifth section provides recommendations and implications for spatial planning policies stemming from the results, while the sixth section provides concluding remarks and future directions of the research.

2. Materials and Methods

This section is structured into three subsections. The study area is described in the first subsection. In the second subsection, the discrete-choice Logit model is used to estimate the relations between land-taking processes and landslide hazard. The third subsection describes input data for the regression model.

2.1. Study Area

In Italy, a sectoral planning tool termed PAI, an acronym of "piano di assetto idrogeologico" (whose word-for-word translation would be "hydrogeological setting plan"), must identify areas prone to natural hazards; i.e., both landslide and flood hazard, and define measures to reduce their magnitude and to prevent or mitigate their impacts. The PAI can be regarded as part of the broader river basin management plan envisaged by the Water Framework Directive (WFD) [32], and its responsibility lies with ad hoc established competent authorities. Accordingly, Italy has been divided into eight river basin districts [33,34], each having its own competent authority.

One of such seven districts coincides with Sardinia, an Italian island around 24,000 km^2 in size and the second-largest island in the Mediterranean Sea Basin. Sardinia is further divided into seven subdistricts [35], one of which, the so-called "Coghinas-Mannu-Temo" subdistrict (hereafter, CMT), is the area chosen for this study (Figure 1).

The reasons for choosing the Sardinian CMT subdistrict for this study are twofold. First, Sardinia is included in the CORINE Land Cover (CLC) inventory coordinated by the European Environment Agency under the Copernicus program of the EU; this makes it possible to retrieve a regularly updated series of land cover maps, of which the most recent one refers to the year 2018 [36]. Second, a comprehensive and detailed spatial assessment of landslide hazard and risk concerning the whole CMT was carried out and officially validated in 2014 [37] and it is publicly available from the regional geoportal.

Located to the north-west of Sardinia, with an area of 5575.5 km^2, CMT stretches over more than one-fifth of the island and it comprises around forty watersheds, of which the largest and most important ones are the four ones from which its name originates, i.e., Coghinas River, Mannu River, Mannu River in Porto Torres, and Temo River (Figure 1, panel C). The prevailing morphology is hilly, heavily marked in the southern border by the Marghine-Goceano mountain chain and by the Mount Limbara rocky granitic massif to the east, with the exception of the Nurra coastal plain to the north-west and the smaller plains around the mouths of the rivers Coghinas and Temo (Figure 2, panel A). As in all of Sardinia, in CMT the climate is typically Mediterranean, mostly lower meso-Mediterranean, but with coastal areas included within the upper thermo-Mediterranean zone and mountain chains in the upper meso-Mediterranean [38] (Figure 2, panel B). As for vegetation, according to the study by Bacchetta et al. [39], more than 41% of the CMT host species belong to the Sardinian thermo-meso-Mediterranean cork tree series, while the Sardinian oak tree series and holm oak tree series occupy around 11% of the CMT each and the other vegetation series take lower percentages (Figure 2, panel C). Geological instability is diffuse in the study area, where landslide events have been recorded for decades: the Italian landslide inventory (IFFI [40,41]) has documented 398 landslide events occurring up to 2007 in the

CMT. The most prominent category is that of diffuse falls or topples (228 events), followed by simple falls or topples (95 events); third comes diffuse superficial instability (42 events), followed by rotational or translational slides (16 events). Very small numbers concern the other categories; due to the geological and geo-lithological characteristics of the study area, no flow events have been reported in the study area (Figure 2, panel D).

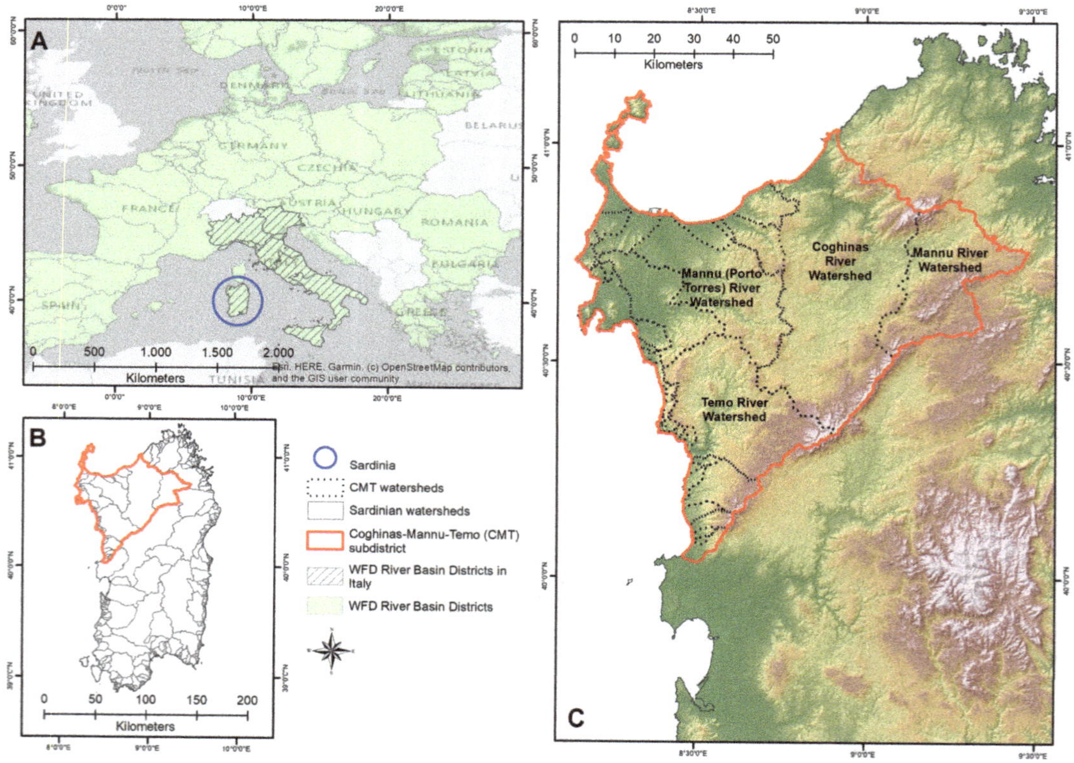

Figure 1. The eight WFD river basin districts in Italy (panel **A**), Sardinian watersheds (panel **B**), and the Coghinas-Mannu-Temo subdistrict with its watersheds (panel **C**).

2.2. Regression Model

The relation between landslide hazard (LH) and the size of land taken up (L_TAKE) is assessed through a linear regression model that uses the LEAC land cover groups as explanatory variables, whose detailed definitions are given in Section 2.3. The covariate representing land take is one of the LEAC groups, namely the variable associated with the artificialized land LEAC group. Dependent and explanatory variables refer to the elements of a 300 m square grid that overlays the study area, and are measured as their percentage share of a grid cell. The model operationalizes as follows:

$$LH = \alpha_0 + \alpha_1 L\_TAKE + \alpha_2 ARA + \alpha_3 PMF + \alpha_4 FOR + \alpha_5 GRSH + \alpha_6 DEPOQ + \alpha_7 VOLSE + \alpha_8 ELEV + \alpha_9 HGLAGGED, \quad (1)$$

where the dependent variable and covariates are identified as shown below:

- LH is for landslide hazard;
- L_TAKE is for artificialized land or land take;
- ARA is for arable land;
- PMF is for pastures and mosaic farmland;

- FOR is for standing forests;
- GRSH is for natural grasslands, sclerophyllous vegetation, and heathlands;
- DEPOQ is for ground substrate characterized by quaternary deposits;
- VOLSE is for ground substrate characterized by volcanic sedimentary rocks;
- ELEV is for the average elevation of a grid element;
- HGLAGGED is the spatially lagged dependent variable that controls for spatial autocorrelation of LH.

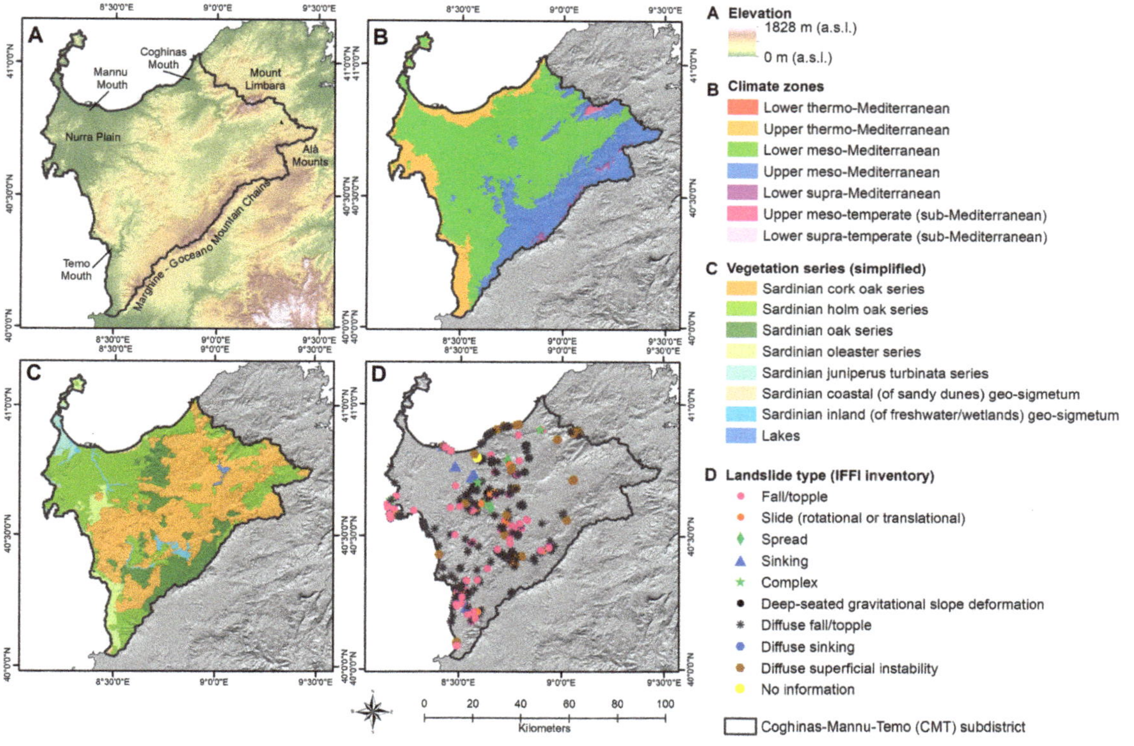

Figure 2. Some features of the Coghinas-Mannu-Temo subdistrict: elevation (panel **A**); climate zones (panel **B**, based on [38]); simplified vegetation series (panel **C**, based on [39]); documented landslides events (panel **D**, based on data from the IFFI project [40]).

The estimates of the coefficients of the multiple linear regression show the correlations between landslide hazard and the land covers of the LEAC taxonomy and, in particular, the interdependence of LH and the size of land take.

The use of a multiple linear regression is motivated by the fact that prior assumptions are not available as regards the functional form of the relations between dependent and explanatory variables, which is consistent with several studies aimed at identifying the interdependence between spatial variables [42–45]. From this point of view, a spatial phenomenon, related to n variables, represented by a surface in an n-dimensional space whose equation is unknown, can be approximately detected, in each of its points, by the tangent hyperplane. The linear equation estimated through the regression model, which relates dependent and explanatory variables, identifies the tangent hyperplane in a small neighborhood of a point of the surface, and in such neighborhood it represents an approximation of the unknown equation of the surface [46,47]. As a consequence, model (1) represents the trace of a hyperplane on a surface in a ten-dimensional space, which reports the correlations between LH and the nine covariates listed above.

The covariates DEPOQ and VOLSE control for the ground substrate, by considering if, and to what extent, landslide hazard is influenced by the conditions of the substrate, which in the study area is mainly featured by cohesive and compact rocks such as volcanic sedimentary successions (VOLSE) and, secondly, by deposits from the quaternary era (DEPOQ), i.e., loose incoherent materials. ELEV controls for the altitude impact on landslide hazard. If the estimates of their coefficients in (1) are significant, this will entail that substrate and elevation are likely to influence LH, at least to some extent. The magnitude of the coefficients will show the size of the impacts, in terms of increase or decrease in the landslide hazard measure.

The sign of ELEV is expected to be negative, since in the study area, on average, landslide hazard conditions are more frequent in lowlands rather than in mountainous locations, as further discussed in Section 3.3, whereas the expected signs of DEPOQ and VOLSE are positive and negative, respectively, since it is intuitively likely that LH will increase as long as the substrate incoherence and looseness increases, and the other way around.

The variable HGLAGGED represents the spatially lagged values of LH, and controls for spatial autocorrelation of the dependent variable in model (1). The HGLAGGED definition is based on the methodology implemented by Zoppi and Lai [48], which builds on Anselin's studies [49,50].

Moreover, a *p*-value test is used to check the level of significance of the estimates of the coefficients of model (1).

2.3. *Input Data for the Regression Model*

The dependent variable and covariates needed to feed into the regression model (1) were calculated with reference to a 300 m vector square grid that covers all of the CMT subdistrict (Figure 3) and comprises a total of 62,231 cells, using three main input spatial datasets listed in Table 1.

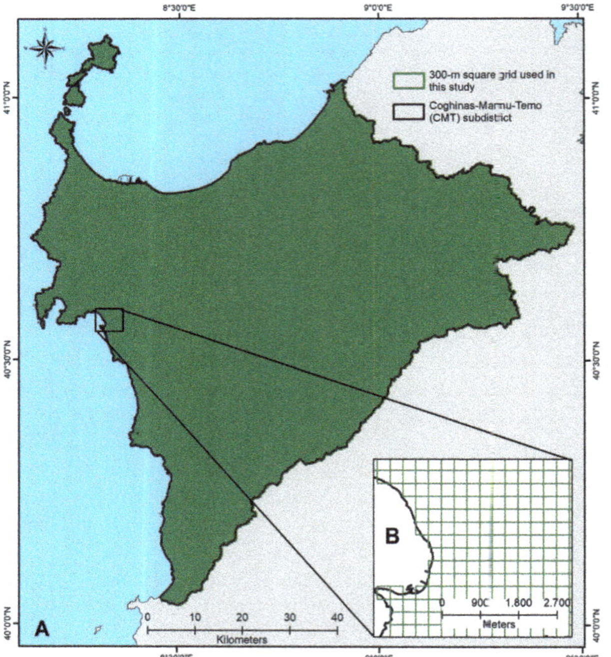

Figure 3. The 300 m vector square grid used in this study: extent of the grid with reference to the Coghinas-Mannu-Temo subdistrict (panel **A**, no. of cells: 62,231), and detail (panel **B**).

Table 1. Input datasets used to compute dependent and explanatory variables.

Input Datasets	Sources	Link
Landslide hazard	Regional geoportal	https://webgis2.regione.sardegna.it/geonetwork/srv/ita/catalog.search#/metadata/R_SARDEG:eb38d6c0-b51f-4df1-acdc-f7a752e7664c (accessed on 17 January 2023)
LEAC land cover groups	Copernicus—Europe's Eyes on Earth program	https://land.copernicus.eu/pan-european/corine-land-cover/clc2018 (accessed on 17 January 2023)
Elevation	Regional geoportal	https://www.sardegnageoportale.it/areetematiche/modellidigitalidielevazione/ (accessed on 17 January 2023)
Geological characteristics	Regional geoportal	https://www.sardegnageoportale.it/index.php?xsl=2420&s=40&v=9&c=14479&es=6603&na=1&n=100&esp=1&tb=14401 (accessed on 17 January 2023)

2.3.1. Landslide Hazard

In compliance with national law no. 183/1989, in Italy each competent authority for a WFD river basin district must approve, as part of the comprehensive river basin management plan, its PAI, which only focuses on landslide and flood risks in the district. The PAI has a dual character: on the one hand, it is a knowledge-oriented and regularly updated tool, which provides for the spatially explicit assessment of flood and landslide risks and hazards, as well as of exposures, hence vulnerable infrastructure, buildings, and land. On the other hand, it is a legally binding plan, which contains provisions that restrict land uses and land transformations in areas prone to landslide or flood hazard: the higher the hazard level, the stricter the restrictions.

The Sardinian Basin Competent Authority approved a first version of its PAI in 2004; since then, the assessment of the hazard level has continuously been updated to integrate new studies in previously non-analyzed areas, or to revise locally the hazard level when a new infrastructure that mitigates natural risks is realized. Accordingly, the landslide hazard map has been revised 42 times so far, and the flood hazard map 59 times.

Landslide hazard levels in the study area were assessed in a study commissioned by the Sardinian Basin Competent Authority in 2011, whose documents are publicly available on the institutional website [37]. The outcomes of the study were approved in 2014 and, as far as the spatially explicit assessment is concerned, integrated within the 36th updated revision of regional PAI spatial dataset available from the regional geoportal.

As per the methodology used in the PAI, landslide hazard (H_L) classes range in the 0–4 interval, as follows: no hazard: $H_L = 0$; moderate hazard: $H_L = 1$; medium hazard: $H_L = 2$; high hazard: $H_L = 3$; very high hazard: $H_L = 4$. For each cell in the 300 m square grid shown in Figure 3, the independent variable LH in model (1) was calculated as the percentage of the cell's area having non-null landslide hazard ($H_L \neq 0$) in the vector data retrieved from the geoportal.

Moreover, LH's spatially lagged variable (HGLAGGED), included in model (1) as a covariate, was calculated using GeoDA (version 1.20) [51], developed by Dr Luc Anselin and his team, based initially at the University of Illinois at Urbana-Champaign and currently at the Center for Spatial Data Science, University of Chicago, United States of America.

2.3.2. LEAC Land Cover Groups

The CORINE (acronym for "Coordination of Information on the Environment") land cover is one of the several spatial datasets made available by the EU through the Copernicus Land Monitoring Services, covering a total of 39 countries, i.e., both members of the European Environment Agency and cooperating countries, and regularly updated every six years using a standardized nomenclature, hence allowing for consistent classifications and measures across time and space, and enabling time-series analyses.

In this study the 2018 CORINE Land Cover vector map (CLC2018) was used. The map provides information on land covers, i.e., on the biophysical characteristics of the Earth's surface, through a hierarchical classification that comprises 44 classes at the third (and lower) level, 17 at the second level, five at the first level, with a minimum mapping unit equaling 25 hectares.

The CLC2018 was next reclassified so as to group the third-level land cover classes following the taxonomy used by the European Environment Agency for land cover accounts [52] and comprising eight groups. Information on how the reclassification was performed is provided in Table 2, whose last column lists the CLC classes that were assembled within a single LEAC group. For the purpose of this study, only five groups out of the eight listed in Table 2 were mapped because three (open space with little or no vegetation; transitional woodland and shrub; wetlands, water bodies and marine waters) are not relevant within CMT. Furthermore, the latter group was not relevant with respect to the aim of this study: indeed, the absence of any relationships between marine or inland waters and landslide hazards is quite straightforward. CLC classes listed in Table 2 that only contain one digit (for instance, "1.") refer to first-level land covers and comprise all of the second- and third-level land covers that detail the first-level one (for instance, 1.1.1, 1.2.1, and so on); likewise, classes containing two digits (for instance, "2.2.") refer to second-level land covers and comprise all of the third-level land covers that detail the second-level one (for instance, 2.2.1, 2.2.2, and so on). For instance, the "standing forests" group includes all of the sub-levels of the 3.1 class, which, in the study area, comprise three third-level land cover classes as follows: 3.1.1 (broad-leaved forests), 3.1.2 (coniferous forests), and 3.1.3 (mixed forests). More specific information on wood types and management can be found in another, and older, land use/land cover map produced in 2008 by the regional administration of Sardinia [53], which further details the CLC taxonomy up to the fifth level. According to this dataset, approximately 28% of the surface covered by the LEAC "standing forest" group in the study area was managed in 2009. Managed forests were almost completely made up of cork oak woods (27%), while negligible percentages concerned other types of managed woods, either broad-leaved (for instance, eucalyptus woods) or coniferous (for instance, pine woods, especially in coastal areas). While many cork oak woods are still managed for production purposes, especially in North-Eastern Sardinia [54], eucalypti and pine trees (both non-native species in the island) were planted mainly for swamp reclamation, slope stability, and erosion control in coastal dunes in the XX century; as of today, they are often unmanaged, to the extent that some have undergone a renaturalization process and have evolved into mixed forests, as a result of successional processes [55] and native species' regaining their spaces.

Table 2. LEAC groups and corresponding CLC classes.

LEAC Groups	CLC Classes
1. Artificialized land (land taken by development)	1.
2. Arable land and permanent crops	2.1. + 2.2. + 2.4.1.
3. Pastures and mosaic farmland	2.3. + 2.4.2 + 2.4.3 + 2.4.4.
4. Standing forests	3.1.
5. Transitional woodland and shrub	3.2.4.
6. Natural grasslands, sclerophyllous vegetation and heathlands	3.2.1 + 3.2.2 + 3.2.3.
7. Open space with little or no vegetation	3.3.
8. Wetlands, water bodies, and marine waters	4. + 5.

Once a vector map of the LEAC groups was retrieved, for each cell in the 300 m square grid shown in Figure 3, the explanatory variables L_TAKE, ARA, PMF, FOR, and GRSH, in model (1) were calculated as the percentage of the cell occupied by LEAC groups listed in Table 2, respectively, as nos. 1, 2, 3, 4, and 6.

2.3.3. Geological Characteristics and Elevation (Control Data)

A 1:25,000 regional geological map of Sardinia was produced at the beginning of the year 2000 building upon geological data collected by the former regional agency for mines and quarries. The spatial dataset, available from the regional geoportal [56], identifies geological characters in compliance with the "CARG" national mapping program initiated in the 1980s by the Italian Geological Society. The taxonomy of the Sardinian geological map is hierarchically structured into five main classes and five levels ([57], pp. 49–108), and a simple reclassification was carried out in this study, whereby i., first-level classes only were considered and, ii., three main groups were retrieved by merging first-level classes. The three groups are as follows: i., quaternary deposits (also comprising lakes); ii., volcanic sedimentary successions; iii., intrusive complexes and metamorphic basements. Finally, for each cell in the 300 m square grid shown in Figure 3, the explanatory variables DEPOQ and VOLSE were calculated as the shares of the cell occupied, respectively, by quaternary deposits and by volcanic sedimentary successions. For any terrestrial cell in the grid, the share occupied by the third group is, fairly obviously, the difference between 100 and the sum of DEPOQ and VOLSE.

Elevation was retrieved from the 10 m resolution digital terrain model (DTM) available "off the shelf" from the regional geoportal [58]. The Sardinian DTM was produced in the early 2010s, based on elevation points and contour lines contained in the 1:10,000 regional technical map (CTR, acronym for the Italian "Carta Tecnica Regionale"). Because the production process was implemented in compliance with the national guidelines issued in 2009 [59], horizontal and vertical accuracy, though not explicitly stated in the DTM metadata, are as follows: horizontal tolerance: 2 m; vertical tolerance in open fields: 2 m; vertical tolerance in densely wooded areas (i.e., in areas where tree canopy cover is over 70% of the surface): $\frac{1}{2}$ of the mean height of the trees. Next, for each cell in the 300 m square grid shown in Figure 3, the explanatory variable ELEV in model (1) was calculated as the average elevation in the cell.

3. Results

This section is organized as follows. The first and second subsections show the spatial framework of, respectively, landslide hazard and the LEAC groups across the study area. The following subsection presents the outcomes of the estimate of regression model (1) implemented into the spatial context identified in Section 2.2.

3.1. *Landslide Hazard in the Study Area*

As Table 3 and Figure 4, panel A, show, in the vast majority of CMT (i.e., 4476.42 km^2, or 80.29% of the CMT land mass) the hazard level was assessed as null by the PAI, while around a fifth of the subdistrict is prone to landslides, mostly of medium (580.01 km^2, or 10.40% of the CMT surface) or high severity (371.84 km^2, or 6.67%); a very small share of the CMT features moderate landslide hazard (107.15 km^2, or 1.92%) and a negligible one is characterized by very high hazard levels (39.80 km^2, i.e., 0.71%). As for the 300 m grid, LH is greater than zero in 30,775 out of the total 62,231 300 m grid cells (Figure 4, panel B); hence, in nearly half of the cells, landslide hazard, of whichever level, affects a certain share of the cell.

Table 3. Landslide hazard levels in the Coghinas-Mannu-Temo (CMT) subdistrict.

Landslide Hazard Level (PAI)		Area (km^2)	Area (% CMT)
Absent	(H_L = 0)	4476.42	80.29
Moderate	(H_L = 1)	107.15	1.92
Medium	(H_L = 2)	580.01	10.40
High	(H_L = 3)	371.84	6.67
Very High	(H_L = 4)	39.80	0.71

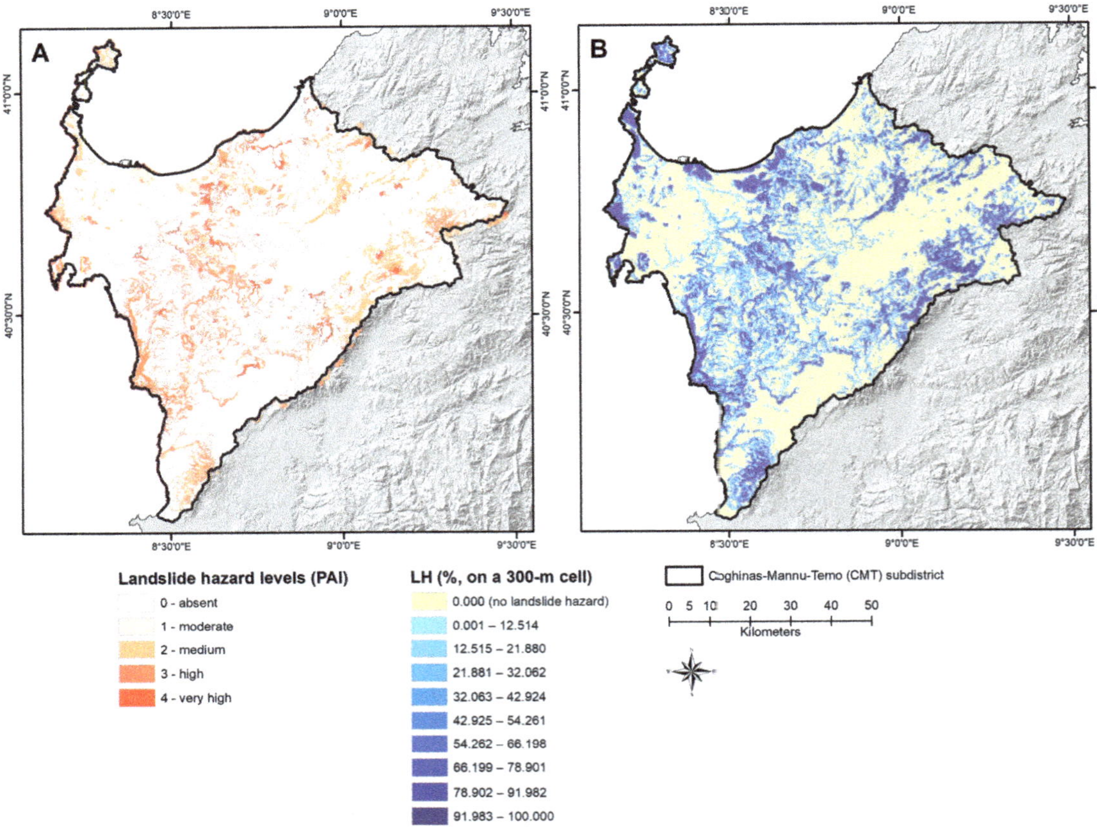

Figure 4. Landslide hazard levels as assessed by the regional PAI in the Coghinas-Mannu-Temo subdistrict (panel **A**), and spatial distribution of the LH variable in the 300 m grid used in this study (panel **B**).

3.2. The Spatial Framework of the LEAC Groups

Three LEAC groups prevail in the CMT subdistrict, as shown in Table 4 and Figure 5, panel A: arable land and permanent crops (32.09%); pastures and mosaic farmland (26.68%); natural grasslands, sclerophyllous vegetation and heathlands (23.19%). Together, they make up 81.96% of the study area. Next come standing forests (14.98%), while artificialized land amounts to 2.37% of the study area, and a negligible share (0.69%) is that of waters, which are not listed in Table 4 because they were not relevant for this study.

Table 4. LEAC groups as share of the Coghinas-Mannu-Temo subdistrict.

LEAC Groups	Area (% CMT)
Artificialized land (land taken by development)	2.37
Arable land and permanent crops	32.09
Pastures and mosaic farmland	26.68
Standing forests	14.98
Natural grasslands, sclerophyllous vegetation, and heathlands	23.19

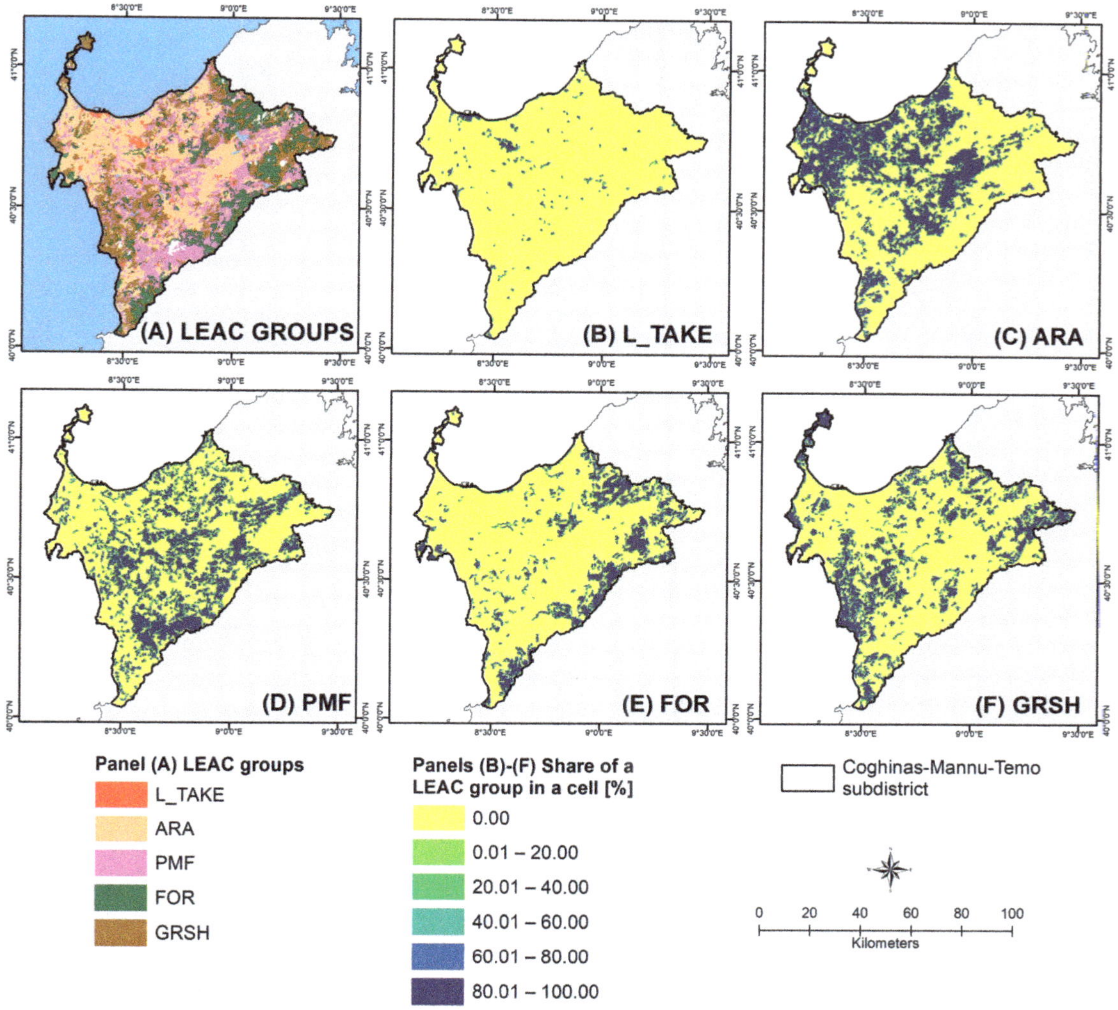

Figure 5. Spatial distribution of land covers in CMT classed through the LEAC groups (panel **A**), and share of each LEAC group within the 300 m grid used in this study (panels **B–F**).

Panels B-F in Figure 5 show the spatial layout of the share of each LEAC group in the 300 m grid cells used within this study. Cells having non-null values of L_TAKE form small and spatially disarticulated bundles. Cells where a share of arable land and permanent crops (ARA) is present cluster especially along the main plains; however, they are spread across the subdistrict, except for the Asinara Island to the north and the mountain areas that delineate the borders of the watersheds. In the latter, clusters of cells hosting standing forests (FOR) are clearly visible in the map, while the Asinara Island is a hotspot for natural grasslands, sclerophyllous vegetation, and heathlands (GRSH), which also feature along the rugged western coast and are scattered across CMT. Finally, cells hosting pasture and mosaic farmland (PMF) are diffuse across the subdistrict, with the larger assemblage along the Marghine mountain chain to the southern border.

3.3. The Outcomes of the Regression Model

The strength and significance of correlations between the explanatory variables in model (1) were preliminarily assessed through the Pearson product–moment correlation coefficient; the outcomes of this assessment, which was carried out on the attribute table of the shapefile containing the 30,775 cells having non-null values of LH, are provided in Appendix A, Table A1. The strongest correlation is that between PMF and GRSH (-0.4033, $p < 0.01$), while $|r| < 0.4$ for the remaining couples of variables. The lack of strong correlations between the explanatory variables highlights the absence of issues of multicollinearity in model (1).

The estimates of the coefficients of DEPOL and VOLSE are significant and show the expected signs, since comparatively higher values of LH are associated with the incoherent and loose substrates that characterize quaternary deposits, and comparatively lower LH values are correlated with the solid and resistant substrates that feature volcanic sedimentary rocks.

Moreover, lower altitudes are associated with higher landslide hazard, and a decrease of 100 m is correlated with an increase of 1.8% in landslide hazard. This outcome may seem rather counterintuitive, since, in general, it is expected that landslide hazard increases with elevation, or the higher the altitude, the higher the probability that landslides may occur. The reason of this finding can be detected from the peculiar spatial taxonomy of landslide hazard in the study area, mapped in Figure 6, which shows the most relevant concentration of high-landslide-hazard cells in locations characterized by comparatively low and medium elevation.

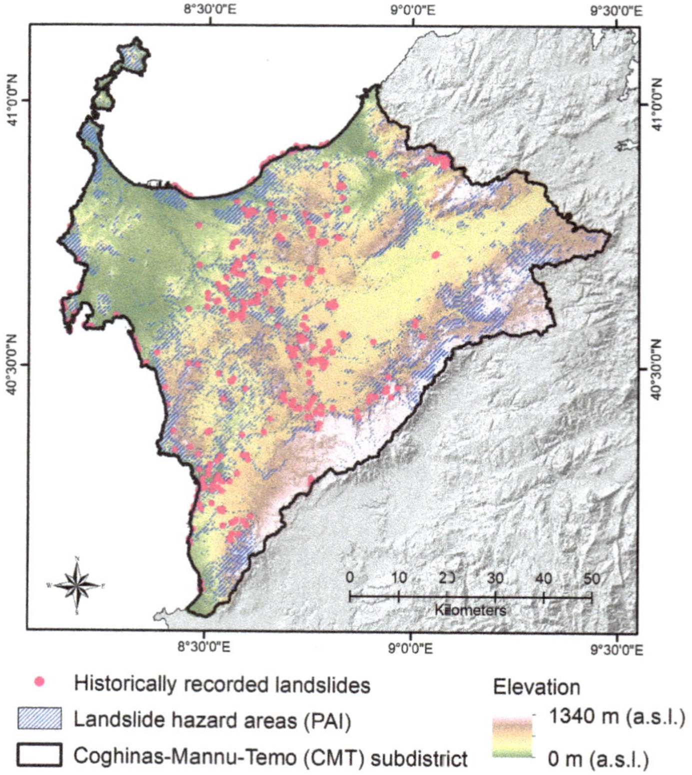

Figure 6. Spatial overlay between historically recorded landslides, landslide hazard areas mapped in the PAI, and regional DTM.

Moreover, the spatially lagged variable shows a positive and significant sign, in terms of *p*-values, which shows an effective control of the spatial autocorrelation of the dependent variable.

That being so, since the estimates related to the control variables DEPOL, VOLSE, and ELEV are statistically significant and consistent with the expectations in terms of the expected signs, whereas the model offers an adequate control for spatial autocorrelation, the impacts of the LEAC covariates and, in particular, the influence of the land take variable on landslide hazard, identified by their estimated coefficients, are reliable and consequential.

The estimated coefficients of the five explanatory variables are significant at 1% and entail the following results, provided that everything else is equal.

Agricultural land, whether characterized by extensive or intensive production, is negatively associated with landslide hazard, showing comparatively low correlations, since, on average, a 10% increase in pastures and mosaic farmland or in arable land corresponds to a 0.7% decrease or to a 1% decrease in landslide hazard.

Positive correlations are shown by the coefficient of FOR, since 10% increases in FOR and GRSH are associated with 1.4% and 0.9% increases in LH, respectively. Increases in forests, natural grasslands, sclerophyllous vegetation, and heathlands are associated with higher values of LH, which entails that such land covers are likely to identify buffer zones with respect to areas characterized by relevant landslide hazard. The spatial contexts featured by these land covers are usually almost totally devoid of human settlements, which highlights a virtuous spatial organization, which aims at protecting urbanized areas from the negative impacts generated by landslides, by preserving natural forests and grasslands from land-taking processes.

All in all, crop production is not associated with increases in LH. On the other hand, forests are the LEAC group that reveals the most relevant positive correlation with landslide hazard, whereas natural grasslands, sclerophyllous vegetation, and heathlands are less relevant in terms of association with LH.

Finally, the regression model identifies the association of LH and L_TAKE as a relevant positive correlation; namely, a 10% increase in L_TAKE is associated with a 0.8% increase in LH. In other words, the higher the size of the land take-related covariate, the higher the size of areas characterized by relevant landslide hazard. This finding highlights that the spatial structure of the study area is characterized by artificialized areas intertwined with areas featured by relevant landslide hazard, or that land-taking processes have taken place in locations that should have been preserved free from urbanization processes due to the magnitude of landslide hazard. Table 5 reports the results of the estimate and relevant statistics of regression model (1).

Table 5. Estimate of regression model (1).

Variable	Coefficient	t-Statistic	*p*-Value
ARA	−0.10613	−11.28802	0.00000
PMF	−0.07235	−7.81156	0.00000
FOR	0.14146	14.72390	0.00000
GRSH	0.09436	10.44108	0.00000
L_TAKE	0.08803	5.85287	0.00000
DEPOL	0.01569	1.68502	0.09201
VOLSE	−0.07308	−15.41213	0.00000
ELEV	−0.00978	−10.60318	0.00000
HGLAGGED	0.96762	23.96984	0.00000
Adjusted R-squared: 0.83247			

4. Discussion

The mapping of landslide hazard in the study area is quite consistent with the taxonomies of similar spatial contexts described and discussed in the current literature. As described in Section 2, the CMT subdistrict features a hilly ground orography (the Marghine-

Goceano Chain and the Mount Limbara), with widespread uphill and downhill stretches, and by a limited coastal plain (the Nurra). As described in Section 3.1, just about one-fifth of the study area is characterized by a more-or-less relevant landslide hazard which, nevertheless, has generated a relevant geological instability, demonstrated by nearly 400 events. Hilly spatial contexts intertwined with plain areas are often associated with limited zones characterized by relevant landslide hazard and by diffused geological instability, as demonstrated by the regional screening of landslide phenomena in the lowlands of Calabria, Southern Italy [60]. The European screening study by Jaedicke et al. [61] identifies the European hotspots concerning landslide hazard, based on the implementation of the models defined by the International Center for Geohazards (ICG) and the Joint Research Center of the European Commission (JRC), which are often located in hilly and plain areas, i.e., with morphological characteristics similar to the CTM subdistrict, sometimes featured by high levels of precipitation and seismic activity. The European screening is quite consistent with what was detected in the case of the landslide inventory implemented by Solís-Castillo et al. [62] as regards the Mexican tropical region of Sierra Costa, characterized by low precipitation rates and landslide hazard diffused over mountainous and plain zones. Analogous findings are shown, among many, in recent studies concerning the Freetown region in Sierra Leone [63], and the Whitsunday Region, located in North Queensland, Australia [64].

The mapping of the LEAC groups in the study area, featuring pastures and mosaic farmland, natural grasslands, sclerophyllous vegetation and heathlands, and arable land, is consistent with the spatial taxonomies reported in other studies concerning landslide hazard in hilly spatial contexts intertwined with plain areas. As in the case of the CTM subdistrict, important relations are identified between landslide hazard and farming production in hilly and plain zones by Rendon et al. [65], which are addressed by several policy tools, aimed at improving the quality of degraded ground and agrosystems, such as the Common agricultural policy [66], the strategy "Farm to Fork" [67] and the Biodiversity Strategy of the EU [68]. According to Borrelli et al. [69] and Panagos et al. [70], landslide hazard and related events in hilly spatial contexts intertwined with plain areas are mainly related to soil erosion phenomena, which should be addressed by increasing soil retention capacity [60], and the endowment of ecosystem services such as ground and superficial water resources quality and recharge, ground and underground biodiversity, and soil resilience to the impacts of climate change and of landslide events [71].

The outcomes of the regression model can be straightforwardly discussed in the theoretical and technical context of the current literature.

Negative correlations are associated with agricultural land, whether it is characterized by intensive or extensive crop farming. This is consistent with the results of several studies which relate the effectiveness of soil conservation practices based on agriculture. For example, Suci et al. [72] highlight the importance of crop farming and crop rotation in improving soil conservation conditions and landslide hazard mitigation in the rural area planning in the Indonesian Cidadap Subdistrict located in Western Java. Extensive and, wherever it is suitable, intensive crop farming are identified as effective approaches to recovering from scars generated by landslide-related events in Mount Elgon, Uganda [73], where such practices are implemented through the direct cooperation of local communities. The extensive mapping of rural areas' exposure to landslide hazard in Central Italy developed by Santangelo et al. [74] shows the association of extensive and intensive crop farming to low-hazard areas as well.

The mapping of the quoted study by Santangelo et al. is consistent with the regression outcomes related to the covariates that identify forests (FOR), and natural grasslands, sclerophyllous vegetation, and heathlands (GRSH). Since the latter two LEAC groups characterize non-urbanized areas, it has to be put in evidence that a virtuous approach to land use planning brings together Central Italy and the Sardinian CMT subdistrict, since areas with relevant landslide hazards have been kept almost totally settlement-free.

The outcomes related to the DEPOQ covariate are consistent with the studies by Sasaky and Sugai concerning the Hachimantai region located in Northeastern Japan [75], and by Akumu et al. [76], where significant landslide hazard is correlated to inland wetlands, whereas coastal wetlands are associated with low LH, which brings together the CMT and the Central Italy coastal wetlands, as characterized by Santangelo et al. [74].

As for forests, woodlands, and shrubs, not only do they almost totally feature the non-urbanized areas of the CMT district, but also they act as spatial contexts whose management is crucial to implement planning policies aimed at decreasing the environmental risk associated with landslide hazard. The association of these LEAC groups with areas characterized by landslide hazard is consistent with their environmental protection function. This issue is widely addressed in the current literature. The enhancement and strengthening of forests and woodlands is basically related to the protection of primary forests, to forest recovery activities, to sustainable management of forests and woodlands, and to tree planting in spatial contexts characterized by different prevailing ecosystems, such as urban and agricultural areas, where these LEAC groups play a decisive role in mitigating the impact of landslide hazard [77]. FAO identifies forest sustainable management as the most important operational category to enhance economic, social, and environmental quality of rural areas, mainly because of its impact on improvement of crop production and productivity connected to protection from flood and landslide effects [78]. Forest and woodlands' recovery and new arboreal plantations are particularly relevant for the definition and implementation of spatial planning policies since the assessment of their economic impact in terms of mitigation and adaptation to climate change is generally recognized as particularly effective in the medium and long runs, especially due to decrease in flood and landslide risk [79,80], as well as for biodiversity protection and enhancement [81–83].

Finally, the regression model shows that landslide hazard is associated with land take in significant and quantitatively relevant terms in the CMT subdistrict. This finding is supported by the fact that the other results of the regression are consistent with the outcomes of several studies available in the current literature, which implies that the definition and implementation of spatial planning policies aimed at addressing landslide hazard in the study area are almost entirely an issue of countering the ongoing land-taking processes and of deurbanizing at least a part of the areas located in landslide-prone zones. This is a key issue in the current scientific and technical debate (among many, [84–87]), and it is widely discussed in the fifth section.

5. Policy Implications

The results show a positive correlation between coefficient of FOR and landslide hazard due to virtuous spatial organization aimed at protecting urbanized areas and at preserving natural forests and grassland from land-taking processes. Forests have positive effects on reducing impacts of landslide. In shallow soils, deep-rooted trees and shrubs may reduce the occurrence of rapid landslide [88] by anchoring and stabilizing superficial soil layer to more sound substrates [89]. Moreover, trees may represent a physical barrier to contrast rocks and debris falls [90]. Therefore, spatial planning policies concerning forest and woodland recovery and plantations of deep-rooted trees and shrubs are particularly significant in order to decrease landslide hazard. From this perspective, two main policy implications can be identified as follows. First, forest management should consider the potential of forests and woodlands for landslide protection by restoring and protecting natural forests [91] and by maintaining forest cover. Health and vitality of forests are two key factors to reduce landslide hazard by strengthening rooting systems of tree in relation to climate change [89]. Secondly, spatial planning should localize forest in high-risk areas in order to support a virtuous spatial organization that locates human settlements and activities in zones characterized by low levels of landslide hazard [89].

The most prominent result is, however, the significant positive correlation concerning L_TAKE; therefore, as far as land covers and their effects on landslide hazard are concerned, controlling land-taking processes is the main road to mitigating the hazard. On this premise, three main groups of policy implications, respectively concerning land densification, land recycling, and strategic environmental assessment, can be identified as follows.

At the international level, governments are using different measures to reduce land-taking processes, such as policy targets [3], financial or fiscal incentives, and environmental assessment of spatial plan and projects [92]. On the other hand, as shown in the introduction, land-taking processes are steadily increasing; therefore, further measures are necessary in order to achieve the EU goal of "no net land take by 2050". According to the EU Soil Strategy for 2030, Member States should integrate the actions defined in the "land take hierarchy"; that is, avoid, reuse, minimize, and compensate, into urban greening plans, and promote the reuse and the recycling of land and high-quality urban soil [3]. In particular, land recycling is defined as "the reuse of abandoned, vacant or underused land for redevelopment. It includes 'grey recycling' and 'green recycling'. Grey recycling is when 'grey' urban objects, such as buildings or transport infrastructures, are built under redevelopment. Green recycling is when 'green' urban objects, such as green urban areas or sport facilities, are built" [93]. Moreover, land recycling includes three components: gray land recycling, green land recycling, and land densification.

As for land densification, it implies that land is developed within existing settlements so as to take advantage of existing infrastructure without using undeveloped land [93]. Therefore, national and regional administrations should promote land recycling strategies within regional plans, to be further downscaled at the local level through municipal masterplans where new development should be allowed only if its impacts on land take are negligible. Moreover, regional strategies should promote a compact urban model based on the land densification concept to reduce demand for undeveloped areas. However, this should not be regarded as a "one size fits all" solution, as local specificities need to be taken into account. Indeed, such measures have been found to be particularly effective in developed countries, whereas in Latin America and Sub-Saharan Africa, as well as in already hyper-dense Asian megacities, further urban densification can bring about negative effects in terms of inequality of spatial distribution and social conditions of the local communities [94].

With regard to land recycling, this is mainly supported through financial and regulatory tools. Concerning the former, financial support through publicly funded programs [95] and subsidies generated through impact fees, soil sealing fees, or improvement levies [96] are among the most common tools to promote land recycling. However, national and regional governments should diversify the set of fiscal tools that usher in creative and innovative ways to manage land uses effectively and efficiently. For example, the transfer of development rights can be used to direct development towards already taken up and well-serviced areas, rather than towards greenfield areas that are poorly connected in terms of transport, infrastructures, facilities, and services. As for regulatory tools, such as zoning schemes and land use regulations, these could successfully promote the participation of the private sector within land recycling projects. Furthermore, flexible and performance-based zoning regulations could be adopted within municipal masterplans [97]: these should pursue strict limits and constraints on land take, while allowing land uses that do not result in artificial land, hence in turn promoting mixed land uses where different functions coexist. However, because in Italy land use plans are drafted and approved by local municipalities independently of each other, monitoring and evaluating the provisions of existing municipal land use plans is necessary in order to understand the cumulative effects of land use regulations in adjacent cities and towns, and their consequences on land-taking processes. In this regard, in Sardinia, the regional administration can play a key role, since the regional planning office actively participates in the approval processes of regional and local plans with a view to ensuring their compliance with both regional planning laws and the regional landscape plan [98]. Due to the complexity of interests at stake, measures concerning land-take prevention and limitation, be they finance-based or regulation-based, call for active involvement of local communities and municipalities [99], and for effective vertical and horizontal cooperation between governments and other public bodies [100].

A third important policy implication concerns strategic environmental assessment (SEA), an appraisal planning tool that is mandatory in EU countries, which ensures that environmental considerations and sustainable-oriented goals are integrated into plan-making processes by assessing their likely effects on the environment, by considering reasonable and more sustainable alternatives, and by taking into account the mutual relations between the environment and the economic and social sectors [100,101]. Through the comparison of alternative land uses, the identification of areas that are more suitable for certain uses, and the evaluation of cumulative, direct, and indirect impacts of land-based investments, SEA can pave the way for the integration, within spatial plans, of measures aimed at preventing or minimizing land take.

In areas prone to landslide hazard (as well as in areas prone to flood hazard, which are not the object of this study), in Sardinia the PAI maps serve as a spatial reference for the PAI regulations, which restrict land uses and prevent land transformations depending on the magnitude of the hazard. In this way, the PAI provides a legally binding framework for municipal masterplans, whose zoning choices must comply with the PAI regulations, contrary to what has been reported in other countries [102], where new development in landslide hazard areas is not prohibited [103]. Hence, this higher-level regional planning tool contributes to limiting land-taking processes in fragile areas, while also providing relevant spatial information to planners in charge of drafting land use plans and appraising them through the SEA. Moreover, because the PAI maps are publicly available through the regional geoportal, they also contribute to raising local authorities' and local communities' awareness of landslide hazard and, by doing so, to granting transparency and legitimacy to restrictions that otherwise would be, in principle, quite conflictual.

6. Conclusions

In this study a novel methodological approach was proposed with a view to analyzing the relations between land-taking processes and landslide hazard in order to understand whether, and to what extent, land-take phenomena are associated with landslide hazard. The outcomes of this analysis were next used to define policy suggestions that, by preventing or minimizing land take, can help mitigating landslide hazard, therefore indirectly preventing the human and economic losses that might result from land mismanagement.

The methodological approach here proposed as a tool to support decision-making processes can be exported to other European contexts, since the CLC inventory is available for 39 European countries [36], among them the 27 EU Member States, provided that a detailed landslide hazard or landslide susceptibility spatial dataset is available, which is not the case for all of the EU Member States, as reported in a recent study by Mateos et al. [102].

Moreover, the methodological approach shows a certain degree of flexibility, allowing for the inclusion of further context-specific spatial or normative variables that might be appropriate or needed in other contexts. In our selected case study, the influence of two LEAC groups (open space with little or no vegetation; transitional woodland and shrub) on landslide hazard could not be assessed because they were not relevant in the study area, hence this might be a matter for future investigation in other contexts. In addition, the impact of specific land cover classes on landslide hazard was here not assessed due to the simplified classification of the LEAC taxonomy, where the 44 CORINE land covers are aggregated into eight groups. Future directions of the research might therefore include a more detailed analysis, where the LEAC groups are (completely or partly) replaced by the CLC classes, which might, however, result in a more complicated implementation of the model and interpretation of its results.

This study has analyzed the relations between land-taking processes and landslide hazards in a cross-section terms; therefore, dynamic issues, such as the evolution process of land covers or the development through time of different forms of urbanization are not within the scope of this work. However, these are relevant future directions related to the research work proposed in this article, which may entail the assessment of the dynamics of the relationship between landslide hazard and land-taking processes. This point is analyzed by Pisano et al. [30], who studied how land cover changes affect landslide hazard in relation to the years 1954, 1981, and 2007. The advantages in analyzing how land cover changes are correlated to landslide hazard are connected with the outstanding dynamicity that characterizes land cover change processes. Although different factors influence landslide hazard in the long run, such as geological and geomorphological phenomena, land covers are also characterized by short-term dynamics, which stresses the relevant added value which may be provided by a time-series-based contextualization of the cross-section assessment here implemented [104]. As regards the dynamic relationships between urbanization processes and landslide hazard, further important research directions are represented by the assessment of landslide phenomena in different cities and towns characterized by diversified urban morphologies, ranging from dense and compact urban fabrics to sparse and widespread urbanization, in order to define and implement planning policies and measures aimed at countering and mitigating landslide hazard.

Author Contributions: F.I., S.L, F.L. and C.Z. collaboratively designed this study. F.L. wrote Section 1. S.L. wrote Section 2.1, Section 2.3, Sections 2.3.1–2.3.3, Section 3.1, Section 3.2 and Appendix A. C.Z. wrote Sections 2.2, 3.3 and 4. F.I. and F.L. jointly wrote Sections 5 and 6. All authors have read and agreed to the published version of the manuscript.

Funding: This study was carried out in the framework of the research project "Implementation and revision of the regional strategy for climate change adaptation (SRACC)—Macro-Action 2, Mainstreaming of the SRACC spatial framework: governance and environmental assessment" (2021–2023), funded by the Autonomous Region of Sardinia.

Institutional Review Board Statement: Ethical review and approval were not applicable, since this study does not involve humans or animals.

Informed Consent Statement: Not applicable.

Data Availability Statement: Data available upon request.

Conflicts of Interest: The authors declare no conflict of interest.

Appendix A

Table A1. Correlations between the explanatory variables used in model (1): Pearson product–moment correlation coefficients and significance levels.

	L_TAKE	ARA	PMF	FOR	GRSH	DEPOQ	VOLSE	ELEV	HGLAGGED
L_TAKE		−0.0541	−0.0715	−0.0798	−0.1096	0.0614	0.0091	−0.0893	0.0415
ARA	*0.0000 ****		−0.2600	−0.2727	−0.3633	0.2006	0.1491	−0.3056	−0.1751
PMF	*0.0000 ****	*0.0000 ****		−0.2750	−0.4033	0.0118	0.1708	−0.0034	−0.1894
FOR	*0.0000 ****	*0.0000 ****	*0.0000 ****		−0.3118	−0.0434	−0.2723	0.3082	0.1759
GRSH	*0.0000 ****	*0.0000 ****	*0.0000 ****	*0.0000 ****		−0.1720	−0.0159	0.0030	0.1699
DEPOQ	*0.0000 ****	*0.0000 ****	*0.0377 ***	*0.0000 ****	*0.0000 ****		−0.3042	−0.2769	0.0186
VOLSE	*(0.1085)*	*0.0000 ****	*0.0000 ****	*0.0000 ****	*0.0052 ****	*0.0000 ****		−0.2143	−0.2093
ELEV	*0.0000 ****	*0.0000 ****	*(0.5523)*	*0.0000 ****	*(0.5982)*	*0.0000 ****	*0.0000 ****		0.0202
HGLAGGED	*0.0000 ****	*0.0000 ****	*0.0000 ****	*0.0000 ****	*0.0000 ****	*0.0011 ****	*0.0000 ****	*0.0004 ****	

Below the diagonal (italicized): p-values and significance levels: ***: $p < 0.01$; **: $p < 0.05$; *: $p < 0.10$; (): $p > 0.10$. $n = 30{,}775$; degrees of freedom = 30,773.

References

1. European Environment Agency. *The European Environment—State and Outlook 2020. Knowledge for Transition to a Sustainable Europe*; Publications Office of the European Union: Luxembourg, 2019; ISBN 978-92-9480-090-9. Available online: https://op.europa.eu/en/publication-detail/-/publication/b312a176-1b69-11ea-8c1f-01aa75ed71a1/language-en (accessed on 17 January 2023).
2. European Environment Agency. *Land Take and Land Degradation in Functional Urban Areas, EEA Report n. 17/2021*; Publications Office of the European Union: Luxembourg, 2021; ISBN 978-92-9480-465-5. Available online: https://www.eea.europa.eu/publications/land-take-and-land-degradation (accessed on 17 January 2023).

3. European Commission. Communication from the Commission to the European Parliament, the Council, the European Economic and Social Committee and the Committee of the Regions. EU Soil Strategy for 2030, Reaping the Benefits of Healthy Soils for People, Food, Nature and Climate. 2021. Available online: https://eur-lex.europa.eu/legal-content/EN/TXT/?uri=CELEX%3A52021DC0699 (accessed on 17 January 2023).
4. European Environment Agency. Indicator Assessment. *Land Take in Europe*. Available online: https://www.eea.europa.eu/data-and-maps/indicators/land-take-3/assessment (accessed on 17 January 2023).
5. Munafò, M. (Ed.) Consumo di Suolo, Dinamiche Territoriali e Servizi Ecosistemici. Edizione 2022. Land-Take, Land Dynamics and Ecosystem Services. 2022 edition Report SNPA 32/22; 2022; ISBN 978-88-448-1124-2. Available online: https://www.snpambiente.it/wp-content/uploads/2022/07/Rapporto_consumo_di_suolo_2022.pdf (accessed on 17 January 2023).
6. Gerten, C.; Fina, S.; Rusche, K. The sprawling planet: Simplifying the measurement of global urbanization trends. *Front. Environ. Sci.* **2019**, *7*, 140. [CrossRef]
7. Seto, K.C.; Güneralp, B.; Hutyra, L.R. Global forecasts of urban expansion to 2030 and direct impacts on biodiversity and carbon pools. *Proc. Natl. Acad. Sci. USA* **2012**, *109*, 16083–16088. [CrossRef] [PubMed]
8. Yuan, Y.J.; Chen, D.X.; Wu, S.H.; Mo, L.J.; Tong, G.J.; Yan, D.H. Urban sprawl decreases the value of ecosystem services and intensifies the supply scarcity of ecosystem services in China. *Sci. Total Environ.* **2019**, *697*, 134170. [CrossRef]
9. Millennium Ecosystem Assessment. *Ecosystems and Human Well-Being: Synthesis*; Island Press: Washington, DC, USA, 2005. Available online: https://www.millenniumassessment.org/documents/document.356.aspx.pdf (accessed on 17 January 2023).
10. Mooney, H.A.; Ehrlich, P.R. Ecosystem services: A fragmentary history. In *Nature's Services. Societal Dependence on Natural Ecosystems*; Daily, G.C., Ed.; Island Press: Washington, DC, USA, 1997; pp. 11–19.
11. Costanza, R. Ecosystem services: Multiple classification systems are needed. *Biol. Conserv.* **2008**, *141*, 350–352. [CrossRef]
12. Holdren, J.P.; Ehrlich, P.R. Human Population and the Global Environment: Population Growth, Rising Per Capita Material Consumption, and Disruptive Technologies Have Made Civilization a Global Ecological Force. *Am. Sci.* **1974**, *62*, 282–292. Available online: http://www.jstor.org/stable/27844882 (accessed on 17 January 2023).
13. Haines-Young, R.; Potschin, M.B. Common International Classification of Ecosystem Services (CICES) V5.1 and Guidance on the Application of the Revised Structure. 2018. Available online: https://cices.eu/content/uploads/sites/8/2018/01/Guidance-V51-01012018.pdf (accessed on 17 January 2023).
14. Costanza, R.; d'Arge, R.; de Groot, R.; Farber, S.; Grasso, M.; Hannon, B.; Limburg, K.; Naeem, S.; Oneill, R.V.; Paruelo, J.; et al. The value of the world's ecosystem services and natural capital. *Nature* **1997**, *387*, 253–260. [CrossRef]
15. Isola, F.; Lai, S.; Leone, F.; Zoppi, C. *Green Infrastructure and Regional Planning—An Operational Framework*; FrancoAngeli: Milan, Italy, 2022; ISBN 9788835141402.
16. Olschewski, R.; Bebi, P.; Teich, P.; Wissen Hayek, U.; Grêt-Regamey, A. Avalanche protection by forests—A choice experiment in the Swiss Alps. *Forest Policy Econ.* **2012**, *17*, 19–24. [CrossRef]
17. Notaro, S.; Paletto, A. The economic valuation of natural hazards in mountain forests: An approach based on the replacement cost method. *J. Forest Econ.* **2012**, *18*, 318–328. [CrossRef]
18. Cruden, D.M.; Varnes, D.J. Landslide types and processes. In *Landslides: Investigation and Mitigation. Transportation Research Board Special Report 247*; Turner, A.K., Schuster, R.L., Eds.; National Academy Press: Washington, DC, USA, 1996; pp. 36–75.
19. Tanveer, I.; Jeffrey, R. *Hazard Mitigation in Emergency Management*; Butterworth-Heinemann: Oxford, UK, 2016; ISBN 978-0-12-420134-7.
20. United State Geological Survey (USGS). Landslide Types and Processes. Fact Sheet 2004–3072. 2004. Available online: https://pubs.usgs.gov/fs/2004/3072/pdf/fs2004-3072.pdf (accessed on 17 January 2023).
21. Huber, S.; Prokop, G.; Arrouays, D.; Banko, G.; Bispo, A.; Jones, R.J.A.; Kibblewhite, M.G.; Lexer, W.; Möller, A.; Rickson, R.J. (Eds.) *Environmental Assessment of Soil for Monitoring. Volume I, Indicators & Criteria. EUR 23490 EN/1*; Office for the Official Publications of the European Communities: Luxembourg, 2008. Available online: http://www.fsp-parrur.irenala.edu.mg/Data-FSP-PARRUR/Les%20activit%C3%A9s/IST/S%C3%A9minaires-Ecoles/Ecole%20th%C3%A9matique/ECOLE%20THEMATIQUE%20SOL_LRI-IRD/Ouvrages/Decline_in_Soil_Biodiversity.pdf (accessed on 17 January 2023).
22. Meneses, B.M.; Pereira, S.; Reis, E. Effects of different land use and land cover data on the landslide susceptibility zonation of road networks. *Nat. Hazards Earth Sys.* **2019**, *19*, 471–487. [CrossRef]
23. Hao, L.; van Westen, C.; Rajaneesh, A.; Sajinkumar, K.S.; Martha, T.R.; Jaiswal, P. Evaluating the relation between land use changes and the 2018 landslide disaster in Kerala, India. *Catena* **2022**, *216*, 106363. [CrossRef]
24. Tasser, E.; Mader, M.; Tappeiner, U. Effects of land use in alpine grasslands on the probability of landslides. *Basic Appl. Ecol.* **2003**, *4*, 271–280. [CrossRef]
25. Vuillez, C.; Tonini, M.; Sudmeier-Rieux, K.; Devkota, S.; Derron, M.H.; Jaboyedoff, M. Land use changes, landslides and roads in the Phewa Watershed, Western Nepal from 1979 to 2016. *Appl. Geogr.* **2018**, *94*, 30–40. [CrossRef]
26. Wasowski, J.; Lamanna, C.; Casarano, D. Influence of land-use change and precipitation patterns on landslide activity in the Daunia Apennines, Italy. *Q. J. Eng. Geol. Hydroge.* **2010**, *43*, 387–401. [CrossRef]
27. Reichenbach, P.; Busca, C.; Mondini, A.C.; Rossi, M. The influence of land use change on landslide susceptibility zonation: The Briga Catchment Test Site (Messina, Italy). *Environ. Manag.* **2014**, *54*, 1372–1384. [CrossRef] [PubMed]
28. García-Ruiz, J.M. The effects of land uses on soil erosion in Spain: A review. *Catena* **2010**, *81*, 1–11. [CrossRef]

29. Glade, T. Landslide occurrence as a response to land use change: A review of evidence from New Zealand. *Catena* **2003**, *51*, 297–314. [CrossRef]
30. Pisano, L.; Zumpano, V.; Malek, Ž.; Rossokopf, C.M.; Parise, M. Variations in the susceptibility to landslides, as a consequence of land cover changes: A look to the past, and another towards the future. *Sci. Total Environ.* **2017**, *601–602*, 1147–1159. [CrossRef] [PubMed]
31. Muñoz-Torrero Manchado, A.; Ballesteros-Cánovas, J.A.; Allen, S.; Stoffel, M. Deforestation controls landslide susceptibility in Far-Western Nepal. *Catena* **2022**, *219*, 106627. [CrossRef]
32. European Parliament and European Council. Directive 2000/60/EC of the European Parliament and of the Council of 23 October 2000 Establishing a Framework for Community Action in the Field of Water Policy. Available online: https://eur-lex.europa.eu/legal-content/EN/TXT/?uri=celex%3A32000L0060 (accessed on 17 January 2023).
33. Ministero dell'Ambiente e della Sicurezza Energetica. Autorità di Bacino. Ministry for the Environment and Energy Security. Competent Authorities for River Basin Districts. Available online: https://www.mite.gov.it/direttive/autorita-di-bacino (accessed on 17 January 2023).
34. European Commission. Environment—Water—Italy. Available online: https://ec.europa.eu/environment/water/participation/map_mc/countries/italy_en.htm (accessed on 17 January 2023).
35. Regione Autonoma della Sardegna, Assessorato Lavori Pubblici. Regional Administration of Sardinia, Office of Public Works. *Piano Stralcio per l'Assetto Idrogeologico—Relazione Generale [Hydrogeological Setting Plan—General Report].* 2004. Available online: https://www.regione.sardegna.it/documenti/1_26_20060913170906.pdf (accessed on 17 January 2023).
36. Copernicus Europe's Eyes on Earth 2022. CORINE Land Cover. Available online: https://land.copernicus.eu/pan-european/corine-land-cover (accessed on 17 January 2023).
37. Sardinian Basin Competent Authority. Studio di Dettaglio e Approfondimento del Quadro Conoscitivo della Pericolosità e del Rischio da Frana nel Sub Bacino n° 3 Coghinas–Mannu–Temo [Detailed Study of the Landslide Hazard and Risk in Sub-District no. 3 Coghinas-Mannu-Temo]. Available online: https://www.regione.sardegna.it/index.php?xsl=509&s=1&v=9&c=12097&tb=8374&st=13 (accessed on 17 January 2023).
38. Canu, S.; Rosati, L.; Fiori, M.; Motroni, A.; Filigheddu, R.; Farris, E. Bioclimate map of Sardinia (Italy). *J. Maps* **2015**, *11*, 711–718. [CrossRef]
39. Bacchetta, G.; Bagella, S.; Biondi, E.; Farris, E.; Filigheddu, R.; Mossa, L. Vegetazione Forestale e Serie di Vegetazione della Sardegna (con Rappresentazione Cartografica alla Scala 1:350.000). Forest Vegetation and Serial Vegetation of Sardinia (with a 1:350,000 Scale Map). *Fitosociologia* **2009**, *46*, 3–82.
40. IFFI—Inventario dei Fenomeni Franosi in Italia. IFFI—Inventory of Landslides Events in Italy. Available online: https://www.progettoiffi.isprambiente.it (accessed on 17 January 2023).
41. Cinus, S.; Demuru, S.; Deriu, M.; Farris, M.; Patteri, G.; Tilocca, G.; Martelli, R. Analisi del Dissesto da Frana in Sardegna. An Analysis of Landslide Phenomena in Sardinia. In *Rapporto sulle Frane in Italia. APAT Rapporti no. 78/2007. Report on Landslides in Italy. APAT Reports no. 78/2007*; Trigila, A., Ed.; APAT: Rome, Italy, 2007; pp. 651–681. Available online: https://www.isprambiente.gov.it/files/pubblicazioni/rapporti/rapporto-frane-2007/Capitolo_25_Sardegna.pdf (accessed on 17 January 2023).
42. Cheshire, P.; Sheppard, S. On the price of land and the value of amenities. *Econ. New Ser.* **1995**, *62*, 247–267. [CrossRef]
43. Stewart, P.A.; Libby, L.W. Determinants of farmland value: The case of DeKalb County Illinois. *Rev. Agr. Econ.* **1998**, *20*, 80–95. [CrossRef]
44. Sklenicka, P.; Molnarova, K.; Pixova, K.C.; Salek, M.E. Factors affecting farmlands in the Czech Republic. *Land Use Policy* **2013**, *30*, 130–136. [CrossRef]
45. Zoppi, C.; Argiolas, M.; Lai, S. Factors influencing the value of houses: Estimates for the city of Cagliari, Italy. *Land Use Policy* **2015**, *42*, 367–380. [CrossRef]
46. Byron, R.P.; Bera, A.K. Linearized estimation of nonlinear single equation functions. *Int. Econ. Rev.* **1983**, *24*, 237–248. [CrossRef]
47. Wolman, A.L.; Couper, E.A. Potential consequences of linear approximation in economics. *Econ. Q.* **2003**, *89*, 51–67.
48. Zoppi, C.; Lai, S. Land-taking processes: An interpretive study concerning an Italian region. *Land Use Policy* **2014**, *36*, 369–380. [CrossRef]
49. Anselin, L. *Spatial Econometrics: Methods and Models*; Kluwer Academic Publishers: Dordrecht, The Netherlands, 1988.
50. Anselin, L. Spatial Econometrics. In *A Companion to Theoretical Econometrics. Spatial Econometrics*; Baltagi, B.H., Ed.; Blackwell Publishing: Oxford, UK, 2003; pp. 310–330.
51. Anselin, L.; Syabri, I.; Kho, Y. GeoDa: An introduction to spatial data analysis. *Geogr. Anal.* **2006**, *38*, 5–22. [CrossRef]
52. European Environment Agency. *Land Accounts for Europe 1990–2000: Towards Integrated Land and Ecosystem Accounting. European Environment Agency Report no. 11*; Office for Official Publications of the European Communities: Luxembourg, 2006. Available online: http://www.eea.europa.eu/publications/eea_report_2006_11 (accessed on 17 January 2023).
53. Regione Autonoma della Sardegna. Sardegna Geoportale. *Carta dell'Uso del Suolo. Regional Administration of Sardinia. Sardinian Geoportal. Land Cover Map.* Available online: https://www.sardegnageoportale.it/index.php?xsl=2420&s=40&v=9&c=14480&es=6603&na=1&n=100&esp=1&tb=14401 (accessed on 11 January 2022).
54. Dettori, S.; Falqui, A.; Filigheddu, M.R.; Sedda, L. Performance di recenti imboschimenti con quercia da sughero in ex-coltivi. Growth performance of cork oak plantations recently established on farmlands in Sardinia, Italy. *Forest* **2006**, *3*, 327–338. [CrossRef]

55. Bianchi, L.; Calamini, G.; Gregori, E.; Paci, M.; Pallanza, S.; Pierguidi, A.; Salbitano, F.; Tani, A.; Vedele, S. Valutazione degli effetti del rimboschimento in zone aride della Sardegna: Risultati preliminari sulla vegetazione. Evaluation of the effects of afforestation in Sardinian arid areas: Preliminary results on vegetation. *L'Italia Forestale e Montana* **2002**, *4*, 353–368.
56. Regione Autonoma della Sardegna. Sardegna Geoportale. *Carta Geologica di base della Sardegna in scala 1:25.000. 1:25,000 Geology Map of Sardinia.* Available online: https://www.sardegnageoportale.it/index.php?xsl=2420&s=40&v=9&c=14479&es=6603&na=1&n=100&esp=1&tb=14401 (accessed on 11 January 2023).
57. Regione Autonoma della Sardegna. Linee Guida per l'Adeguamento dei Piani Urbanistici Comunali al PPR e al PAI. *Assetto ambientale. Bozza 2008. Guidelines for the Elaboration of Municipal Land-Use Plans Compliant with the Regional Landscape Plan and the Hydrogeological Setting Plan. Environmental Setting. 2008 Draft.* Available online: https://www.sardegnaterritorio.it/documenti/6_83_20081007095546.zip (accessed on 11 January 2022).
58. Regione Autonoma della Sardegna. Sardegna Geoportale. *Modelli Digitali del Terreno e delle Superfici. Regional Administration of Sardinia. Sardinian Geoportal. Digital Terrain and Digital Surface Models.* Available online: https://www.sardegnageoportale.it/areetematiche/modellidigitalidielevazione (accessed on 17 January 2023).
59. Brovelli, M.; Crespi, M.; Cilloccu, F.; Dequal, S.; Lingua, A.M. *Ortoimmagini 1:10000 e Modelli Altimetrici—Linee Guida*; CISIS: Rome, Italy, 2009; 1:10,000 Orthoimages and Elevation Models—Guidelines. Available online: https://www.sardegnageoportale.it/index.php?xsl=2425&s=330919&v=2&c=14471&t=1&tb=14401 (accessed on 11 January 2022).
60. Vennari, C.; Gariano, S.L.; Antronico, L.; Brunetti, M.T.; Iovine, G.; Peruccacci, S.; Terranova, O.; Guzzetti, F. Rainfall thresholds for shallow landslides occurrence in Calabria, Southern Italy. *Nat. Hazards Earth Syst. Sci. Discuss.* **2013**, *1*, 5141–5179. [CrossRef]
61. Jaedicke, C.; Van Den Eeckhaut, M.; Nadim, F.; Hervás, J.; Kalsnes, B.; Vangelsten, B.V.; Smith, J.T.; Tofani, V.; Ciurean, R.; Winter, M.G.; et al. Identification of landslide hazard and risk 'hotspots' in Europe. *Bull. Eng. Geol. Environ.* **2014**, *73*, 325–339. [CrossRef]
62. Solís-Castillo, B.; Mendoza, M.E.; Vázquez Castro, G.; Bocco, G. Landslide inventory map of the tropical dry Sierra Costa Region, Michoacán México. *Phys. Geog.* **2019**, *40*, 554–577. [CrossRef]
63. Lahai, Y.A.; Anderson, K.F.E.; Jalloh, Y.; Rogers, I.; Kamara, M. A comparative geological, tectonic and geomorphological assessment of the Charlotte, Regent and Madina landslides, Western area, Sierra Leone. *Geoenv. Dis.* **2021**, *8*, 1–17. [CrossRef]
64. Bradbury, T. *Whitsunday Landslide Study—Landslide Susceptibility Investigation and Mapping*; Cardno: Paget, Australia, 2019. Available online: https://www.whitsundayrc.qld.gov.au/downloads/file/367/08-whitsunday-landslide-study-2019 (accessed on 17 January 2023).
65. Rendon, P.; Steinhoff-Knopp, B.; Burkhard, B. Linking ecosystem condition and ecosystem services: A methodological approach applied to European agroecosystems. *Ecos. Serv.* **2022**, *53*, 101387. [CrossRef]
66. European Commission. The New Common Agricultural Policy: 2023–27. Available online: https://agriculture.ec.europa.eu/common-agricultural-policy/cap-overview/new-cap-2023-27_en#documents (accessed on 17 January 2023).
67. European Commission. A Farm to Fork Strategy for a fair, healthy and environmentally-friendly food system. *Communication from the Commission to the European Parliament, The Council, The European Economic and Social Committee and the Committee of the Regions (COM(2020) 381 Final).* Available online: https://eur-lex.europa.eu/resource.html?uri=cellar:ea0f9f73-9ab2-11ea-9d2d-01aa75ed71a1.0001.02/DOC_1&format=PDF (accessed on 17 January 2023).
68. European Commission. EU Biodiversity Strategy for 2030. *Bringing Nature Back into Our Lives. Communication from the Commission to the European Parliament, the Council, the European Economic and Social Committee and the Committee of the Regions (COM(2020) 380 Final).* Available online: https://eur-lex.europa.eu/resource.html?uri=cellar:a3c806a6-9ab3-11ea-9d2d-01aa75ed71a1.0001.02/DOC_1&format=PDF (accessed on 17 January 2023).
69. Borrelli, P.; Robinson, D.A.; Panagos, P.; Lugatod, E.; Yang, J.E.; Alewella, C.; Wuepper, D.; Montanarella, L.; Ballabio, C. Land use and climate change impacts on global soil erosion by water (2015–2070). *Proc. Natl. Sci. Acad. USA* **2020**, *117*, 21994–22001. [CrossRef]
70. Panagos, P.; Ballabio, C.; Himics, M.; Scarpa, S.; Matthews, F.; Bogonos, M.; Poesen, J.; Borrelli, P. Projections of soil loss by water erosion in Europe by 2050. *Environ. Sci. Pol.* **2021**, *124*, 380–392. [CrossRef]
71. Lal, R. Soil conservation and ecosystem services. *Int. Soil Water Conserv. Res.* **2014**, *2*, 36–47. [CrossRef]
72. Suchi, M.I.; Supriatna, S.; Rustanto, A. Vegetative conservation of landslide prone areas in the Cidadap Watershed Area, Sukabumi Regency. In *The 4th Life and Environmental Sciences Academics Forum 2020: IOP Conference Series; Earth and Environmental Science*; IOP Publishing: Bristol, UK, 2021; Volume 846, p. 012024. [CrossRef]
73. Nakileza, B.R.; Majaliwa, M.J.; Wandera, A.; Nantumbwe, M.C. Enhancing resilience to landslide disaster risks through rehabilitation of slide scars by local communities in Mt Elgon, Uganda. *J. Disaster Risk Stud.* **2017**, *9*, 1–11. [CrossRef]
74. Santangelo, M.; Marchesini, I.; Bucci, F.; Cardinali, M.; Cavalli, M.; Crema, S.; Marchi, L.; Alvioli, M.; Guzzetti, F. Exposure to landslides in rural areas in Central Italy. *J. Maps* **2021**, *17*, 124–132. [CrossRef]
75. Sasaki, N.; Sugai, T. Distribution and development processes of wetlands on landslides in the Hachimantai volcanic group, NE Japan. *Geogr. Rev. Jap. Ser. B* **2015**, *87*, 103–114. [CrossRef]
76. Akumu, C.E.; Henry, J.; Gala, T.; Dennis, S.; Reddy, C.; Teggene, F.; Haile, S.; Archer, R.S. Inland wetlands mapping and vulnerability assessment using an integrated geographic information system and remote sensing techniques. *Global J. Environ. Sci. Manag.* **2018**, *4*, 387–400. [CrossRef]

77. Cariñanos, P.; Calaza, P.; Hiemstra, J.; Pearlmutter, D.; Vilhar, U. The role of urban and peri-urban forests in reducing risks and managing disasters. *Unasylva* **2018**, *69*, 53–58. Available online: https://www.researchgate.net/profile/Paloma-Carinanos/publication/324014447_The_role_of_urban_and_peri-urban_forests_in_reducing_risks_and_managing_disasters/links/5abcee9b45851584fa6e9702/The-role-of-urban-and-peri-urban-forests-in-reducing-risks-and-managing-disasters.pdf (accessed on 17 January 2023).
78. FAO (Food and Agriculture Organization of the United Nations). *Global Forest Resources Assessment 2020. Main Report*; FAO: Rome, Italy, 2020. [CrossRef]
79. Mansourian, S.; Doncheva, N.; Valchev, K.; Vallauri, D. *Lessons Learnt from 20 Years of Floodplain Forest Restoration: The Lower Danube Landscape*; WWF Field Series 2019—WWF France: Paris, France, 2019. Available online: https://wwfint.awsassets.panda.org/downloads/lessons_learnt_fromthe_lower_danube_landscape.pdf (accessed on 17 January 2023).
80. de Groot, R.S.; Blignaut, J.; Van der Ploeg, S.; Aronson, J.; Elmqvist, T.; Farley, J. Benefits of investing in ecosystem restoration. *Conserv. Biol.* **2013**, *27*, 1286–1293. [CrossRef] [PubMed]
81. Morecroft, M.D.; Duffield, S.; Harley, M.; Pearce-Higgins, J.W.; Stevens, N.; Wattsand, O.; Whitake, J. Measuring the success of climate change adaptation and mitigation in terrestrial ecosystems. *Science* **2019**, *366*, eaaw9256. [CrossRef]
82. Watson, J.E.M.; Evans, T.; Venter, O.; Williams, B.; Tulloch, A.; Stewart, C.; Thompson, I.; Ray, J.C.; Murray, K.; Salazar, A.; et al. The exceptional value of intact forest ecosystems. *Nat. Ecol. Evol.* **2018**, *2*, 599–610. [CrossRef]
83. Forest Europe. *State of Europe's Forests 2020*; Ministerial Conference on the Protection of Forests in Europe: Bonn, Germany, 2020. Available online: https://foresteurope.org/wp-content/uploads/2016/08/SoEF_2020.pdf (accessed on 23 December 2022).
84. Segoni, S.; Caleca, F. Definition of environmental indicators for a fast estimation of landslide risk at national scale. *Land* **2021**, *10*, 621. [CrossRef]
85. Fiorini, L.; Zullo, F.; Marucci, A.; Romano, B. Land take and landscape loss: Effect of uncontrolled urbanization in Southern Italy. *J. Urban Manag.* **2019**, *8*, 42–56. [CrossRef]
86. Shu, H.; Hürlimann, M.; Molowny-Horas, R.; González, M.; Pinyol, J.; Abancó, C.; Ma, J. Relation between land cover and landslide susceptibility in Val d'Aran, Pyrenees (Spain): Historical aspects, present situation and forward prediction. *Sci. Total Environ.* **2019**, *693*, 133557. [CrossRef]
87. Garcia, R.A.C.; Oliveira, S.C.; Zêzere, J.L. Assessing population exposure for landslide risk analysis using dasymetric cartography. *Nat. Hazard Earth Sys.* **2016**, *16*, 2769–2782. [CrossRef]
88. Sidle, R.C.; Ziegler, A.D.; Negishi, J.N.; Nik, A.R.; Siew, R.; Turkelboom, F. Erosion processes in steep terrain—Truths, myths, and uncertainties related to forest management in Southeast Asia. *Forest. Ecol. Manag.* **2006**, *224*, 199–225. [CrossRef]
89. RECOFTC; FAO. Forests and Landslides. Available online: https://www.recoftc.org/sites/default/files/publications/resources/recoftc-0000049-0001-en.pdf (accessed on 17 January 2023).
90. Guthrie, R.H.; Hockin, A.; Colquhoun, L.; Nagy, T.; Evans, S.G.; Ayles, C. An examination of controls on debris flow mobility: Evidence from coastal British Columbia. *Geomorphology* **2010**, *114*, 601–613. [CrossRef]
91. Zhang, Y.; Shen, C.; Zhou, S.; Luo, X. Analysis of the influence of forests on landslides in the Bijie Area of Guizhou. *Forests* **2022**, *13*, 1136. [CrossRef]
92. Schatz, E.-M.; Bovet, J.; Lieder, S.; Schroeter-Schlaack, C.; Strunz, S.; Marquard, E. Land take in environmental assessments: Recent advances and persisting challenges in selected EU countries. *Land Use Policy* **2021**, *111*, 105730. [CrossRef]
93. European Environment Agency. Indicator Assessment. Land Recycling and Densification. Available online: https://www.eea.europa.eu/data-and-maps/indicators/land-recycling-and-densification/assessment-1 (accessed on 17 January 2023).
94. Li, M.; Verburg, P.H.; van Vliet, J. Global trends and local variations in land take per person. *Landsc. Urban Plan.* **2022**, *218*, 104308. [CrossRef]
95. Dull, M.; Wernstedt, K. Land recycling, community revitalization, and distributive politics: An analysis of EPA brownfields program support. *Policy Stud. J.* **2010**, *38*, 119–141. [CrossRef]
96. Tobias, S.; Conen, F.; Duss, A.; Wenzel, L.M.; Buser, C.; Alewell, C. Soil sealing and unsealing: State of the art and examples. *Land Degrad. Dev.* **2018**, *29*, 2015–2024. [CrossRef]
97. Voghera, A.; Giudice, B. Defining a social-ecological performance to prioritize compensatory actions for environmental regeneration. The experimentation of the environmental compensation plan. *Sustain. Cities Soc.* **2020**, *61*, 102357. [CrossRef]
98. Zoppi, C.; Lai, S. Assessment of the regional landscape plan of Sardinia (Italy): A participatory-action-research case study type. *Land Use Policy* **2010**, *27*, 690–705. [CrossRef]
99. Kovács, E.; Kelemen, K.; Kalóczkai, A.; Margóczi, K.; Pataki, G.; Gébert, J.; Málovics, G.; Balázs, B.; Roboz, A.; Krasznai Kovács, E.; et al. Understanding the links between ecosystem service trade-offs and conflicts in protected areas. *Ecosyst. Serv.* **2015**, *12*, 117–127. [CrossRef]
100. Floris, M.; Zoppi, C. Ecosystem services and spatial planning: A study on the relationship between carbon sequestration and land-taking processes. *Arch. Studi Urbani Reg.* **2020**, 127. [CrossRef]
101. Organisation for Economic Cooperation and Development (OECD). Applying Strategic Environmental Assessment. Good Practice Guidance for Development Co-Operation. DAC Guidelines and Reference Series. OECD Publications. 2006. Available online: https://read.oecd-ilibrary.org/development/applying-strategic-environmental-assessment_9789264026582-en#page1 (accessed on 17 January 2023).

102. Mateos, R.M.; López-Vinielles, J.; Poyiadji, E.; Tsagkas, D.; Sheehy, M.; Hadjicharalambous, K.; Liscák, P.; Podolski, L.; Laskowicz, I.; Iadanza, C.; et al. Integration of landslide hazard into urban planning across Europe. *Landsc. Urban Plan.* **2020**, *196*, 103740. [CrossRef]
103. Bucała-Hrabia, A.; Kijowska-Strugała, M.; Śleszyński, P.; Rączkowska, Z.; Izdebski, W.; Malinowski, Z. Evaluating the use of the landslide database in spatial planning in mountain communes (the Polish Carpathians). *Land Use Policy* **2022**, *112*, 105842. [CrossRef]
104. Promper, C.; Puissant, A.; Malet, J.-P.; Glade, T. Analysis of land cover changes in the past and the future as contribution to landslide risk scenario. *Appl. Geogr.* **2014**, *53*, 11–19. [CrossRef]

land

MDPI

Article

Spatiotemporal Change Analysis and Prediction of the Great Yellow River Region (GYRR) Land Cover and the Relationship Analysis with Mountain Hazards

Chunliu Gao [1,2,3], Deqiang Cheng [2,*], Javed Iqbal [4] and Shunyu Yao [5,6]

1 School of Cultural Industry & Tourism Management, Henan University, Kaifeng 475001, China
2 Key Research Institute of Yellow River Civilization and Sustainable Development & Collaborative Innovation Center on Yellow River Civilization Jointly Built by Henan Province and Ministry of Education, Henan University, Kaifeng 475001, China
3 Laboratory of the Yellow River Cultural Heritage, Henan University, Kaifeng 475001, China
4 Department of Earth Sciences, The University of Haripur, Haripur 22620, Pakistan
5 China Institute of Water Resources and Hydropower Research, Beijing 100038, China
6 Research Center on Flood and Drought Disaster Reduction of the Ministry of Water Resources, Beijing 100038, China
* Correspondence: chengdq90@henu.edu.cn; Tel.: +86-166-0280-6361

Abstract: The study of land use/land cover (LULC) changes plays an important guiding role in regional ecological protection and sustainable development policy formulation. Especially, the simulation study of the future scenarios may provide a hypothetical prospect which could help to determine the rationality of current and future development policies. In order to support the ecological protection and high-quality development strategy of the Yellow River Basin proposed by the Chinese government, the Great Yellow River Region (GYRR) is taken as the research area. The multi-period land cover data are used to carry out the analysis of land cover changes. The MOLUSCE (Modules for Land Use Change Simulations) plugin of QGIS software is used to carry out a land cover simulation and prediction study for 2030 on a large regional scale. Finally, the land cover status in the mountainous areas of the GYRR is analyzed thoroughly. The results show a decrease in agricultural land and increase in forest land during the past 25 years from 1995 to 2020, and that this trend would continue to 2030. The landscape pattern index analysis indicates that the land cover in the GYRR has become more and more abundant, and the degree of fragmentation has become higher and higher, while landscape patches were more evenly distributed in the GYRR until 2020. On the other hand, the landscape pattern would tend to achieve a certain degree of stability in 2030. The decrease in farmland and the increase in forest land illustrate the efforts made by the GYRR residents and governments in improving the ecological environment under the policy of returning farmland to forests and grasslands. On the other hand, although the residential areas in the mountainous areas are far away from the mountain hazard historical points because of consideration during construction with the help of the development of disaster prevention and mitigation over the years, there could be problem of rapid and haphazard urbanization. It is worth mentioning here that the harmonious and sustainable development of people and land in the GYRR mountainous areas still requires a large amount of effort.

Keywords: land cover; QGIS; MOLUSCE; Great Yellow River Region; mountain hazards

Citation: Gao, C.; Cheng, D.; Iqbal, J.; Yao, S. Spatiotemporal Change Analysis and Prediction of the Great Yellow River Region (GYRR) Land Cover and the Relationship Analysis with Mountain Hazards. *Land* **2023**, *12*, 340. https://doi.org/10.3390/land12020340

Academic Editors: Matej Vojtek, Andrea Petroselli and Raffaele Pelorosso

Received: 29 December 2022
Revised: 25 January 2023
Accepted: 25 January 2023
Published: 27 January 2023

1. Introduction

All lives on the earth depend on land, which is the material basis for human survival and development. Land use refers to the activities related to the focused development and utilization of land resources by human beings, such as industrial land, agricultural land, residential land, transportation land, etc. Land cover refers to the natural or man-made coverage of the land surface. The material coverage related to various land uses mentioned above includes crops, forests, grasslands, houses, and so on. Therefore, the land use is a process occurring on the earth's surface, while the land cover is the result of various surface processes. Whether at the regional scale, national scale, or even global scale, change in land use is constantly causing the accelerated change of land cover [1,2].

Land use/land cover (LULC) changes affect the natural basis of human survival and development. Climate, soil, vegetation, water resources, and biodiversity are deeply affected. They are closely related to global climate change, biodiversity reduction, ecological environment evolution, and the sustainability of human–environment interaction [3]. The research on land use and land cover changes could provide some reference for policy formulation, land planning, and many other aspects. Nowadays, LULC change research has become one of the core topics of global change research [4]. Many national government agencies, scientific research departments, and social groups are paying attention to land use and land cover change research, which involves a series of major issues such as the protection and management of the ecological environment [5,6], the effective development and rational protection of regional resources [7], the protection of arable land and food security [8], and the sustainable development of the social economy [9,10].

At present, there are many models that analyze and simulate land use and land cover change, such as the Markov chain model [11,12], cellular automata model [13], the future land use simulation (FLUS) model [14], cellular automata Markov (CA–Markov) model [15], SLEUTH [16,17], etc. Every model has its own specialty for addressing the composite issues of land use and land cover changes. Now, various LULC prediction models have also been applied to different regional scales. Han et al. [18] simulated future land use scenarios for Beijing from 2010 to 2020 by combining the Conversion of Land Use and its Effects at Small regional extent (CLUE-S) model with a Markov model. Arsanjani et al. [19] used a hybrid model consisting of the logistic regression model, Markov chain (MC), and cellular automata (CA) to improve the performance of the standard logistic regression model, and predicted the future land use for 2016 and 2026 in the metropolitan area of Tehran, Iran. Kafy et al. [20] used the Cellular Automata (CA) and the Artificial Neural Network (ANN) machine learning algorithms to simulate the LULC and seasonal land surface temperature (LST) scenarios of Chattogram, Bangladesh for 2029 and 2039. Puangkaew and Ongsomwang [21] simulated the LULC data of Phuket Island using the CLUE-S model. Based on the CA–Markov model, Chen et al. obtained a predicted land use map of a hilly area, Jiangle, China, for 2014. Li et al. [22] presented a Future Land-Use Simulation (FLUS) system to simulate global LUCC in relation to human–environment interactions from 2010 to 2100. In general, people may pay more attention to the simulation of land use and land cover on medium and small scales. However, with the deepening of cross regional economic and cultural exchanges, the simulation of land use and land cover on a large regional scale is receiving more and more attention [23]. The improvement of computer computing ability also provides conditions for the simulation of land use and land cover on a large regional scale.

In order to achieve long-term peace and stability in the Yellow River Basin, the Chinese government has set the ecological protection and high-quality development of the Yellow River Basin national strategies that are equally as important as the coordinated development of Beijing, Tianjin, and Hebei, the development of the Yangtze River economic belt, the construction of the Great Bay area of Guangdong, Hong Kong, and Macao, and the integrated development of the Yangtze River Delta [24]. In this study, we performed the analysis of land cover changes and modeled the future scenario of Land cover with the help of the Modules for Land Use Change Simulation (MOLUSCE) plugin within QGIS software [25]. As compared with other land cover simulation tools, MOLUSCE has the advantages of being open source, free of charge, and simple to operate. We used land cover data from 1995 to 2020 with a five-year interval, along with spatial variables, such as elevation, relief, slope, monthly average temperature, annual precipitation, river network density, Gross Domestic Product (GDP), population, road network density, and city density.

The logistic regression was used to construct transition potential modeling, and the Cellular Automata was used to do the future land cover simulation of 2030. On the other hand, we analyzed the land cover changes between different years, especially the land cover changes in the mountainous areas of the Great Yellow River Region (GYRR), and comprehensively discussed relationships between land cover and the mountain hazards in this region. This study confirms that the MOLUSCE plug-in could be effectively applied to the simulation of land cover on a large regional scale, and it is also an attempt to explore the relationship between land cover change and mountain hazards on a large regional scale.

2. Materials and Methods

2.1. Study Area

The Yellow River, located in the north-central part of China (Figure 1), is the second-longest river in China, with a total length of 5464 km [26]. It flows through the Qinghai Tibet Plateau, Inner Mongolia Plateau, Loess Plateau, and Huang-Huai-Hai Plain [27], and goes through nine provinces, including Qinghai, Sichuan, Gansu, Ningxia, Inner Mongolia, Shaanxi, Shanxi, Henan, and Shandong [28,29]. The terrain of the Yellow River Basin is high in the West and low in the East [30]. According to statistics, the total area of the Yellow River Basin is 795,000 km^2 [31,32]. The annual average temperature of the basin is about 7 °C and the annual average precipitation is about 440 mm [33]. Now, the Yellow River basin has become one of the most vulnerable areas of ecological environment in China due to its complex landforms and climate differences. Serious water pollution, land desertification, gradual reduction of runoff, intensified soil erosion, and vegetation degradation [34] have become the focus of sustainable development of the Yellow River Basin. On 18 September 2019, the "Ecological protection and high-quality development in the Yellow River River Basin" was upgraded to a major national strategy by the China's government on a forum in Zhengzhou, Henan, China [35,36].

It should be noted that the Yellow River is a special river which exists in the form of suspended river on the ground in the lower reaches. According to statistics, thousands of years before, and until, 1946, the Yellow River burst 1593 times, and 26 major river diversions occurred [37–39]. Among them, the northernmost diversion occupied the Hai River and flowed into the Bohai Sea; the southernmost diversion passed through the Huai River (Figure 1). Considering the particularity of the Yellow River, we believe that the relevant research on the Yellow River cannot be limited to the existing basin, because its lower reaches are bounded by artificial levees and do not show a natural state. Therefore, we selected the Yellow River Basin, the Huai River Basin, and the Hai River Basin, which all are greatly affected by the Yellow River, to form the GYRR (Figure 1), and used them as the research area in response to "ecological protection and high-quality development of the Yellow River Basin". For the GYRR, relevant scholars have put forward similar concepts, such as the "Great Yellow River theory" of Guo [40], which defines a similar research area to guide relevant researchers to explore the development, evolution, generation, watershed size, source, rheology, estuary, river length, disaster, and contribution of the Yellow River. Mostern [41], in his book "The Yellow River-A Natural and Unnatural history", also selected a similar study area to introduce many research aspects of the Yellow River, such as history, loess, levies, and levees.

The GYRR is bounded by the Yanshan and Yinshan Mountains in the north, Helan and Qilian Mountains in the west, Qinling and Dabie Mountains in the South, and Bohai and Yellow Sea in the East. The division of the surrounding mountains causes the GYRR to become an independent geographical unit. The Yellow River, which has changed its course for many times, has become the tie linking different parts of the geographical unit. This area has become the main and core area of the Yellow River civilization. In terms of administrative divisions, the GYRR occupies all of Shandong, Shanxi, and Ningxia, most places in Henan and Hebei, the east part of Qinghai, the middle and north parts of Shaanxi, the north part of Jiangsu and Anhui, the south part of Gansu, the northwest corner of Sichuan, and the middle part of Inner Mongolia. A total of 12 provinces are involved.

In terms of geomorphology, the western areas of the GYRR are the mountainous areas, while the eastern part is a large area of alluvial plains. The area percentage of plains and platforms is about 34.96%, and that of mountainous areas is 65.04% (Figure 2).

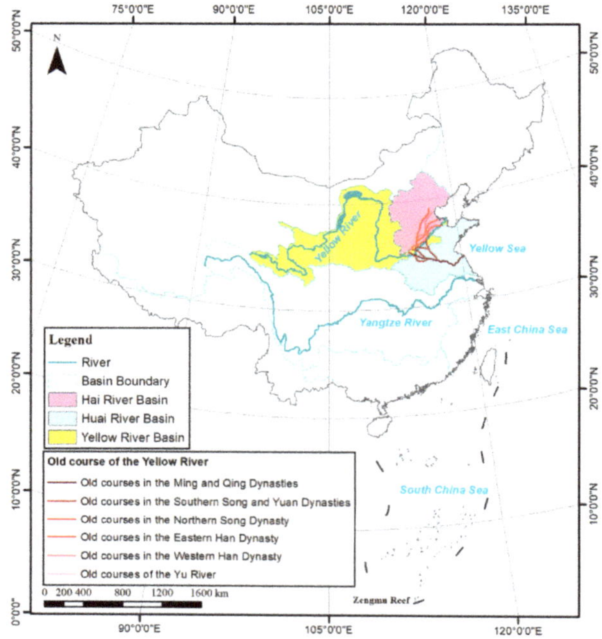

Figure 1. GYRR extent, including the Yellow River Basin, the Huai River Basin, and the Hai River Basin. A similar region concept has been recognized and mentioned by many scholars [40,41]. Historically, the Yellow River has burst and changed its course many times, affecting a wide area. At present, the lower reaches of the Yellow River are overland rivers, which are not natural rivers, but are significantly affected by human activities. Therefore, the study of the Yellow River should consider the history and river characteristics. It is more reasonable to take the area affected by the Yellow River as the study area of the Yellow River. In particular, we propose that historical archaeologists may take this area as the research area for Yellow River civilization archaeology.

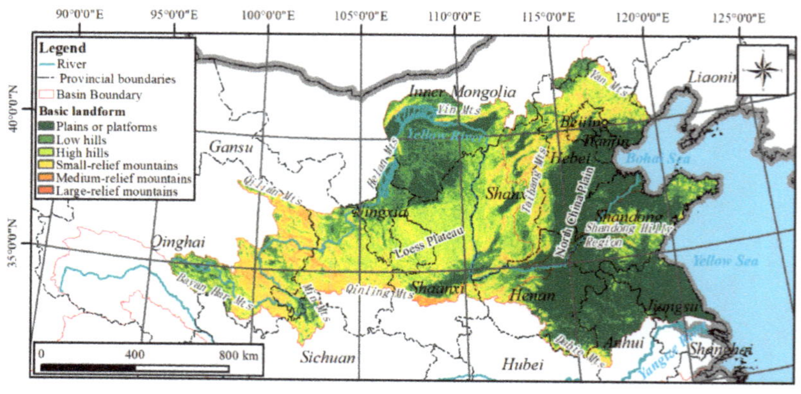

Figure 2. Landforms of the GYRR. The mountainous areas occupy about two-thirds of the GYRR. The cultural exchange in the GYRR is convenient, and forms the unique Yellow River civilization.

2.2. *Materials*

2.2.1. Land Cover Data

The land cover data used in this study (1995, 2000, 2005, 2010, 2015, and 2020) were downloaded from the land cover classification data set released by the European Space Agency (ESA) climate change initiative [42]. The spatial resolution is 300 m. Using the international Intergovernmental Panel on Climate Change (IPCC) land categories, the land cover types were divided into 10 categories: (i) agriculture, (ii) forest, (iii) grassland, (iv) wetland, (v) settlement, (vi) permanent snow and ice, (vii) shrubland, (viii) sparse vegetation, (ix) bare area, and (x) water (Figure 3). We resampled these land cover data and obtained multi-temporal 1000-m resolution land cover data.

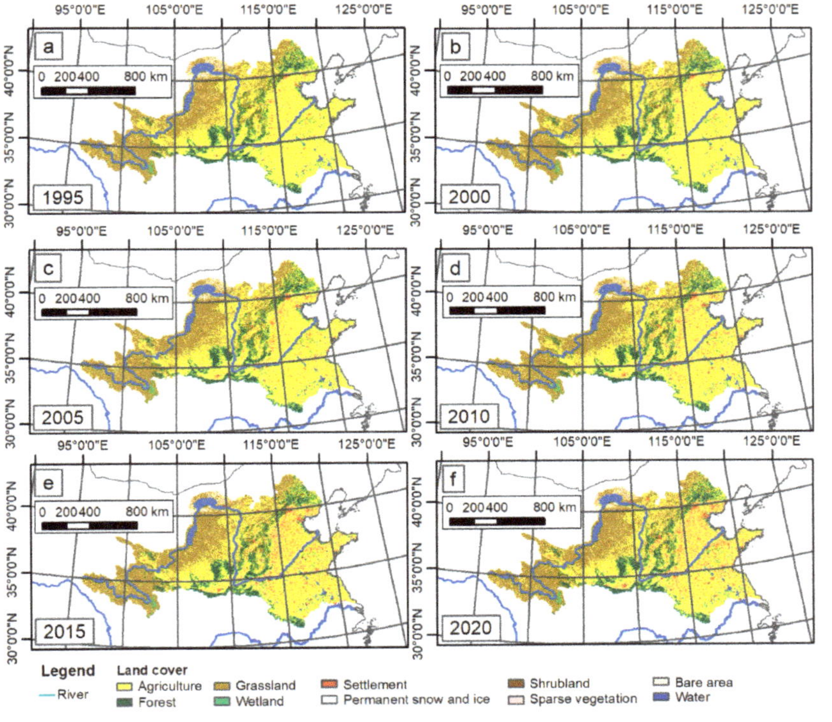

Figure 3. Land cover of the GYRR from 1995 to 2020. (**a**) Land cover in 1995; (**b**) land cover in 2000; (**c**) land cover in 2005; (**d**) land cover in 2010; (**e**) land cover in 2015; (**f**) land cover in 2020. Although the size of the pictures was limited, we could still find the subtle differences between them. For example, the settlement areas displayed in red shows an obvious increasing trend.

2.2.2. Spatial Variables Affecting the Land Cover Change

Physical and socioeconomic elements may cause alterations in land cover. For the selection of spatial variables affecting land cover change, we mainly referred to the relevant literature [14,43–49]. After comparison and analysis, we employed a variety of physical and socioeconomic elements (Table 1), including the elevation, topographic relief, slope, annual average temperature, annual average precipitation, river network density, GDP, population, road network density, and city density.

The elevation data (Figure 4a) were downloaded from the EarthEnv website (https://www.earthenv.org/topography). We found that the landform of the whole GYRR is high in the west region and low in the east region. There are many mountains in the west, and alluvial plains and hills in the east. The highest altitude of the whole area is 6018 m. The topographic relief was calculated from elevation, and the maximum relief in this region is 1112 m. The high value of relief is mainly distributed in the Taihang Mountains (Figure 4b). The slope data (Figure 4c) were also downloaded from the EarthEnv website (https://www.earthenv.org/topography). The high value distribution of the slope is similar to the relief high value distribution. The maximum value of the slope is 38.36°. The temperature and precipitation data were downloaded from the WorldClim website. WorldClim version 2.1 climate data for 1970–2000 was released in January 2020. They provide monthly climate data for minimum, mean, and maximum temperature, precipitation, solar radiation, wind speed, water vapor pressure, and total precipitation at the four spatial resolutions, between 30 s and 10 min. Each download is a "zip" file that contains 12 GeoTiff (.tif) files, one for each month of the year (January is 1; December is 12). We obtained the annual average temperature by averaging the 12-monthly mean temperature data. It was found that the maximum annual average temperature in this area is 16.18°C and the minimum is −13.68 °C (Figure 4d). Due to the influence of monsoons, the temperature in the East is higher, while the influence of ocean in the West is weak, and the temperature is lower. The precipitation data were also taken from the WorldClim website. We summed up the 12-monthly precipitation data to obtain the annual average precipitation. The precipitation in the GYRR decreases from Southeast to Northwest. The annual maximum precipitation can reach 1723 mm (Figure 4e). The river network density was calculated using the river network data (Figure 4f). In addition, data related to human activities mainly include GDP, population, road density, and city density. Due to the accumulation of human beings in the plain area, the four above-mentioned factors show the characteristics of high density in the plain area (Figure 4g–j).

Table 1. Data sources.

Data	Source	Access Date
Elevation	https://www.earthenv.org/topography [50]	20 May 2022
Relief	Calculated from Elevation	20 May 2022
Slope	https://www.earthenv.org/topography [50]	20 May 2022
Temperature	https://www.worldclim.org/data/index.html [51]	22 May 2022
Precipitation	https://www.worldclim.org/data/index.html [51]	22 May 2022
River	https://www.hydrosheds.org/products/hydrorivers [52]	28 May 2022
GDP	https://www.nies.go.jp/link/population-and-gdp.html [53]	6 June 2022
Population	https://landscan.ornl.gov/ [54]	10 June 2022
Road	https://sedac.ciesin.columbia.edu/data/set/groads-global-roads-open-access-v1/data-download [55]	11 June 2022
City	https://www.resdc.cn/data.aspx?DATAID=211 [56]	15 June 2022

2.2.3. Mountain Hazards in the GYRR

Mountain hazards generally refer to the hazards that can threaten human beings and their living environment in mountainous areas [57,58]. Tang et al. [59] discussed and defined "mountain hazards" in the 1980s, and considered that landslides, collapses, mudslides, soil erosion, ice avalanches, frozen soil hazards, earthquakes, hail, and other hazards in the mountainous areas could all be classified as mountain hazards. As compared with the above-mentioned broad categories, mountain hazards, in a narrow sense, could be understood as the phenomenon through which the water and soil materials move along the slope under the driving force of gravity and have a certain destructive capacity [60]. Debris flows, landslides, collapses, and mountain torrents are the representatives of common typical mountain hazards. In this study, we collected data on landslides, mountain torrents, and debris flows in the GYRR. For the collection of landslide and debris flow data, the global landslide catalog (GLC) from 2007 to 2017 produced by the National Aeronautics and Space Administration (NASA) of the United States was downloaded to collect rainfall-induced landslide and debris flow events. The data sources of the GLC include media, disaster databases, scientific reports, etc. [61]. On the other hand, the Dartmouth flood Observatory was established in 1993, mainly recording major global flood events from January 1985 [62].

For the mountain torrents, as a special flood occurring in the mountainous areas, the mountainous areas of the GYRR were used to screen the above flood event points and to obtain the mountain torrent points.

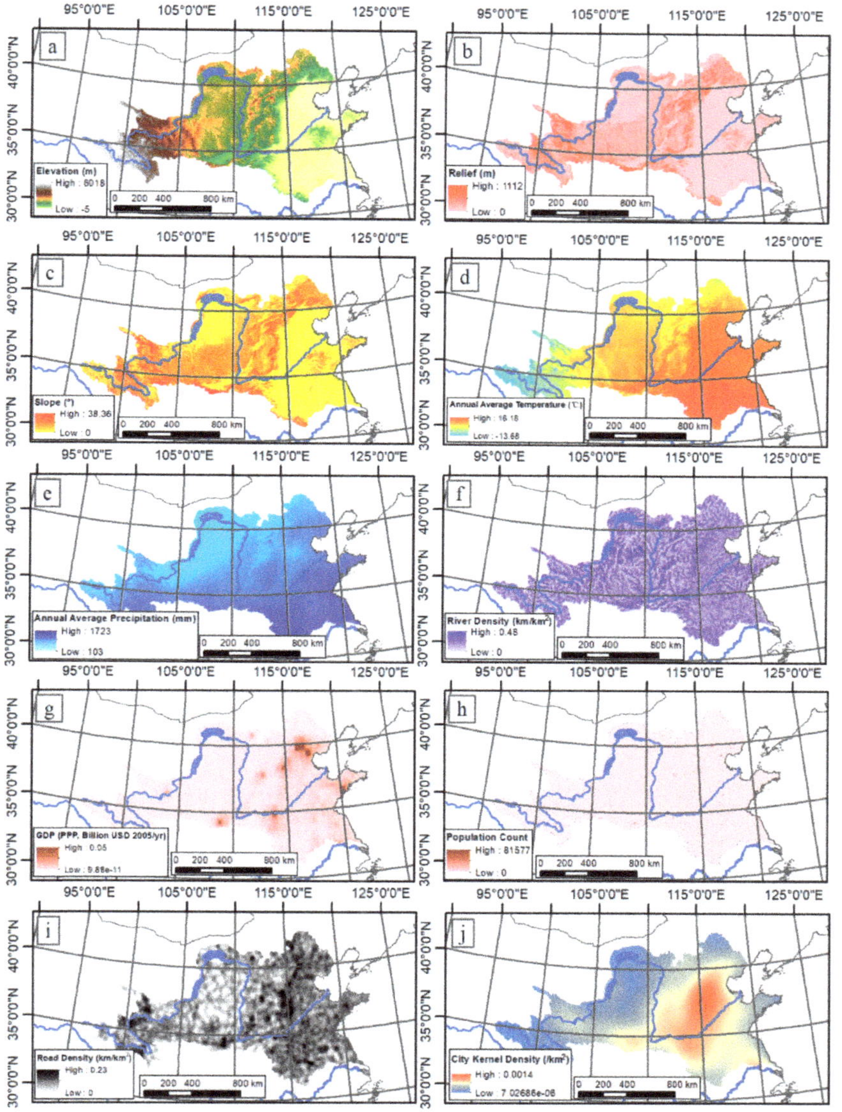

Figure 4. Impact factors considered in land cover prediction. (**a**) Elevation; (**b**) relief; (**c**) slope; (**d**) annual average temperature; (**e**) annual average precipitation; (**f**) river density; (**g**) GDP; (**h**) population; (**i**) road density; (**j**) city kernel density. We have considered as many natural and socio-economic factors as possible based on the availability of the data.

2.3. Methods

2.3.1. MOLUSCE Plugin

Asia Air Survey Co., Ltd. (AAS) released MOLUSCE (Modules for Land Use Change Evaluation) at FOSS4G 2013, which was a conference for people working with open-source tools. As a user-friendly plug-in for QGIS 2.0 and above, MOLUSE is designed to analyze, model, and simulate land use/cover changes. MOLUSCE is well suited to analyze land use and forest cover changes between different time periods, model land use/cover transition potential or areas at risk of deforestation, and simulate future land use and forest cover changes [63].

2.3.2. Correlation Analysis

Correlation analysis refers to the analysis of two or more variable elements with correlation, to measure the closeness of the correlation between the variable factors. The measurement of the closeness of the relationship between geographical elements is mainly realized through the calculation and interpretation of the correlation coefficient. Pearson's correlation and Cramer's coefficient are the main correlation analysis methods in the MOLUSCE plugin of QGIS. Among them, Pearson correlation analysis is a measurement method of vector similarity [64]. The output range is from −1 to + 1, where 0 represents no correlation, negative value represents negative correlation, and positive value represents positive correlation.

The correlation degree is usually judged by the following value ranges:

- 0.8–1.0: extreme correlation;
- 0.6–0.8: strong correlation;
- 0.4–0.6: moderate correlation;
- 0.2–0.4: weak correlation;
- 0.0–0.2: very weak correlation or no correlation.

2.3.3. Change Analysis and Transition Potential Modeling

We used the MOLUSCE plugin inside QGIS to compute the land cover change between the research intervals. For transition potential modeling, we used the logistic regression approach. The elevation, relief, slope, monthly average temperature, annual average precipitation, river density, GDP, population count, road density, and city kernel density were used as the explanatory factors.

2.3.4. Prediction and Model Validation

The MOLUSCE plugin can not only efficiently compute land cover change analyses, but is also well-suited for simulating future scenarios of land cover. We used the CA Simulation tool [65–67] of the MOLUSCE plugin inside QGIS to simulate the future land cover after we finished the transition potential modeling operation using the logistic regression approach. Next, we entered the reference map and simulated map for comparison and verification on the Validation page of the MOLUSCE plugin, and obtained the relevant Kappa coefficient values as a reference to check the accuracy of the simulation results.

2.3.5. Annual Rate of Change Analysis

The annual rate of change (ARC) could be used to represent the magnitude of change between corresponding years. In order to obtain the annual rate of change for each land cover type, the area difference between the final year and initial year was divided by the area of initial year and time (year) period. We used Equation (1) to assess the annual rate of change in land cover categories [25,68]:

$$\mathrm{ARC} = \frac{\mathrm{Area_{Final}} - \mathrm{Area_{Initial}}}{\mathrm{Area_{Initial}} * \mathrm{t}} \times 100\% \tag{1}$$

where ARC is the annual rate of change in land cover categories. $\mathrm{Area_{Final}}$ and $\mathrm{Area_{Initial}}$ are the areas of final and initial year, and t is the interval of years between the final year and initial year.

2.3.6. Landscape Pattern Index Analysis

The landscape pattern is the arrangement of landscape blocks of different sizes and shapes formed naturally or artificially in landscape space [69]. The landscape pattern indexes, as the sub-classification of landscape indexes, reflect the structural characteristics of land use/land cover types [70]. There are many types of landscape pattern indexes, and because of the application of new theories in landscape ecology, they are constantly being pushed forward [71,72]. Researchers often extend this part of the functions of the Geographic Information System (GIS) to form a unique landscape index software package based on GIS, such as Fragstats software package.

In order to study the spatial structure characteristics of different land cover types in the GYRR, this study first introduced the landscape diversity index to characterize them. Shannon's Diversity Index (SHDI) is a measurement index which is widely used in ecology based on information theory, and it is equal to the negative sum of the area ratio of each patch type multiplied by the natural logarithm of its value at the landscape level:

$$\mathrm{SHDI} = -\sum_{i=1}^{s} P_i ln P_i \quad (2)$$

where s is the amount of patches, and P_i the area ratio of each patch type. When SHDI = 0, it indicates that the whole landscape is composed of only one patch, and an increase in SHDI indicates that the patch types increase or distribute equally in the landscape space. In a landscape system, the richer the land use/land cover is, the higher the degree of fragmentation is, and more uncertain information content leads to a higher calculated SDHI value. The diversity depends on two factors: the number of types and the evenness of area combination; therefore, the diversity index is the comprehensive embodiment of type richness and combination complexity [73].

Shannon's Evenness Index (SHEI) equals the SHDI divided by the maximum possible diversity under a given landscape abundance (all patch types are equally distributed). The smaller the SHEI value is, the more likely it is that some patch types may dominate the landscape, and a value that is close to 1 indicates that there is no obvious dominant type in the landscape while patch types are evenly distributed. Therefore, when SHEI = 0, it indicates that the landscape is composed of only one type of patch without diversity, and SHEI = 1 indicates that the patches are evenly distributed and have the greatest diversity.

$$\mathrm{SHEI} = \frac{SHDI}{SHDI_{max}} \quad (3)$$

where SHDI is Shannon's Diversity Index, and SHDI_{max} is the maximum possible diversity under a given landscape abundance (all patch types are equally distributed) [74].

2.4. Technology Roadmap

In the process of this research, our work includes the following steps (Figure 5):

(1) We downloaded the land cover data for six years, including 1995, 2000, 2005, 2010, 2015, and 2020, and then, by comparing and analyzing the data of the first year (1995) and the last year (2020), we gained general insight into the land cover change in the GYRR in the past 25 years.

(2) The data of 10 geographical elements from different data sources, including elevation, relief, slope, annual average temperature, annual average precipitation, river network density, GDP, population, road density, and city density were collected.

(3) The MOLUSCE plugin was found in the plugin installation window and installed.

(4) We opened the MOLUSCE tool in the Raster menu drop-down list. Initial (2000) and final (2010) land cover data were used as input. Geographic impact factors, such as spatial variable, were used as input in the "inputs" tab of the MOLUSCE tool. Then, the subsequent operations were carried out step by step. The data generated in the previous

step were the basis for the next operation. In particular, we carried out the prediction of land cover in 2030 after verifying the effectiveness of the prediction model.

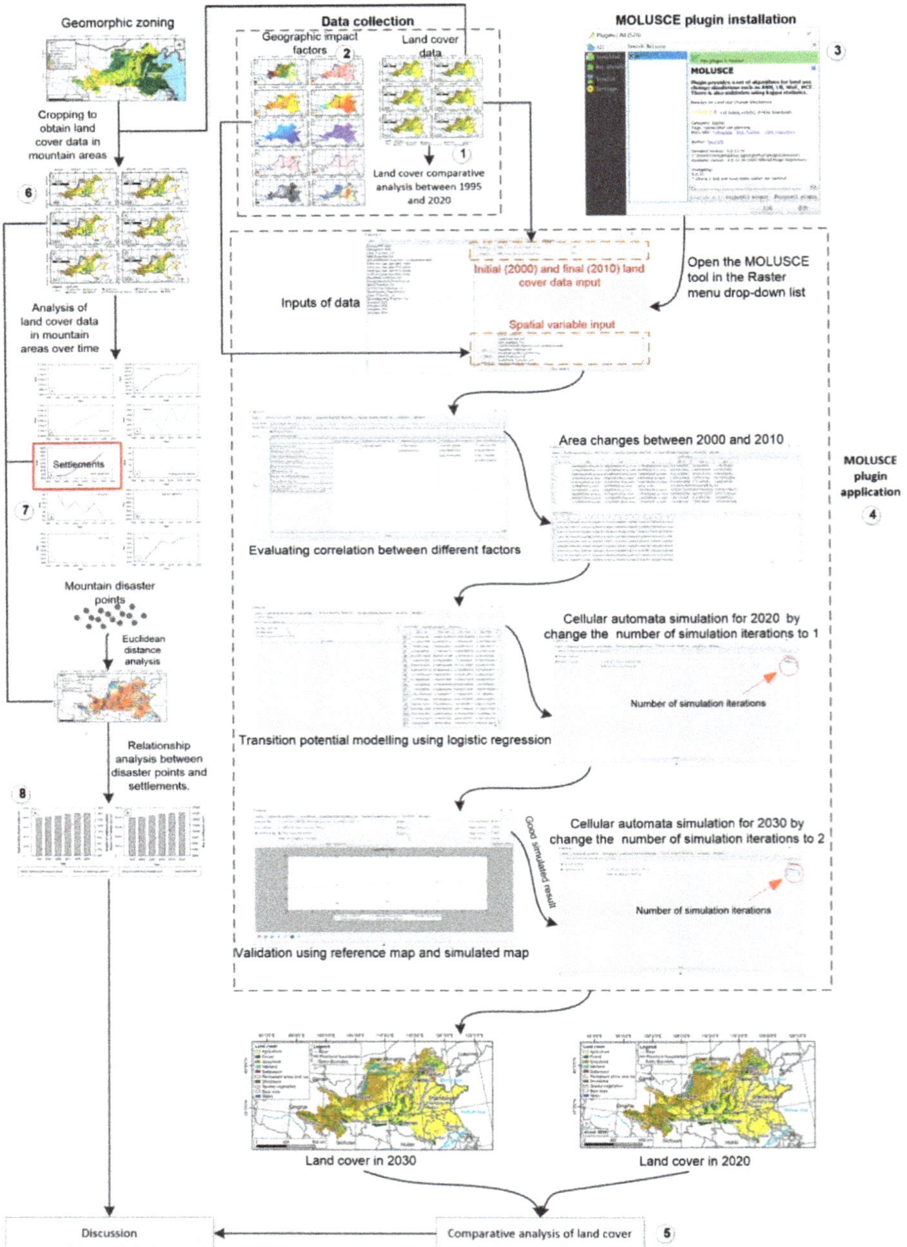

Figure 5. Technology roadmap. In this technical route, we not only paid attention to the land cover change in the whole GYRR, but also paid attention to the land cover change in mountainous areas which account for two-thirds of the whole GYRR, especially the change in residential areas in the mountainous areas.

(5) By comparing the land cover data in 2020 and 2030, we analyzed the land cover change over the next 10 years.

(6) On the other hand, we used the mountainous areas of the GYRR to cut out the land cover data of six periods from 1995 to 2020.

(7) The area changes in different land cover types in the mountainous areas of the GYRR were analyzed over time.

(8) Mountain hazard data points were downloaded and sorted to conduct the Euclidean distance analysis. The relationship analyses include distance from hazard points, number of settlement patches, distance from hazard points, and area of settlements.

3. Results

3.1. Correlation between Geographical Variables

We calculated the Pearson correlation coefficient, as shown in Table 2. After comparison, it was found that the variables having strong correlation with each other include temperature and elevation, city density and temperature, city density and elevation, and relief and slope.

Table 2. Pearson correlation coefficient between different variables.

	Temperature	Road Density	Elevation	GDP	City Density	Slope	Population	Relief	River Density	Precipitation
Temperature		0.26	−0.95	0.35	0.68	−0.48	0.19	−0.48	0.09	0.48
Road Density			−0.27	0.30	0.32	−0.20	0.15	−0.15	0.09	0.19
Elevation				−0.37	−0.64	0.49	−0.17	0.49	−0.07	−0.34
GDP					0.37	−0.24	0.19	−0.22	0.06	0.17
City Density						−0.19	0.17	−0.19	0	0.35
Slope							−0.15	0.87	−0.21	−0.05
Population								−0.14	0.04	0.12
Relief									−0.19	−0.07
River Density										−0.12
Precipitation										

3.2. Area Changes and Landscape Pattern Features

The statistical analysis was done on various land cover areas between 1995 and 2020. The area change and the ARC of the same land cover were calculated (Table 3). It was noted that the land cover with the largest change was agricultural land, with a decrease of −16,437 km^2. The increase in settlement area is the largest one, with an area of +27,364 km^2 and an ARC of 223.69%. The increase in settlement shows the enhancement of human activities in the past 25 years.

The area transfer analysis was also performed between different land cover types. According to the area transfer matrix (Table A1), between 1995 and 2020, large change situations include: 11,171 km^2 agricultural land was transformed into forest land, 53366 km^2 agricultural land was transformed into grassland, and 20,047 km^2 agricultural land was transformed into settlement land. In terms of forest land, 10,233 km^2 forest land was transformed into agricultural land, and 101,506 km^2 forest land was transformed into grassland. On the other hand, 52,088 km^2 grassland was transformed into agricultural land, 12,597 km^2 grassland was transformed into forest land, and 7645 km^2 grassland was transformed into settlement land. The Chord diagram (Figure 6) was used to express the land cover change. It was found that agriculture, grassland, and forest are the main land cover types, and account for most of the land studied.

We analyzed the landscape pattern indexes SHDI and SHEI in the GYRR, and the values of the two indexes increased gradually with time (Figure 7). The continuous increase in the SHDI value indicate that the land cover in the GYRR had become more and more abundant, and the higher the degree of fragmentation was, the greater the uncertain information content became. SHEI was getting bigger and bigger, approaching 1, which indicated that there was no obvious dominant type in the GYRR, and landscape patches were more and more evenly distributed in the GYRR.

Table 3. Land cover change from 1995 to 2020.

Land cover type	Area in 1995 (km^2)	Area in 2020 (km^2)	Area change (km^2)	ARC
Agriculture	788,382	771,945	−16,437	−0.08%
Forest	116,638	126,528	+9890	+8.48%
Grassland	476,334	474,525	−1809	−0.38%
Wetland	7005	5410	−1595	−22.77%
Settlement	12,233	39,597	+27,364	+223.69%
Permanent snow and ice	178	169	−9	−5.06%
Shrubland	1257	646	−611	−48.61%
Sparse vegetation	8662	6681	−1981	−22.87%
Bare area	25,027	15,683	−9344	−37.34%
Water	16,126	15,024	−1102	-6.83%

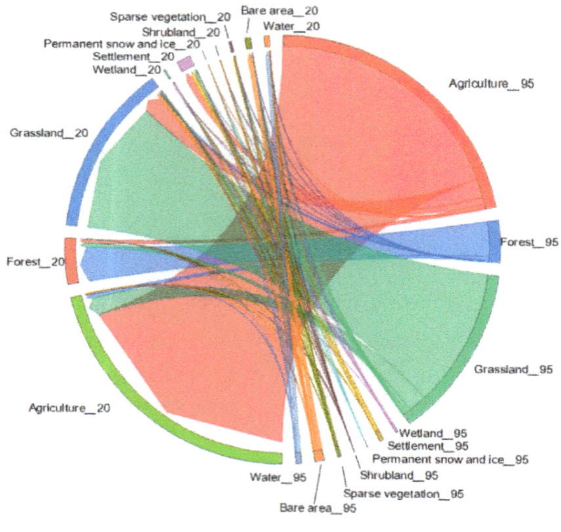

Figure 6. Land cover change in GYRR from 1995 to 2020 using a Chord diagram expression. The right semicircle shows the proportions of different land covers in 1995, and the left semicircle shows the proportions of different land covers in 2020. The arrows in the circle indicate the land cover change.

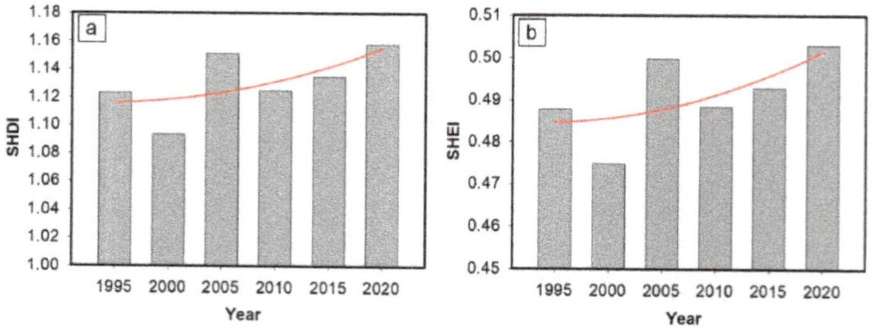

Figure 7. Landscape pattern index analysis. (**a**) SHDI of GYRR; (**b**) SHEI of GYRR. The red line in the figure is a trend line added by the authors.

3.3. Land Cover Prediction in 2020 and Validation

We used the MOLUSCE plugin for the simulation of land cover in 2020. Using the projected 2020 data (Figure 8a) for comparison with the actual land cover data in 2020 (Figure 8b), the percentage of correctness was calculated as 96.42%, the Kappa (overall) was 0.94, the Kappa (histo) was 0.98, and the Kappa (loc) was 0.95. The results show that the 10-year interval prediction model has a good result on land cover simulation and prediction.

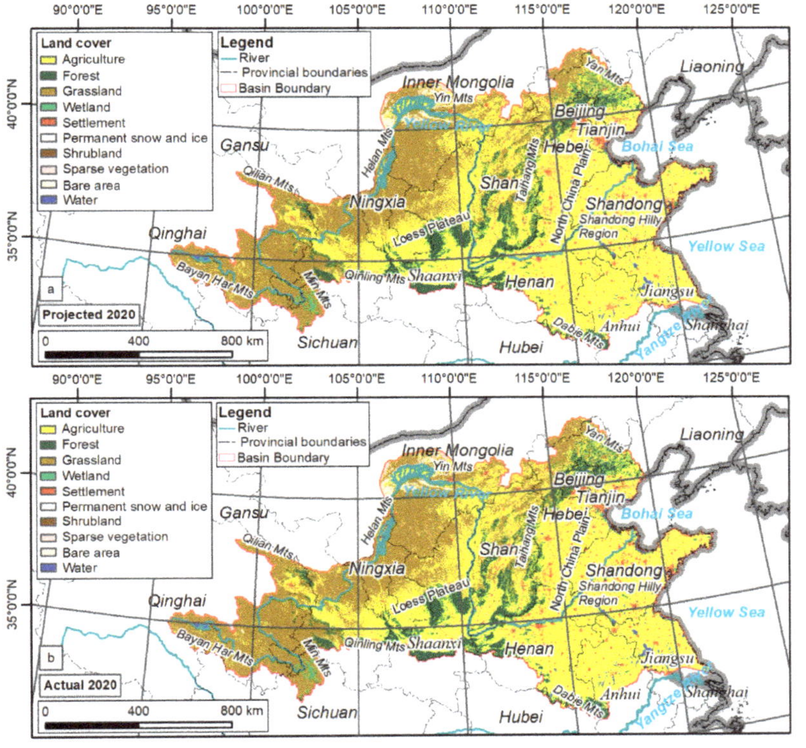

Figure 8. Actual and projected land cover in 2020. (**a**) Projected land cover in 2020; (**b**) actual land cover in 2020. We know that the more similar the two above maps are, the better the simulation results will be. However, there are still some subtle differences between the two maps. For example, the expansion trend of the simulated settlements was still conservative compared with that of the real settlements, and the real settlements expanded more rapidly, for example, in cities in Henan Province.

3.4. Land Cover Prediction in 2030

The above experimental results show that the 10-year interval land cover prediction model has good results. At the windows "Cellular Automata Simulation", the results show the option "Number of Simulation iterations". This means that, if only 1 is entered, it will be projected into the future only once. For example, if the land cover data are for 2000 and 2010, the land cover of 2020 will be projected when entering 1, and 2030 will be projected in the case of changing the "Number of Simulation iterations" to 2. In this study, the actual land cover in 2020 was used as the input of the model to simulate and predict the land cover in 2030 (Figure 9).

According to the statistical results of various land cover types, agricultural land, wetland, permanent snow and ice, shrubland, and sparse vegetation would be further reduced. The area of forest, grassland, settlement, bare area, and water would increase (Table 4).

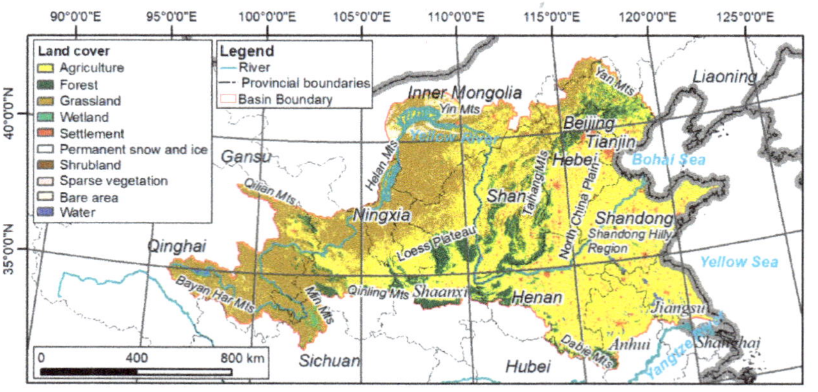

Figure 9. Land cover prediction of the GYRR in 2030.

Table 4. Land cover change from 2020 to expected 2030.

Land Cover Type	Area in 2020 (km^2)	Area in 2030 (km^2)	Area Change (km^2)	ARC
Agriculture	771,945	767,522	−4423	−0.57%
Forest	126,528	129,153	2625	2.07%
Grassland	474,525	483,604	9079	1.91%
Wetland	5410	4533	−877	−16.21%
Settlement	39,597	36,983	−2614	6.60%
Permanent snow and ice	169	83	−86	−50.89
Shrubland	646	533	−113	−17.49%
Sparse vegetation	6681	4282	−2399	−35.91
Bare area	15,683	16,287	604	3.85%
Water	15,024	13,228	−1796	11.95%

In particular, it can be noted that a large amount of agricultural land would still turn into grassland and settlement. Some settlement land would be converted into agricultural land and grassland (Table A2). On the other hand, we analyzed the change in landscape pattern index (SHDI and SHEI) and found that the two indexes did not change much (Figure 10). This result shows that the land cover change in the GYRR may enter a stable development stage when it reaches a certain degree in the future.

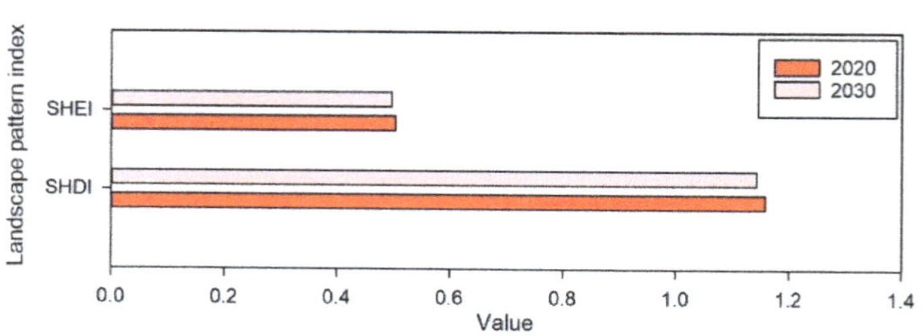

Figure 10. Landscape pattern index (SHDI and SHEI) contrast analysis between 2020 and 2030.

3.5. Land Cover Change in Mountainous Areas

For the whole GYRR, the area percentage of plains and platforms is about 34.96%, and that of mountainous areas is 65.04% (Figure 2). The unique energy gradient causes the mountains to become an area of natural hazards development, such as debris flows, landslides, collapses, avalanches, soil erosion, and mountain torrents. These mountain hazards may destroy urban and rural settlements, damage roads, bridges, and engineering facilities, bury farmlands and forests, and block rivers and reservoirs. They may cause huge casualties, property losses, and ecological damage, seriously threaten the lives and property of the people in the mountainous areas and the safety of engineering construction, and restrict the development of resources and economy in the mountainous areas [60]. We cropped out the land cover of the mountainous areas of the GYRR in different years (Figure 11).

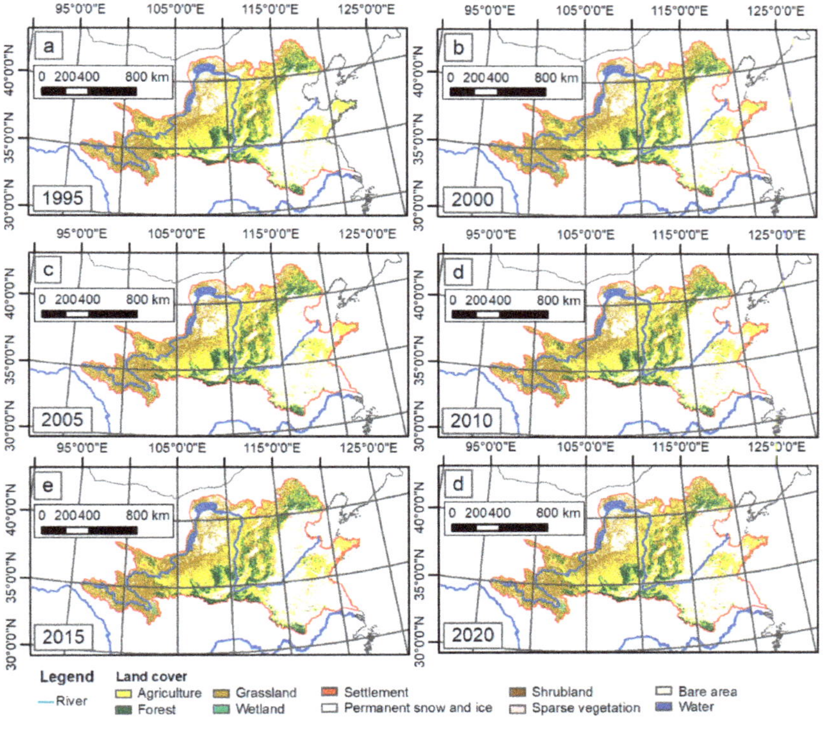

Figure 11. Land cover in mountainous areas of the GYRR in different years.

Blind expansion of cities in mountainous areas can easily cause mountain hazards. It would cause huge losses to the life, property, and safety of urban residents. For example, on 14 August 2017, a devastating geo-hazard chain—debris slide, debris flow, and sediment-laden flood—occurred in Freetown, Sierra Leone, resulting in at least 500 deaths, more than 600 missing, and hundreds of houses destroyed. Although rainfall was a trigger factor for the Sierra Leone disaster, rapid and haphazard urbanization increased the hazard and vulnerability [75]. The development of mountain towns is generally affected by many factors such as social economy, topography, and geomorphology. Compared with plain towns, their infrastructure is relatively weak. In particular, poor urban planning and inadequate consideration of risks could lead to the construction of housing in dangerous areas. On the other hand, the removal of hillside vegetation increases erosion potential; low cost buildings using fragile building materials and methods could lack resilience; inadequate risk management leads to weak emergency response.

We have also made area statistics for 10 land cover types (Figure 12). It can be seen that the settlement area in the mountainous areas had been increasing continuously in the past 25 years (Figure 12e), with an ARC value of +14.97%. Thanks to the policy of returning farmland to forests and grasslands implemented by the Chinese government in mountainous areas, the area of ecological land such as forest land and grassland had been continuously increased and, at the same time, the ecological environment had been improved as a gratifying result (Figure 12b,c).

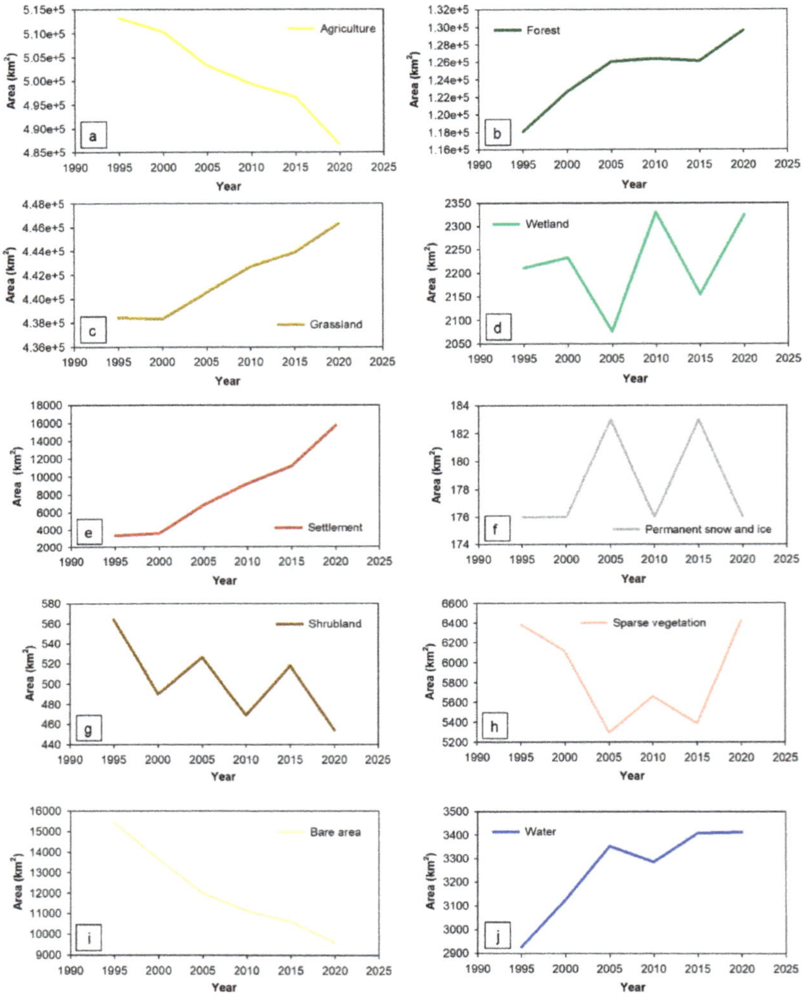

Figure 12. Land cover in mountainous areas of the GYRR in different years. We tried to use colors similar to the colors of different land cover types.

We superimposed landslide, debris flow, and mountain torrent points on the base map of the GYRR (Figure 13). It was found that the mountain-hazard points are mainly distributed in the Central and Western regions of the GYRR. Next, the Euclidean distance is calculated using these mountain-hazard point data in order to represent the distance from the hazard point (Figure 13).

We made the statistics on the number of settlement patches, area of settlements in the mountainous areas of the GYRR, and the average Euclidean distance from the hazard points during the period from 1995 to 2020 with a time interval of five years (Figure 14). According to the statistical results, when the number of settlement patches in mountainous areas continued to increase, the distance between settlements in the GYRR and hazard sites was also increased (Figure 14a). On the other hand, the mountainous residential areas in the GYRR also increased; however, the distance from the hazard point was also increasing (Figure 14b). The above two situations show that, although the intensity of human development in the mountains of the GYRR had been increasing, the awareness of avoiding hazards was also improved.

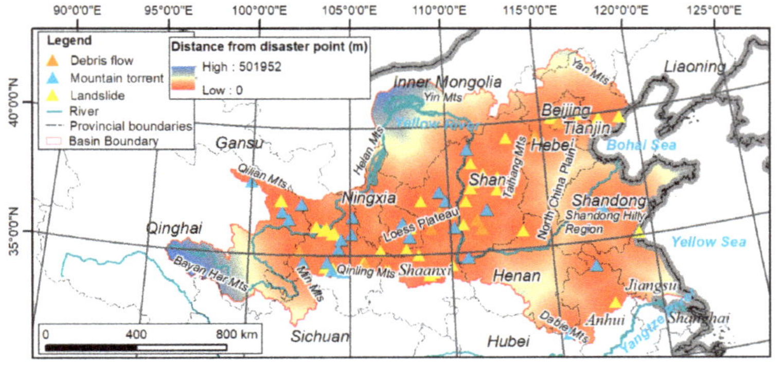

Figure 13. Mountain-hazard distribution and calculation of Euclidean distance around them.

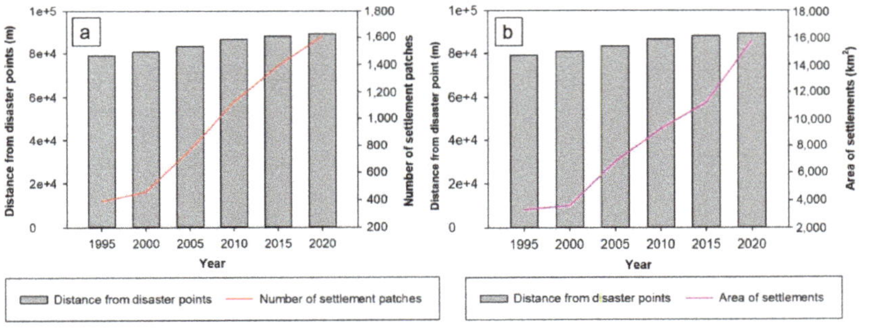

Figure 14. Relationship between hazard points and settlements. (**a**) Distance from hazard points and number of settlement patches; (**b**) distance from hazard points and area of settlements.

4. Discussion

4.1. Decrease of Farmland and the Increase of Woodland under Returning Farmland to Forest and Grassland

In 1998, Wuqi County of Yan'an, Shaanxi, China began to forbid grazing on the mountains [76]. After that, this small county, located in the northwest of the GYRR, began to take the lead in implementing the policy of returning farmland to forests [77]. Since 1999, Yan'an has reached a forest coverage rate of more than 50% and a vegetation coverage rate of more than 80% by returning farmland to forest over more than 20 years [78]. This is only a microcosm of China's project of returning farmland to forest and grassland. From 1999 to 2013, a total of 298,000 km^2 farmland in China was returned to forest. The project covers 2279 counties, with 32 million farmers and 124 million farmers directly benefiting. The central government of China has invested 64.7 billion dollars in the project of returning farmland to forest [79]. From 2014 to 2018, the new round of returning farmland to forest and grassland involved 25 provinces (regions) including Hebei, Shanxi, Inner Mongolia, and others [80]. The Loess Plateau of the GYRR was one of the earliest regions to implement the project of returning farmland to forest and grassland which has made great contributions to the improvement of forest coverage in China [81]. Returning farmland to forest and grassland provides a reasonable explanation for the reduction in farmland and the increase of forest land in the GYRR.

4.2. Urbanization in Mountainous Areas of China

Mountainous areas account for 24% of the earth's land area, and more than 12% of the world's population lives in mountainous areas [82,83]. China has a large number of mountains. The mountainous areas accounts for about 70% of the land area, and the population accounts for about 45% of the total population of the country. Due to the geographical and economic marginality of mountainous areas, the overall level of urbanization development in mountainous areas is far lower than the average level of China. The low level of urbanization and the slow urbanization process in mountainous areas, and a large number of agricultural population gathered in mountainous areas, will certainly bring great pressure to the ecological environment in mountainous areas. Among China's 1429 county-level administrative units in mountainous areas, 54% of the counties have an urbanization rate of less than 20%, and only 10% have an urbanization rate of more than 40% [84]. The development of natural resources, especially mineral resources, has played a significant role in promoting the urbanization of mountainous areas, and a number of resource-based cities have emerged. Tourism is a potential tool to promote the diversified development of mountain economy, increase the employment of mountainous residents, alleviate the poverty in mountainous areas, promote the participation of mountainous areas in economic globalization activities, and correct the development gap in mountainous areas. In recent years, tourism has become a new driving force to promote the urbanization of mountainous areas, such as Emeishan City, Wuyishan City, Tai'an City, and Jiuzhaigou County, and other counties and cities have developed rapidly through tourism. At the same time, due to the lack of management and the lag in planning, some mountainous tourism cities have also experienced excessive urbanization [85]. Compared with plain towns, the urban planning in mountainous areas of China seriously lags behind the urban construction. At present, low-level spread of built-up areas and inefficient use of land are common in urban construction in mountainous areas. The level of urban functional layout is not clear. Especially with the increase in population and the shortage of construction land, the important functional layout of mountain towns basically ignores the avoidance of mountain hazards, and the ability of disaster prevention and mitigation is weak. For example, the area which was most seriously affected by the huge debris flow in Zhouqu, Gansu Province happened to be the most densely-populated and prosperous area [86,87]. The development of mountain towns in the GYRR also faces the above problems, accompanied by the increase in the area of residential areas and the number of residential patches.

4.3. Harmonious and Sustainable Development of People and Land in Mountainous Areas

The development of urbanization in mountainous areas should also be compatible with the resources and be coordinated with the land space and environmental capacity [88]. People should adhere to the fundamental support of ecological industry and form an intensive and ecological development model in order to improve the quality of urbanization. It is necessary to change the traditional direction and mode of urbanization development, gradually lead the urbanization construction to the road of new urbanization, implement the green development strategy, intensively utilize resources, improve resource efficiency, promote the intensive utilization of water, soil, and energy resources, and accelerate the construction of resource-saving cities and towns [89]. The development of urbanization in mountainous areas cannot ignore the restrictions of and close relationship with the mountainous environmental factors. We must grasp the basic characteristics and laws of the mountainous environment from different scales and regional differences, and deeply analyze the typical examples of the pattern, resources, and environmental characteristics, as well as the process of urbanization in mountainous areas. At present, the GYRR is carrying out the urbanization of the mountainous areas with an ARC value of +14.97%. There are many problems that need to be seriously considered to minimize the ecological problems caused by the rapid urbanization of the mountainous areas and realize the harmonious and sustainable development of the relationship between people and land in the mountainous areas.

4.4. Keeping Away from Hazards Benefitting from Disaster Prevention and Mitigation in Mountainous Areas of China

China's mountains account for more than two-thirds of the total land area. With the rapid growth of population and inappropriate production activities in mountainous areas, mountain hazards occur frequently, and people have been disturbed and destroyed by mountain hazards in the process of utilization in mountainous areas [90]. Since the 1960s, China's relevant departments have begun to carry out the investigation and control of mountain hazards. For example, the color scientific and educational film "debris flow," released in 1965, brought the debris flow phenomenon onto the screen, which played a very strong role in publicizing and popularizing debris flow knowledge [91]. At present, many departments and colleges in China have trained many professional scientific and technological workers in theoretical research and disaster mitigation-prevention practice regarding mountain-hazards, and laid down a solid foundation in theoretical and technical reserves, becoming a very active scientific and technological force in China [92,93]. On the other hand, China has integrated the study of debris flows, landslides, floods, and other hazards, combined the construction of the large environment with the management of small watersheds, carried out comprehensive research on the process of various hazards, implemented comprehensive disaster mitigation, scientifically assessed the current situation and trend of hazards, and put forward quantitative indicators [94]. With the increasing awareness of disaster prevention and mitigation in mountainous areas, although the proportion of residential areas and the number of residential patches in mountainous areas continue to increase, the safety of residential areas in mountainous areas has been continuously improved due to the conscious distance from mountain hazard points of urban construction [95].

5. Conclusions

In this study, the GYRR was selected as the research area. The land cover change analysis, as well as simulation and prediction of future land cover, was performed, focusing especially on the analysis of the relationship between land cover in mountainous areas and mountain hazards. This work verifies that the MOLUSCE plug-in could be effectively applied to land cover simulation on a large regional scale. Based on the analysis in the current study, the following conclusions are drawn:

(1) Based on multi-period land cover data and physical and socioeconomic factors, the logistic regression and CA model within the MOLUSCE plugin in QGIS software was used to perform the future simulation of land cover in the GYRR. This could provide a reference for related research, especially for large regional-scale land cover simulation.

(2) The decrease in farmland and the increase in forest land illustrate the efforts made by the government and residents of the GYRR in improving the ecological environment during the past 25 years.

(3) According to the simulation and prediction results for land cover in 2030, the agricultural land will decrease, and the forest land will increase. At the same time, the increase in land cover in residential areas could not be ignored, which indicates the continuous development of urbanization in the GYRR. On the other hand, landscape pattern index analysis shows that the land cover in the GYRR may enter a roughly stable development stage when it reaches a certain degree in 2030.

(4) Returning farmland to forest and grassland in the GYRR is conducive to ecological improvement. On the other hand, although the residential areas in mountainous areas were built as far away as possible from the mountain hazard points during construction, there could be a problem of rapid and haphazard urbanization, which should also be paid attention to.

Author Contributions: C.G., J.I. and D.C. designed the method, conceived the experiments. C.G. and D.C. analyzed the data; C.G., D.C., J.I. and S.Y. wrote the paper. All authors have read and agreed to the published version of the manuscript.

Funding: This research was supported by the Research Topics of Henan Social Science Federation (SKL-2022-2717), the National Natural Science Foundation of China: (grant No. 42171186), the Major Project of China National Social Science Fund in Art (grant No. 21ZD03), and the Research Start-up Fund of Henan University (No. CX3050A0250560, Higher Education Commission of Pakistan, NRPU project No.15732).

Data Availability Statement: Publicly available datasets were used in this study. We have added the relevant data URL in the article.

Acknowledgments: The authors would like to thank all colleagues who gave us help during this study. We hope that the relevant research on the Yellow River basin could consider the same research area of GYRR, especially the research related to the archaeology and cultural heritage of the Yellow River.

Conflicts of Interest: The authors declare no conflict of interest.

Appendix A

Table A1. Area transfer matrix between land cover in 1995 and land cover in 2020.

Land Cover Type		2020 (km²)										
		Agriculture	Forest	Grassland	Wetland	Settlement	Permanent Snow and Ice	Shrubland	Sparse Vegetation	Bare Area	Water	SUM
1995 (km²)	Agriculture	701,727	11,171	53,366	287	20,047	2	36	174	118	1454	788,382
	Forest	10,233	101,506	4562	8	123	0	12	1	5	188	116,638
	Grassland	52,088	12,597	399,627	478	7645	10	61	943	1744	1141	476,334
	Wetland	1224	0	1083	4015	206	0	0	0	6	471	7005
	Settlement	1675	7	62	3	10,473	0	0	0	0	13	12,233
	Permanent snow and ice	0	0	32	0	0	145	0	0	1	0	178
	Shrubland	140	456	94	0	9	0	420	106	26	6	1257
	Sparse vegetation	683	2	3358	2	64	0	77	4215	223	38	8662
	Bare area	481	2	9681	9	246	5	40	1101	13,437	25	25,027
	Water	2710	81	945	546	592	0	0	67	16	11,169	16,126
	SUM	771,945	126,528	474,525	5410	39,597	169	646	6681	15,683	15,024	1,456,208

Table A2. Area transfer matrix between land cover in 2020 and land cover in 2030.

Land Cover Type		2030 (km²)										
		Agriculture	Forest	Grassland	Wetland	Settlement	Permanent Snow and Ice	Shrubland	Sparse Vegetation	Bare Area	Water	SUM
2020 (km²)	Agriculture	747,492	1457	15,042	4	7718	0	5	60	70	97	771,945
	Forest	1911	123,088	1421	0	97	0	3	0	0	8	126,528
	Grassland	8075	4273	459,873	9	966	0	1	162	1057	109	474,525
	Wetland	76	17	193	4342	622	0	0	2	0	158	5410
	Settlement	9490	61	3273	132	26,431	0	6	29	50	125	39,597
	Permanent snow and ice	86	0	0	0	0	83	0	0	0	0	169
	Shrubland	3	52	71	0	0	0	510	3	7	0	646
	Sparse vegetation	64	45	1789	0	3	0	8	4014	707	51	6681
	Bare area	36	9	1240	0	4	0	0	5	14,388	1	15,683
	Water	289	151	702	46	1142	0	0	7	8	12,679	15,024
	SUM	767,522	129,153	483,604	4533	36,983	83	533	4282	16,287	13,228	1,456,208

References

1. Houghton, R.A. The worldwide extent of land-use change. *BioScience* **1994**, *44*, 305–313. [CrossRef]
2. Tumer, B.; Meyer, W.; Skole, D. Global land use/land cover change: Toward an integrated program of study. *Ambio* **1994**, *23*, 91–95.
3. Anderson, J.R. Land use and land cover changes. A framework for monitoring. *J. Res. By Geol. Surv.* **1977**, *5*, 143–153.
4. Meyer, W.B.; Turner, B. Land-use/land-cover change: Challenges for geographers. *GeoJournal* **1996**, *39*, 237–240. [CrossRef]
5. Wang, H.; Liu, X.; Zhao, C.; Chang, Y.; Liu, Y.; Zang, F. Spatial-temporal pattern analysis of landscape ecological risk assessment based on land use/land cover change in Baishuijiang National Nature Reserve in Gansu Province, China. *Ecol. Indic.* **2021**, *124*, 107454. [CrossRef]
6. Muchová, Z.; Tárníková, M. Land cover change and its influence on the assessment of the ecological stability. *Appl. Ecol. Environ. Res* **2018**, *16*, 2169–2182. [CrossRef]
7. Liu, Y.; Wang, D.; Gao, J.; Deng, W. Land use/cover changes, the environment and water resources in Northeast China. *Env. Manag.* **2005**, *36*, 691–701. [CrossRef]
8. Jianchu, X.; Fox, J.; Vogler, J.B.; Yongshou, Z.P.F.; Lixin, Y.; Jie, Q.; Leisz, S. Land-use and land-cover change and farmer vulnerability in Xishuangbanna prefecture in Southwestern China. *Env. Manag.* **2005**, *36*, 404–413. [CrossRef]
9. Caldas, M.; Walker, R.; Arima, E.; Perz, S.; Aldrich, S.; Simmons, C. Theorizing land cover and land use change: The peasant economy of amazonian deforestation. *Ann. Assoc. Am. Geogr.* **2007**, *97*, 86–110. [CrossRef]
10. Lambin, E.F. Modelling and monitoring land-cover change processes in tropical regions. *Prog. Phys. Geogr.* **1997**, *21*, 375–393. [CrossRef]
11. Cabral, P.; Zamyatin, A. Markov processes in modeling land use and land cover changes in Sintra-cascais, Portugal. *Dyna* **2009**, *76*, 191–198.
12. Paul, S.; Li, J.; Wheate, R.; Li, Y. Application of object oriented image classification and markov chain modeling for land use and land cover change analysis. *J. Environ. Inform.* **2018**, *31*, 30–40. [CrossRef]
13. Qiang, Y.; Lam, N.S. Modeling land use and land cover changes in a vulnerable coastal region using artificial neural networks and cellular automata. *Environ. Monit. Assess.* **2015**, *187*, 1–16. [CrossRef] [PubMed]
14. Liu, X.; Liang, X.; Li, X.; Xu, X.; Ou, J.; Chen, Y.; Li, S.; Wang, S.; Pei, F. A future land use simulation model (flus) for simulating multiple land use scenarios by coupling human and natural effects. *Landsc. Urban Plan.* **2017**, *168*, 94–116. [CrossRef]
15. Huang, Y.; Yang, B.; Wang, M.; Liu, B.; Yang, X. Analysis of the future land cover change in Beijing using CA–Markov chain model. *Environ. Earth Sci.* **2020**, *79*, 1–12. [CrossRef]
16. Liu, X.; Wei, M.; Zeng, J. Simulating urban growth scenarios based on ecological security pattern: A case study in Quanzhou, China. *Int. J. Environ. Res. Public Health* **2020**, *17*, 7282. [CrossRef] [PubMed]
17. Chaudhuri, G.; Clarke, K. The sleuth land use change model: A review. *Environ. Resour. Res.* **2013**, *1*, 88–105.
18. Han, H.; Yang, C.; Song, J. Scenario simulation and the prediction of land use and land cover change in Beijing, China. *Sustainability* **2015**, *7*, 4260–4279. [CrossRef]
19. Arsanjani, J.J.; Helbich, M.; Kainz, W.; Boloorani, A.D. Integration of logistic regression, markov chain and cellular automata models to simulate urban expansion. *Int. J. Appl. Earth Obs. Geoinf.* **2013**, *21*, 265–275. [CrossRef]
20. Kafy, A.-A.; Shuvo, R.M.; Naim, M.N.H.; Sikdar, M.S.; Chowdhury, R.R.; Islam, M.A.; Sarker, M.H.S.; Khan, M.H.H.; Kona, M.A. Remote sensing approach to simulate the land use/land cover and seasonal land surface temperature change using machine learning algorithms in a fastest-growing megacity of Bangladesh. *Remote Sens. Appl. Soc. Environ.* **2021**, *21*, 100463. [CrossRef]
21. Puangkaew, N.; Ongsomwang, S. Remote sensing and geospatial models to simulate land use and land cover and estimate water supply and demand for water balancing in Phuket island, Thailand. *Appl. Sci.* **2021**, *11*, 10553. [CrossRef]
22. Li, X.; Chen, G.; Liu, X.; Liang, X.; Wang, S.; Chen, Y.; Pei, F.; Xu, X. A new global land-use and land-cover change product at a 1-km resolution for 2010 to 2100 based on human–environment interactions. *Ann. Am. Assoc. Geogr.* **2017**, *107*, 1040–1059. [CrossRef]
23. Ren, Y.; Lü, Y.; Comber, A.; Fu, B.; Harris, P.; Wu, L. Spatially explicit simulation of land use/land cover changes: Current coverage and future prospects. *Earth-Sci. Rev.* **2019**, *190*, 398–415. [CrossRef]
24. Yong, X.; Chuansheng, W. Ecological protection and high-quality development in the yellow river basin: Framework, path, and countermeasure. *Bull. Chin. Acad. Sci.* **2020**, *35*, 875–883.
25. Abbas, Z.; Yang, G.; Zhong, Y.; Zhao, Y. Spatiotemporal change analysis and future scenario of LULC using the CA-ANN approach: A case study of the greater bay area, china. *Land* **2021**, *10*, 584. [CrossRef]
26. Xiao, C.; Song, L. Climate Services for the Development Plan of the Yellow River Basin in China. *EGU Gen. Assem.* **2020**, 6231.
27. Yuan, Z.; Yan, D.-H.; Yang, Z.-Y.; Yin, J.; Yuan, Y. Temporal and spatial variability of drought in Huang-Huai-Hai river basin, China. *Theor. Appl. Climatol.* **2015**, *122*, 755–769. [CrossRef]
28. Shao, W.; Yang, D.; Hu, H.; Sanbongi, K. Water resources allocation considering the water use flexible limit to water shortage—A case study in the Yellow River Basin of China. *Water Resour. Manag.* **2009**, *23*, 869–880. [CrossRef]
29. Lan, H.; Peng, J.; Zhu, Y.; Li, L.; Pan, B.; Huang, Q.; Li, J.; Zhang, Q. Research on geological and surfacial processes and major disaster effects in the Yellow River Basin. *Sci. China Earth Sci.* **2022**, *65*, 234–256. [CrossRef]
30. Jiang, W.; Yuan, L.; Wang, W.; Cao, R.; Zhang, Y.; Shen, W. Spatio-temporal analysis of vegetation variation in the Yellow River Basin. *Ecol. Indic.* **2015**, *51*, 117–126. [CrossRef]

31. Yang, T.; Xu, C.Y.; Shao, Q.; Chen, X.; Lu, G.H.; Hao, Z.C. Temporal and spatial patterns of low-flow changes in the Yellow River in the last half century. *Stoch. Environ. Res. Risk Assess.* **2010**, *24*, 297–309. [CrossRef]
32. Künzer, C.; Ottinger, M.; Liu, G.; Sun, B.; Baumhauer, R.; Dech, S. Earth observation-based coastal zone monitoring of the Yellow River delta: Dynamics in China's second largest oil producing region over four decades. *Appl. Geogr.* **2014**, *55*, 92–107. [CrossRef]
33. Tang, Y.; Tang, Q.; Tian, F.; Zhang, Z.; Liu, G. Responses of natural runoff to recent climatic changes in the Yellow River Basin, China. *Hydrol. Earth Syst. Sci. Discuss.* **2013**, *10*, 4489–4514.
34. Wang, S.; Liu, J.; Yang, C. Eco-environmental vulnerability evaluation in the Yellow River Basin, China. *Pedosphere* **2008**, *18*, 171–182. [CrossRef]
35. Zhang, X.; Zhou, Y.; Han, C. Research on high-quality development evaluation and regulation model: A case study of the Yellow River water supply area in Henan Province. *Water* **2023**, *15*, 261. [CrossRef]
36. Chen, Y.; Zhu, M.; Lu, J.; Zhou, Q.; Ma, W. Evaluation of ecological city and analysis of obstacle factors under the background of high-quality development: Taking cities in the Yellow River Basin as examples. *Ecol. Indic.* **2020**, *118*, 106771. [CrossRef]
37. Shu, L.; Finlayson, B. Flood management on the lower Yellow River: Hydrological and geomorphological perspectives. *Sediment. Geol.* **1993**, *85*, 285–296. [CrossRef]
38. Wu, H.; Li, X.; Qian, H. Detection of anomalies and changes of rainfall in the Yellow River Basin, China, through two graphical methods. *Water* **2018**, *10*, 15. [CrossRef]
39. Zhang, R. Morphologic evolution of north china plain and causes of channel changes and overflows of the Yellow River. *Geol. Miner. Resour. South China* **2000**, *4*, 52–57.
40. Guo, L. A conception about innovating the new Yellow River theory: Dedicated to the centenary of academician Huang Bingwei's birth and the ninety-five years of academician Wu Chuanjun's birth. *Areal Res. Dev.* **2013**, *32*, 1–5.
41. Mostern, R. *The Yellow River: A Natural and Unnatural History*; Yale University Press: New Haven, CN, USA, 2021.
42. ECMWF. Land Cover Classification Gridded Maps from 1992 to Present Derived from Satellite Observations. Available online: https://cds.climate.copernicus.eu/cdsapp#!/dataset/satellite-land-cover?tab=form (accessed on 16 August 2022).
43. Pervez, M.S.; Henebry, G.M. Assessing the impacts of climate and land use and land cover change on the freshwater availability in the Brahmaputra River Basin. *J. Hydrol. Reg. Stud.* **2015**, *3*, 285–311. [CrossRef]
44. Rutherford, G.N.; Bebi, P.; Edwards, P.J.; Zimmermann, N.E. Assessing land-use statistics to model land cover change in a mountainous landscape in the European Alps. *Ecol. Model.* **2008**, *212*, 460–471. [CrossRef]
45. Rahman, M.T.U.; Esha, E.J. Prediction of land cover change based on CA-ANN model to assess its local impacts on Bagerhat, southwestern coastal Bangladesh. *Geocarto Int.* **2022**, *37*, 2604–2626. [CrossRef]
46. Rodríguez Eraso, N.; Armenteras-Pascual, D.; Alumbreros, J.R. Land use and land cover change in the Colombian Andes: Dynamics and future scenarios. *J. Land Use Sci.* **2013**, *8*, 154–174. [CrossRef]
47. Buğday, E.; Buğday, S.E. Modeling and simulating land use/cover change using artificial neural network from remotely sensing data. *Cerne* **2019**, *25*, 246–254. [CrossRef]
48. Saputra, M.H.; Lee, H.S. Prediction of land use and land cover changes for north Sumatra, Indonesia, using an artificial-neural-network-based cellular automaton. *Sustainability* **2019**, *11*, 3024. [CrossRef]
49. Chen, Y.; Li, X.; Liu, X.; Ai, B.; Li, S. Capturing the varying effects of driving forces over time for the simulation of urban growth by using survival analysis and cellular automata. *Landsc. Urban Plan.* **2016**, *152*, 59–71. [CrossRef]
50. Amatulli, G.; Domisch, S.; Tuanmu, M.-N.; Parmentier, B.; Ranipeta, A.; Malczyk, J.; Jetz, W. A suite of global, cross-scale topographic variables for environmental and biodiversity modeling. *Sci. Data* **2018**, *5*, 1–15. [CrossRef]
51. Fick, S.E.; Hijmans, R.J. Worldclim 2: New 1-km spatial resolution climate surfaces for global land areas. *Int. J. Climatol.* **2017**, *37*, 4302–4315. [CrossRef]
52. Lehner, B.; Grill, G. Global river hydrography and network routing: Baseline data and new approaches to study the world's large river systems. *Hydrol. Process.* **2013**, *27*, 2171–2186. [CrossRef]
53. NIES. Global Dataset of Gridded Population and gdp Scenarios. Available online: https://www.nies.go.jp/link/population-and-gdp.html (accessed on 6 June 2022).
54. Bright, E.; Coleman, P. *Landscan Global 2000*, 2000 ed; Oak Ridge National Laboratory: Oak Ridge, TN, USA, 2001.
55. Center for International Earth Science Information Network—CIESIN—Columbia University; Information Technology Outreach Services—ITOS—University of Georgia. *Global Roads Open Access Data Set, Version 1 (groadsv1)*; NASA Socioeconomic Data and Applications Center (SEDAC): Palisades, New York, NY, USA, 2013.
56. Resource and Environment Science and Data Center. Global Residential Distribution Data. Available online: https://www.resdc.cn/data.aspx?DATAID=211 (accessed on 15 June 2022).
57. Cui, P.; Jia, Y. Mountain hazards in the tibetan plateau: Research status and prospects. *Natl. Sci. Rev.* **2015**, *2*, 397–399. [CrossRef]
58. Peng, C.; Rong, C.; Lingzhi, X.; Fenghuan, S. Risk analysis of mountain hazards in Tibetan Plateau under global warming. *Adv. Clim. Change Res.* **2014**, *10*, 103.
59. Tang, B.; Liu, S.; Liu, S. Study on mountain calamities in China. *Mt. Res.* **1984**, *2*, 1–7.
60. Cui, P. Progress and prospects in research on mountain hazards in China. *Prog. Geogr.* **2014**, *33*, 145–152.
61. NASA. Global Landslide Catalog Export. Available online: https://data.nasa.gov/Earth-Science/Global-Landslide-Catalog-Export/dd9e-wu2v (accessed on 20 January 2020).

62. Brakenridge, G.R. Global Active Archive of Large Flood Events. Available online: http://floodobservatory.colorado.edu/Archives/ (accessed on 20 December 2019).
63. AAS. Molusce—Quick and Convenient Analysis of Land Cover Changes. Available online: https://nextgis.com/blog/molusce/ (accessed on 20 March 2022).
64. Benesty, J.; Chen, J.; Huang, Y.; Cohen, I. Pearson correlation coefficient. In *Noise Reduction in Speech Processing*; Springer: Berlin/Heidelberg, Germany, 2009; pp. 1–4.
65. Codd, E.F. *Cellular Automata*; Academic Press: Cambridge, MA, USA, 2014.
66. White, R.; Engelen, G. Cellular automata and fractal urban form: A cellular modelling approach to the evolution of urban land-use patterns. *Environ. Plan. A* **1993**, *25*, 1175–1199. [CrossRef]
67. Guan, D.; Li, H.; Inohae, T.; Su, W.; Nagaie, T.; Hokao, K. Modeling urban land use change by the integration of cellular automaton and markov model. *Ecol. Model.* **2011**, *222*, 3761–3772. [CrossRef]
68. Muhammad, R.; Zhang, W.; Abbas, Z.; Guo, F.; Gwiazdzinski, L. Spatiotemporal change analysis and prediction of future land use and land cover changes using QGIS MOLUSCE plugin and remote sensing big data: A case study of Linyi, China. *Land* **2022**, *11*, 419. [CrossRef]
69. Hulshoff, R.M. Landscape indices describing a dutch landscape. *Landsc. Ecol.* **1995**, *10*, 101–111. [CrossRef]
70. Zhang, Q.; Fu, B.; Chen, L. Several problems about landscape pattern change research. *Sci. Geogr. Sin.* **2003**, *23*, 270–275.
71. Leitao, A.B.; Ahern, J. Applying landscape ecological concepts and metrics in sustainable landscape planning. *Landsc. Urban Plan.* **2002**, *59*, 65–93. [CrossRef]
72. Naveh, Z.; Lieberman, A.S. *Landscape Ecology: Theory and Application*; Springer Science & Business Media: Berlin/Heidelberg, Germany, 2013.
73. Yue, W.; Xu, J.; Xu, L.H. An analysis on eco-environmental effect of urban land use based on remote sensing images: A case study of urban thermal environment and NDVI. *Acta Ecol. Sin.* **2006**, *26*, 1450–1460.
74. Gao, C.; Cheng, L. Tourism-driven rural spatial restructuring in the metropolitan fringe: An empirical observation. *Land Use Policy* **2020**, *95*, 104609. [CrossRef]
75. Cui, Y.; Cheng, D.; Choi, C.E.; Jin, W.; Lei, Y.; Kargel, J.S. The cost of rapid and haphazard urbanization: Lessons learned from the Freetown landslide disaster. *Landslides* **2019**, *16*, 1167–1176. [CrossRef]
76. Xin, S.U.; Wang, J.J.; Hui, L.I.; Niu, Y.L. Evolutionary processes in agricultural eco-economic system of Wuqi County after converting slope farmland into forest and grassland. *Bull. Soil Water Conserv.* **2010**, *30*, 186–190. [CrossRef]
77. Du, Y.; Zhang, D.; Yao, S. Analysis of the driving forces of SLCP based on the weights of evidence model—A case study of Wuqi, Shaanxi Province. *Res. Soil Water Conserv.* **2017**, *24*, 325–332.
78. Yifan, X.; Shunbo, Y.; Yuanjie, D.; Lei, J.; Yuanyuan, L.; Qing, G. Impact of the 'grain for green'project on the spatial and temporal pattern of habitat quality in Yan'an city, China. *Chin. J. Eco-Agric.* **2020**, *28*, 575–586.
79. Zhou, H. Nothing short of a miracle: The two decades' Chinese restoration of forests and grasslands from farmland. *Ecol. Civiliz. World* **2019**, 10–19.
80. Li, S.; Liu, M. The development process, current situation and prospects of the conversion of farmland to forests and grasses project in China. *J. Resour. Ecol.* **2022**, *13*, 120–128.
81. Ning, J.; Liu, J.; Kuang, W.; Xu, X.; Zhang, S.; Yan, C.; Li, R.; Wu, S.; Hu, Y.; Du, G. Spatiotemporal patterns and characteristics of land-use change in China during 2010–2015. *J. Geogr. Sci.* **2018**, *28*, 547–562. [CrossRef]
82. Price, M.F. Forests in sustainable mountain development. In *Global Change and Mountain Regions*; Springer: Berlin/Heidelberg, Germany, 2005; pp. 521–529.
83. Körner, C.; Jetz, W.; Paulsen, J.; Payne, D.; Rudmann-Maurer, K.; Spehn, E.M. A global inventory of mountains for bio-geographical applications. *Alp Botany* **2017**, *127*, 1–15. [CrossRef]
84. Deng, W.; Tang, W. General directions and countermeasures for urbanization development in mountainous areas of China. *J. Mt. Sci.* **2013**, *31*, 168–173.
85. Baiping, Z.; Shenguo, M.; Ya, T.; Fei, X.; Hongzhi, W. Urbanization and de-urbanization in mountain regions of China. *Mt. Res. Dev.* **2004**, *24*, 206–209. [CrossRef]
86. Wang, G. Lessons learned from protective measures associated with the 2010 Zhouqu debris flow disaster in China. *Nat. Hazards* **2013**, *69*, 1835–1847. [CrossRef]
87. Hu, K.; Cui, P.; Zhang, J. Characteristics of damage to buildings by debris flows on 7 august 2010 in Zhouqu, Western China. *Nat. Hazards Earth Syst. Sci.* **2012**, *12*, 2209–2217. [CrossRef]
88. Yu, X.; Ma, S.; Cheng, K.; Kyriakopoulos, G.L. An evaluation system for sustainable urban space development based in green urbanism principles—a case study based on the Qinba mountainous area in China. *Sustainability* **2020**, *12*, 5703. [CrossRef]
89. Liu, J.; Raven, P.H. China's environmental challenges and implications for the world. *Crit. Rev. Environ. Sci. Technol.* **2010**, *40*, 823–851. [CrossRef]
90. Chang, J. Formative causes of landslide and debris flow in Lanzhou city and preventives. *Res. Soil Water Conserv.* **2003**, *10*, 250–252.
91. Du, R.; Li, H.; Tang, B.; Zhang, S. Research on debris flow for thirty years in China. *J. Nat. Disasters* **1995**, *4*, 64–73.
92. Cui, P. Advances in debris flow prevention in China. *Sci. Soil Water Conserv.* **2009**, *7*, 7–13.

93. Liu, C.; Guo, L.; Ye, L.; Zhang, S.; Zhao, Y.; Song, T. A review of advances in China's flash flood early-warning system. *Nat. Hazards* **2018**, *92*, 619–634. [CrossRef]
94. Peng, C.; Fenghuan, S.; Qiang, Z.; NingSheng, C.; ZHANG, Y. Risk assessment and disaster reduction strategies for mountainous and meteorological hazards in Tibetan Plateau. *Chin. Sci. Bull.* **2015**, *60*, 3067–3077.
95. Dong, Y.; Jinlong, R.; Guangze, Z.; Zhengxuan, X.; Tao, F. Application of remote sensing technology in site selection of a station on Sichuan-Tibet railway. *Bull. Surv. Mapp.* **2021**, *12*, 83–87.

land

MDPI

Article

Planform Changes in the Lower Mahaweli River, Sri Lanka Using Landsat Satellite Data

Vindhya Basnayaka [1], Jayanga T. Samarasinghe [2], Miyuru B. Gunathilake [3,4], Nitin Muttil [5,6] and Upaka Rathnayake [1,*]

1 Department of Civil Engineering, Faculty of Engineering, Sri Lanka Institute of Information Technology, Malabe 10115, Sri Lanka
2 Department of Earth Environmental and Resource Sciences, University of Texas, El Paso, TX 79968, USA
3 Hydrology and Aquatic Environment, Environment and Natural Resources, Norwegian Institute of Bioeconomy and Research, 1433 Ås, Norway
4 Water, Energy and Environmental Engineering Research Unit, Faculty of Technology, University of Oulu, P.O. Box 8000, FI-90014 Oulu, Finland
5 Institute for Sustainable Industries & Liveable Cities, Victoria University, P.O. Box 14428, Melbourne, VIC 8001, Australia
6 College of Engineering and Science, Victoria University, Melbourne, VIC 8001, Australia
* Correspondence: upaka.r@sliit.lk or upakasanjeewa@gmail.com

Citation: Basnayaka, V.; Samarasinghe, J.T.; Gunathilake, M.B.; Muttil, N.; Rathnayake, U. Planform Changes in the Lower Mahaweli River, Sri Lanka Using Landsat Satellite Data. *Land* **2022**, *11*, 1716. https://doi.org/10.3390/land11101716

Academic Editors: Matej Vojtek, Andrea Petroselli and Raffaele Pelorosso

Received: 11 September 2022
Accepted: 30 September 2022
Published: 3 October 2022

Abstract: Major development projects along rivers, like reservoirs and other hydraulic structures, have changed not only river discharges but also sediment transport. Thus, changes in river planforms can be observed in such rivers. In addition, river centerline migrations can be witnessed. The Mahaweli River is the longest in Sri Lanka, having the largest catchment area among the 103 major river basins in the country. The river has been subjected to many development projects over the last 50 years, causing significant changes in the river discharge and sediment transport. However, no research has been carried out to evaluate the temporal and spatial changes in planforms. The current seeks to qualitatively analyze the river planform changes of the Lower Mahaweli River (downstream to Damanewewa) over the past 30 years (from 1991 to 2021) and identify the major planform features and their spatiotemporal changes in the lower Mahaweli River. Analyzing the changes in rivers requires long-term data with high spatial resolution. Therefore, in this research, remotely sensed Landsat satellite data were used to analyze the planform changes of Lower Mahaweli River with a considerably high resolution (30 m). These Landsat satellite images were processed and analyzed using the QGIS mapping tool and a semi-automated digitizing tool. The results show that major changes in river Mahaweli occurred mainly in the most downstream sections of the selected river segment. Further, the river curvature was also comparatively high downstream of the river. An oxbow lake formation was observed over time in the most downstream part of the Mahaweli River after 2011. Centerline migration rates were also calculated with the generated river centerlines. It was found that the rates were generally lower than about 30 m per year, except for at locations where river meandering was observed. The main limitations of this study were the possible misclassifications due to the resolution of images and obstructions caused by cloud cover in the Landsat images. To achieve more accurate estimates, this study could be developed further with quantitative mathematical analysis by also considering the sediment dynamics of the Mahaweli River.

Keywords: river morphodynamics; centerline migration; Landsat data; planform changes; remote sensing

1. Introduction

The geomorphology of rivers is highly dynamic and has to be closely monitored. Most of the large meandering rivers in the world, which carry significant sediment loads, are subjected to change over time [1]. In fact, braided rivers exhibit high variability within short periods [2]. These changes are often due to sediment movement, river hydraulic and hydrological processes, topography, and the vegetation properties of the riverine environments [3]. Studying the geomorphological changes in rivers is paramount in understanding the natural dynamics and will be helpful in predicting the patterns of future planform changes [4]. The knowledge gained through these analyses will be significant in the decision-making stage of river and floodplain management.

According to the literature, many investigations have been carried out for in recent decades on river planform changes, although these analyses represent different scales and have adopted different methods [3–8]. In many historical studies (during the 1990s and early-2000s), researchers used field data, analytical models, or numerical models to analyze river planform dynamics [9–12]. With the massive development of technology in recent years, more sophisticated computers with high computing power and satellite remote sensing data have become available. Therefore, most recent studies have taken advantage of remotely sensed satellite data (geospatial imagery) and tools such as GIS and algorithms, developed with programming languages such as MATLAB and Python [12]. In some cases, the spatial scale of river planform changes were only a few meters, while in others, there were migrations of up to a few hundred meters. Hence, the need arises for data with high spatial and temporal resolution [13]. Further, this will be crucial when analyzing highly dynamic rivers that contain large numbers of meandering bends.

To analyze the planform changes in rivers, long-term records of the river geomorphology are required. However, in many countries, there are no such databases that contain constant and long-term records of river planforms. This limitation is addressed by using remotely sensed data [14]. With the recent advancements in technology, the availability of free data has been hugely increased. Therefore, these remotely sensed data are used in many projects and research, not only in the water sector but also in many other fields. The importance of remotely sensed satellite data are their free and global availability and considerably high resolution (about 30 m). Further, some of these data are available for more than 30 years (ex: Landsat satellite data is available since 1984) [15].

Many researchers all over the world have used Landsat data to investigate the temporal variation of river meandering [16]. As it was stated, these techniques are chosen ahead of onsite surveying due to logistic and financial issues. Wray [17] has used multispectral Landsat imagery to observe the changes in Palaeochannels of the Namoi River, New South Wales, Australia. Landsat 7 ETM+ data were used in this analysis and the historical changes in Namoi River were presented. Schwenk et al. [18] have incorporated the RivMAP toolbox with Landsat images to detect the physics in meandering migrations in multi-decades for the Ucayali River (which is the headstream of the Amazon River). Therefore, the usage of Landsat data by linking them to the analyzer revealed the significance of such data for the evaluation of river meandering. In addition, Nagel et al. [19] used cloud computing with Landsat data to understand the migration of communities along the Amazon River due to the meandering course of the river. Therefore, the indirect impacts were also assessed based on Landsat data.

However, other than by Basnayaka et al. [15], these recent advances have not been applied in Sri Lanka to understand the meandering behavior of rivers. The River Mahaweli is important due to its flat bathymetry downstream. In addition, the sediment flow is obstructed by several reservoirs along the river. Therefore, assessments of the temporal variation of the river, in terms of its meandering and planform changes, are essential. However, survey data along the river section are limited. Thus, this research addresses these research gaps by analyzing Landsat images of the Mahaweli River to identify the major planform features and their spatial and temporal variation over a 30 year period from 1991 to 2021. The findings of this research will be helpful for managing the riverine environment of the Lower Mahaweli River. Additionally, the proposed method is also applicable for narrow rivers (widths of less than 100 m and greater than the spatial resolution of the imagery, i.e., 30 m) as the method considers only river centerline variations.

2. Study Area and Problem Statement

Among the 103 major river basins in Sri Lanka, Mahaweli is the largest river basin with a catchment area of about 10,448 km^2 [20], which is nearly one-sixth of the country. Further, the Mahaweli River is the longest in Sri Lanka, with a length of about 335 km. The annual precipitation received by the whole basin amounts to 28 $\times$ 10^9 m^3 and the river has an annual discharge of 8.8 $\times$ 10^9 m^3 [20]. According to the Köppen-Geiger climate classification [21], the upper Mahaweli basin belongs to the equatorial fully humid climate zone, while the lower Mahaweli basin lies in the equatorial monsoonal climate zone. River Mahaweli has been subjected to many development projects including hydropower generation projects, supplying drinking water, and agriculture development projects since the beginning of the 1970s. The river has been subjected to many pieces of research which are mostly related to water quality, hydro-meteorological characteristics, sediments geochemistry, water demand management, etc. [22–24]. However, the river morphodynamics of the Mahaweli River is a subject that still needs attention.

Figure 1 clearly showcases the complex river paths of the Mahaweli River (refer to Figure 1a,c) and its meandering behavior of Mahaweli River (refer to Figure 1b). In addition, the river planforms are highly observable (refer to Figure 1c) and they are dynamic based on the flow characteristics depending on the monsoons. Therefore, these observations re-state the importance of having a scientific analysis of the temporal variation of meandering behavior and the river planforms. Thus, a segment of the Mahaweli River to perform the analysis of planform variation over time was selected in this study. The selected river segment is the lower Mahaweli River starting from Damanewewa up to the sea outfall in Trincomalee and has a total length of about 107 km (around 1/3 of the river). The selected river segment is illustrated in Figure 2. The red color box shows the region of interest (ROI) which is Lower Mahaweli River from Damanewewa to Trincomalee.

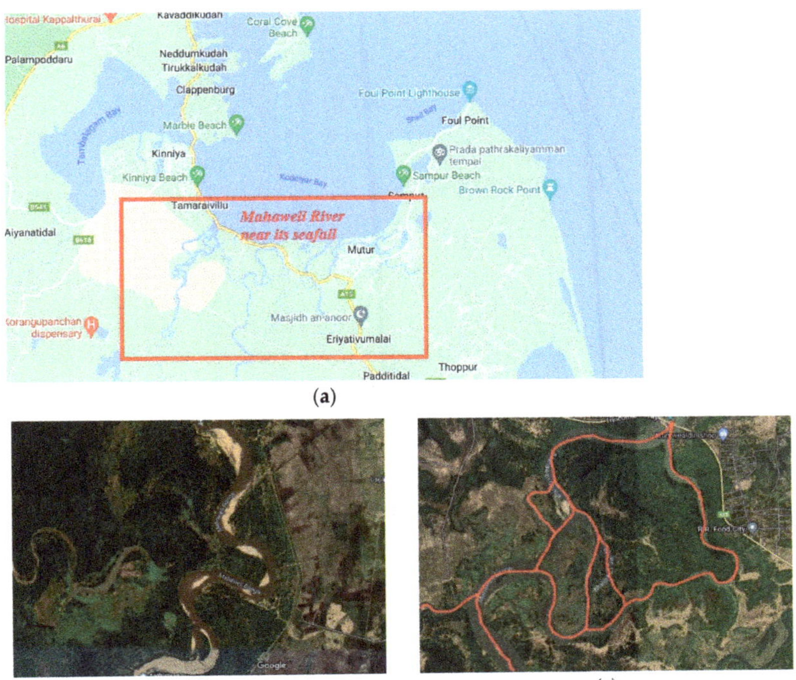

Figure 1. River meandering behavior: (**a**) Mahaweli River near its sea outfall; (**b**) sediment deposition along the river; (**c**) complex river path with river planforms (source: Google maps).

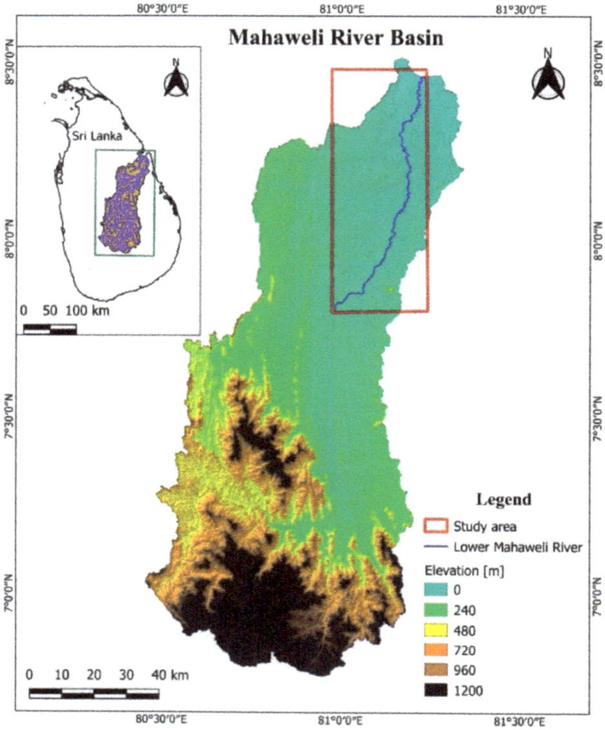

Figure 2. Map of the Mahaweli River basin, Sri Lanka and the selected river segment for the analysis.

3. Data and Methods

3.1. Use of Landsat Imagery

Landsat satellite images provide the most comprehensive record (more than 35 years) for medium resolution (10–50 m) global scale Earth Observation data [25]; images are captured at approximately 16 day intervals. The Landsat mission started in 1984 and continues to today. Currently, there are four Landsat products: Landsat 1–5 MSS, Landsat 4–5 TM, Landsat 7, and Landsat 8–9. The application of Landsat data is widespread because of their free and global scale availability for a considerably long period (about 35 years). Yet, one major issue of these data is obstructions caused by the clouds. Further, in some of the L7 images, there are missing data stripes due to a failure that occurred in the scan line corrector (SLC) [18]. Therefore, in our study, these data were exempted in selecting the imagery. The current study uses only two types of Landsat data, i.e., Landsat 4–5 Thematic Mapper (TM), and Landsat 8 Operational Land Imager (OLI) and Thermal Infrared Sensors (TIRS) products, to estimate the spatiotemporal variability of river planform of Mahaweli River.

There are several methods to download or access Landsat data, including the United States Geological Survey (USGS) website, other environmental data catalogues such as USGS EarthExplorer, LandsatLook Viewer, USGS GloVis, and data catalogues available in cloud-based platforms like Google Earth Engine [26]. To analyze the river planform changes in the Mahaweli River, we downloaded Landsat images from the USGS Earthexplorer website for 30 years spanning from 1991 to 2021. To have a more accurate classification, Landsat images were selected based on the cloud cover criterion. Images with the least cloud cover on the river were selected by visually inspecting the available images. Since the selected three river segments are distributed over two Landsat tiles, (path -141, row–54, 55; refer to Figure 3) both tiles were downloaded and merged using version 3.16 of QGIS software. A tile has 5000 × 5000 pixels with a resolution of 30 m. Before starting the image classification, all the data were cropped to the interested area to reduce the processing time.

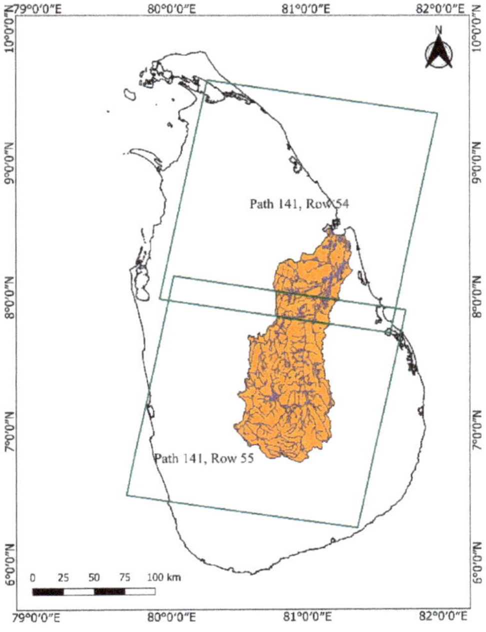

Figure 3. Arrangement of Landsat scenes covering the study area.

3.2. Generating the River Masks

The first step in evaluating the planform changes in the Mahaweli River was to extract the river masks for each selected year using Landsat imagery. Different techniques are used in the literature [27–34] to extract surface water pixels from Landsat images. Index-based classification is one such method that is used to identify water pixels in satellite remote sensing data. Commonly used water indexes are Automated Water Extraction Index for images with shadows ($AWEI_{shadow}$) [27], Tasselled Cap Wetness (TCW_{Crist}) index [28], Normalized Difference Water Index (NDWI) [29], Modified Normalized Difference Water Index (MNDWI) [30], and Automated Water Extraction Index for images without shadows ($AWEI_{no\ shadow}$) [31], Water Index (WI_{2006}) [32], Water Index (WI_{2015}) [33] and Water Index (WI_{2018}) [34].

In the current study, first, we used three water indices to generate the water masks from Landsat images, considering the accuracy of each index in classifying the water pixels. According to Fisher et al. [31], the accuracy of classifying water pixels has slightly increased in WI_{2015} compared to the WI_{2006}. The water index introduced by Raheem and Alwan (WI_{2018}) [34] outperforms most of the previously defined water indices in the classification of water pixels and is comparable to the MNDWI. The mathematical expressions for these indices are given in literature and these are presented in Table 1.

Table 1. Different water indexes defined in literature for water pixel classification using multi-spectral Landsat satellite data. (Surface reflectance values (ρ) of the seven bands are indicated as $\rho_{B1} - \rho_{B7}$ and are the inputs for the equations).

Index	Equation	Source
NDWI	$\frac{\rho_{B2} - \rho_{B4}}{\rho_{B2} + \rho_{B4}}$	[29]
MNDWI	$\frac{\rho_{B2} - \rho_{B5}}{\rho_{B2} + \rho_{B5}}$	[30]
TCW_{Crist}	$0.0315\rho_{B1} + 0.2021\rho_{B2} + 0.3102\rho_{B3} + 0.1594\,\rho_{B4} - 0.6806\rho_{B5} - 0.6109\rho_{B7}$	[28]
$AWEI_{shadow}$	$\rho_{B1} + 2.5 \times \rho_{B2} - 1.5 \times (\rho_{B4} + \rho_{B5}) - 0.25 \times \rho_{B7}$	[27]
$AWEI_{no\ shadow}$	$4 \times (\rho_{B2} - \rho_{B5}) - (0.25 \times \rho_{B4} + 2.75 \times \rho_{B5})$	[30]
WI_{2015}	$1.7204 + 171\rho_{B2} + 3\rho_{B3} - 70\rho_{B4} - 45\rho_{B5} - 71\rho_{B7}$	[33]
WI_{2018}	$\frac{0.12\rho_{B1} + 0.231\rho_{B4}}{0.752\rho_{B6} + 0.223\rho_{B5}}$	[34]

However, the use of these indices alone did not allow us to accurately classify surface water. One major reason for these classification errors may be the presence of mixed water areas such as wetlands. Therefore, it was decided to use Modified Normalized Difference Water Index (MNDWI) together with two vegetation indices; Normalized Difference Vegetation Index (NDVI) and Enhanced Vegetation Index (EVI), as suggested in the literature [35–38] to reduce the misclassification of water areas. The NDVI is used to identify the vegetation percentages by accounting for the greenness of each pixel, while EVI has the capability to correct the atmospheric disturbances and noise from canopy backgrounds [39]. The relationship of MNDWI > NDVI and MNDWI > EVI was used to find out the pixels where the water signals are greater than the vegetation signals and to remove the effect of mixed water and vegetation. The thresholds of EVI were selected following a trial-and-error method by comparing the water masks with Google Earth images. The reflectance values of the images can be affected by the atmospheric condition that exists at the time of acquiring the images. Therefore, the use of different thresholds for different images can be justifiable considering the above fact. These resulting raster images were then used in delineating the river centerlines with the use of the WebPlotDigitizer tool [40].

3.3. Cleaning the River Masks

Having a well-defined river centerline is necessary for the estimation of river direction, curvature, and migration rates and patterns. Therefore, the created watermasks were cleaned before digitizing the centerline to remove the misclassified pixels and have a continuous connection to the river. Further, it removes the unwanted areas such as the water pixels that are not hydraulically connected to the main river (i.e., tributaries, remainders of cut-off channels, and riverside cities).

3.4. Delineating River Centerlines

WebPlotDigitizer is a semi-automated digitizing tool that can be used to automatically digitize images, plots, or maps. In the current study, version 4.5 of WebPlotDigitizer was used [40] to delineate the centerline of the river using the binary watermask. As the first step, images of watermask map were loaded into the WebPlotDigitizer. Then the image was georeferenced using four known points (two points on the X axis and two points on the Y axis). WebPlotDigitizer software allows users to select the method of digitizing; either to digitize the plots/images automatically or manually. When digitizing, the software creates points along the river within a user-specified distance. To have a continuous connection to the river, the gaps of channel masks need to be filled manually by adding additional points for the centerline. The coordinates of these points can be then saved as a comma-separated values (.csv) file or directly copy the data to an Excel or a text file. When saving the (X, Y) coordinate data were arranged in the order from upstream to downstream.

3.5. Estimation of River Planform Geometry

Three parameters were calculated to identify the spatial and temporal variations of the river centerline in the Mahaweli River. The distance between two points of the river centerline (ds_i), inflection angle (θ_i) and the curvature (c_i) between two successive points (x_i, y_i) and (x_{i+1}, y_{i+1}) along the centreline was computed by the Equations (1)–(3).

$$ds_i = \sqrt{(x_{i+1} - x_i)^2 + (y_{i+1} - y_i)^2} \quad (1)$$

$$\theta_i = arctan(\frac{y_{i+1} - y_i}{x_{i+1} - x_i}) \quad (2)$$

$$c_i = \frac{\theta_{i+1} - \theta_{i-1}}{ds_{i+1} - ds_{i-1}} \quad (3)$$

The two parameters, i.e., inflection angle and curvature, can be used to identify the bends of a river. The zero crossing points of the curvature represent meandering bends [8]. Therefore, the above three parameters were calculated and plotted to study the spatial variation of the Mahaweli River from upstream to downstream. A schematic diagram that illustrates the relationship between curvature and inflection points are shown in Figure 4. The points 1–8 represent the inflection points where the bends are separating.

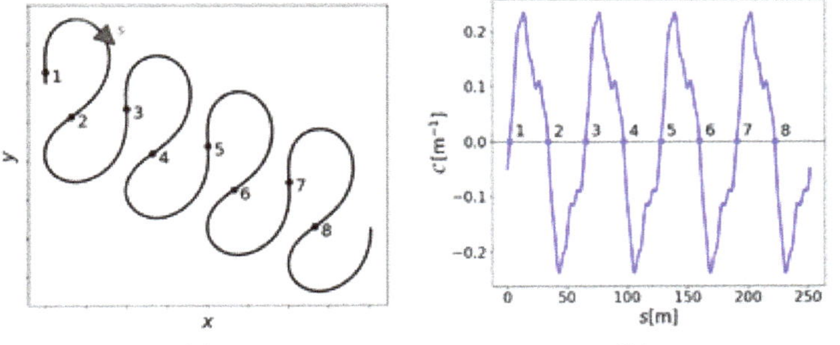

Figure 4. Schematic diagram of bend separation using curvature of the channel: (**a**) Bends; (**b**) inflection points [8].

3.6. Calculating the Centerline Migration

The centerline coordinates extracted from the WebPlotDigitizer tool were used to map the river centerlines for the selected nine years. These point data were loaded into the QGIS software to create the polylines for centerlines. Then the geometric attribute tools of QGIS were utilized in calculating the centerline migration between two adjacent years. The centerline migration distances were calculated with an interval of 200 m along the centerline, with the help of geometric attribute tools in QGIS. To compute the migration rates, these migrated distances were then divided by the number of years between the considered two scenarios and obtained the annual migration rates.

3.7. Overall Methodology

The overall methodology carried out in this research is summarized in Figure 5. The extracted Landsat images spanning from 1991 to 2021 were processed as shown in the following flow chart to understand the temporal variation of river planforms in the lower Mahaweli River.

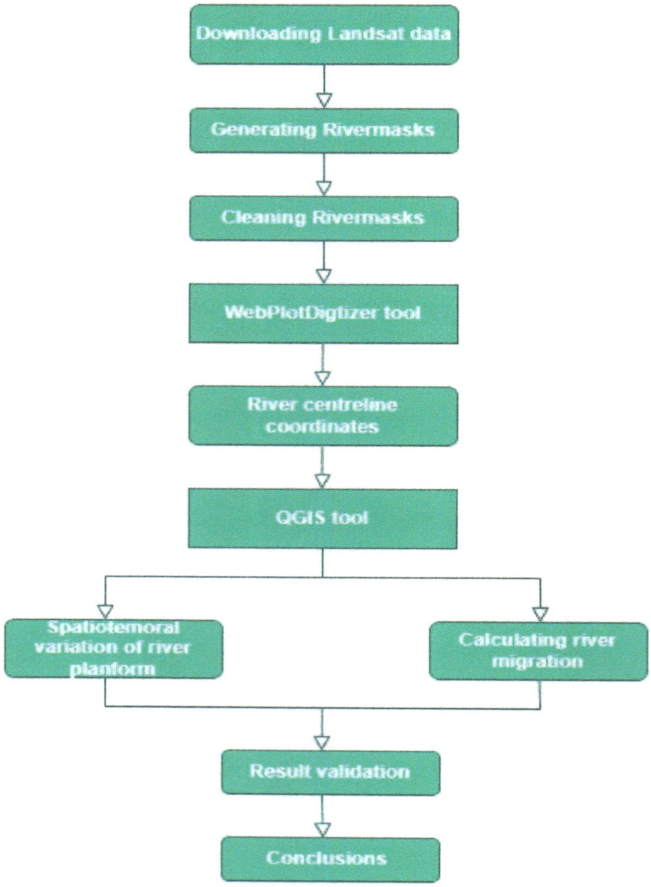

Figure 5. Flowchart of the overall methodology of the study.

4. Results and Discussion

4.1. River Planform

The annual river masks of the Mahaweli River were analyzed to quantify the planform variations from 1991 up to 2021. As the selected river stretch has a total length of 107 km, it was divided into four sections. The generated river masks are presented in Figure 6. Considering the availability of cloud-free images and the selected time span, nine years were selected for our analysis.

It can be seen from river masks that there are disconnections in the main river in a few places. These can occur due to the misclassifications of pixels. These disconnections are indicated in red color boxes (refer to Figure 6b). One of the major reasons for these errors is the cloud cover in the selected images. Additionally, errors can be attributed to the lowered resolution of images compared with the river width. Sometimes, the Landsat images may not be able to capture the river in narrow areas, i.e., where the width of the river is smaller or close to the pixel size (30 m). The river braiding/anabranching areas are highlighted in yellow. Along the selected river segment of the Mahaweli River, there were four clearly visible braided areas, as marked in Figure 6. According to the extracted river masks, an oxbow lake formed near Trincomalee at the most downstream point of the Mahaweli River. This lake formation is illustrated in Figure 6a with a green box.

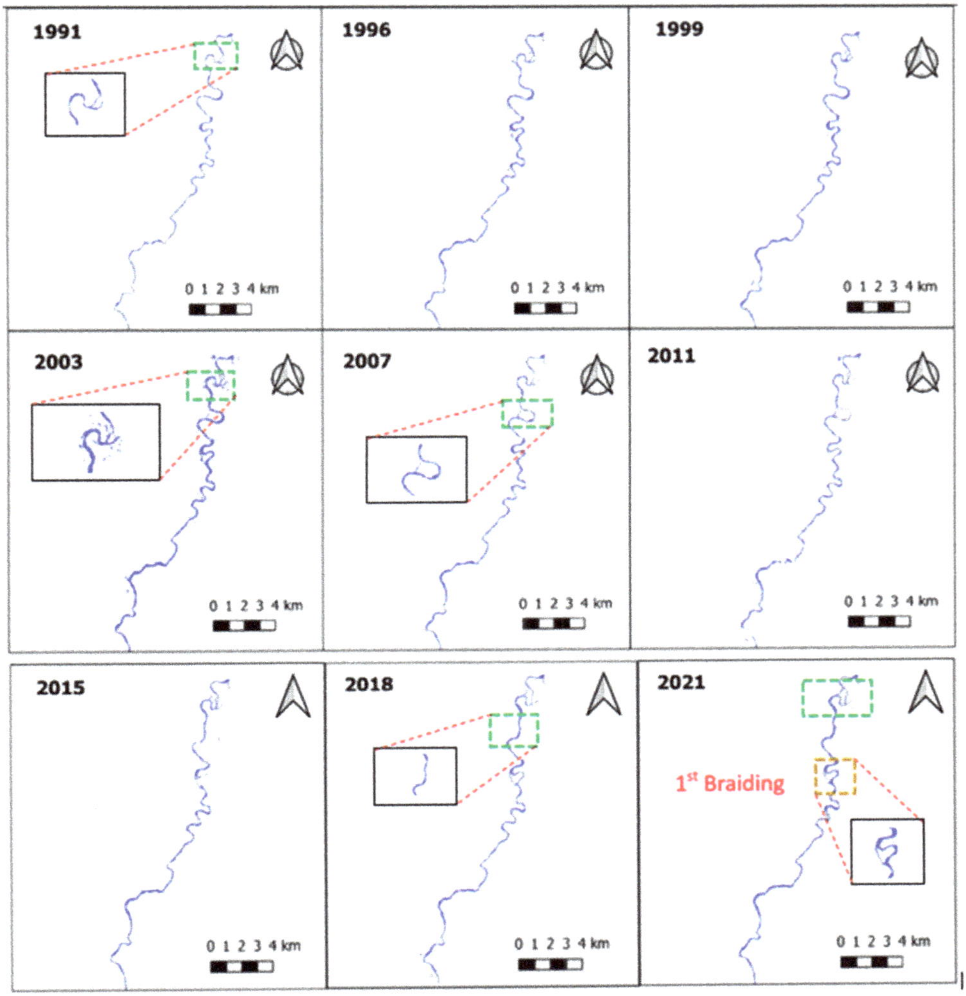

(**a**)

Figure 6. *Cont.*

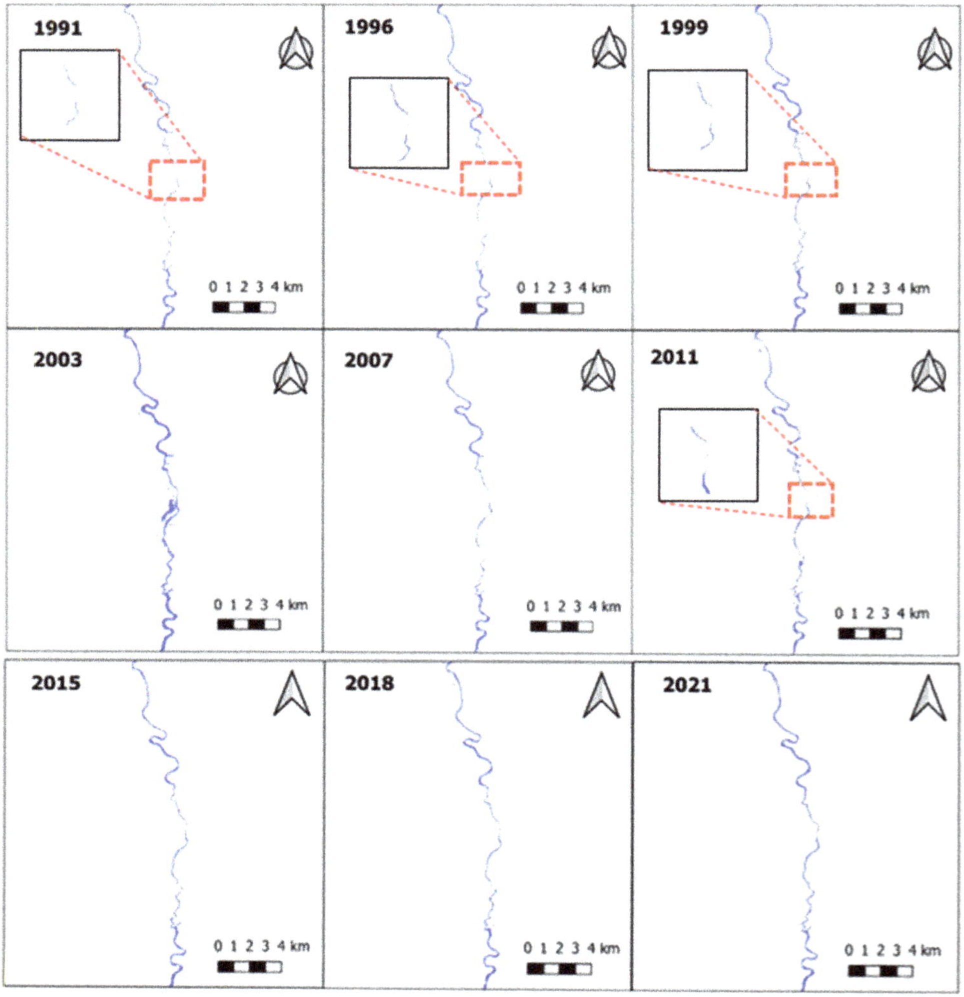

(**b**)

Figure 6. *Cont.*

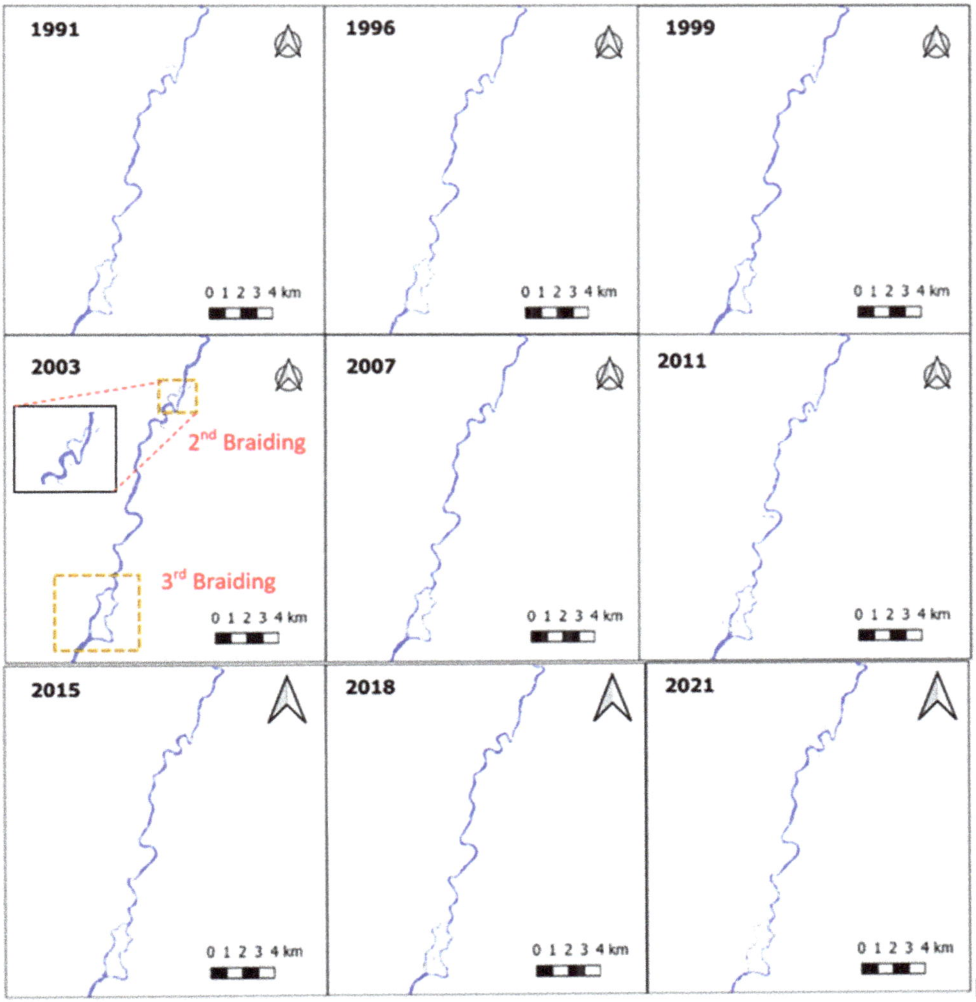

(c)

Figure 6. *Cont.*

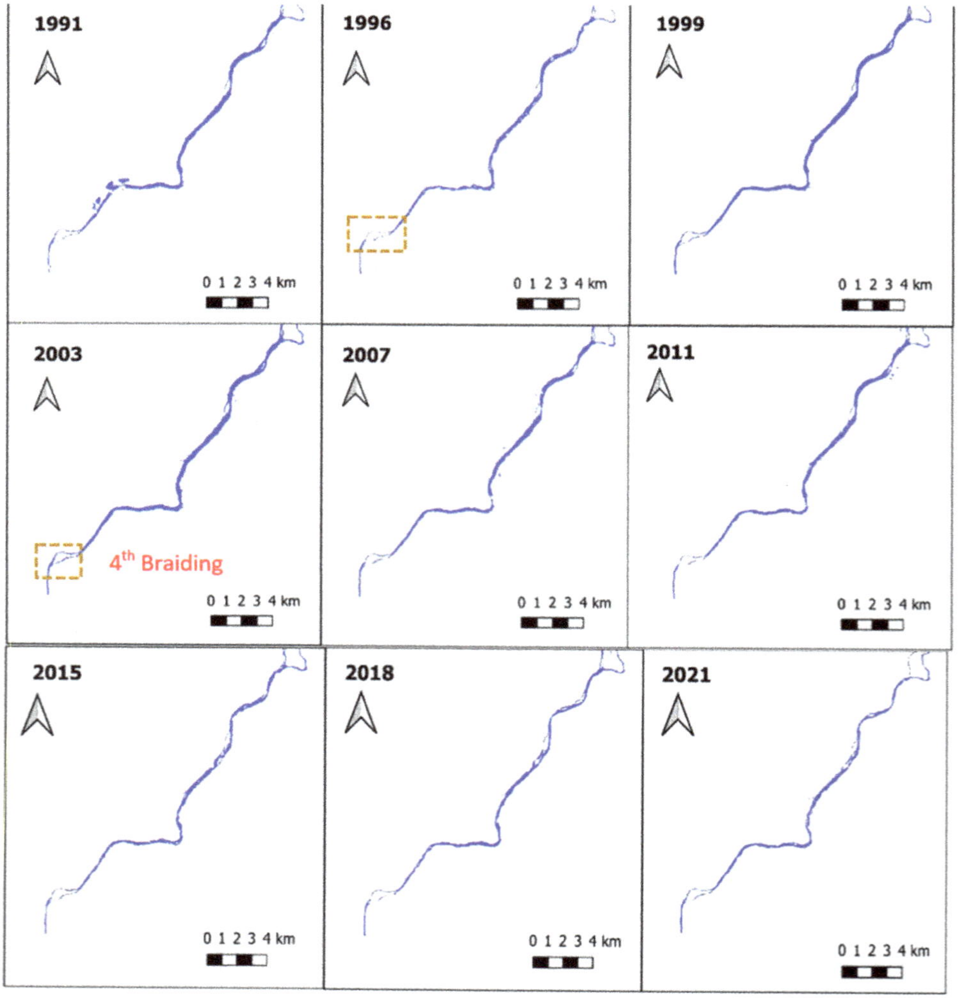

(**d**)

Figure 6. River masks generated for the Mahaweli River from Damanewewa to Trincomalee: (**a**) Most downstream river part of the selected Mahaweli River stretch near to Trincomalee; (**b**,**c**) middle stretches of the selected river segment; (**d**) most upstream part of the selected river stretch starts from Damanewewa.

4.2. Spatiotemporal Variation of River Centreline

The river centerlines points extracted from the watermasks were processed in QGIS to create the centerlines and compare the variations. The centerlines of the main river are presented in Figure 6. Since centerline migrations in the considered river stretch were mainly observed at the most downstream sections of Mahaweli River, here we have presented only the stretch (a) described in the previous sub-section. The figure shows the formation of an oxbow lake (highlighted in red color) over time during these 30 years. The location of the oxbow lake observed downstream of the Mahaweli River is highlighted in Figure 7.

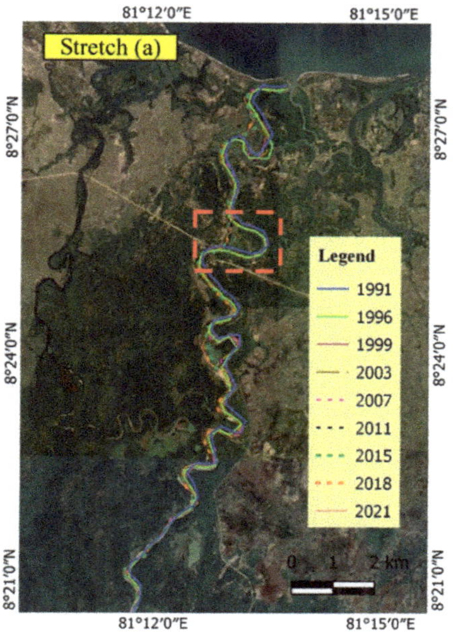

Figure 7. Temporal variation of river centerline.

4.3. Variation of River Curvature

The river curvature for each year was computed and plotted against the distance along the river centerline. These high curvatures represent the locations of bends along the river. As shown in Figure 8, the frequency of high curvature bends increases in the downstream direction. The high curvature of rivers drives the meandering process by causing erosion in the outer bank and accretion in the inner bank. The river curvature values are comparable with the values obtained in a similar study done for Deduru Oya in Sri Lanka [15]. However, the maximum curvature of the selected segment of the Mahaweli River is higher than that of the Deduru Oya River. According to the literature [41], the average bend curvature is related to the average migration rate of a certain river bend. Therefore, these river curvature values are indications of possible meandering bends in the selected river segment.

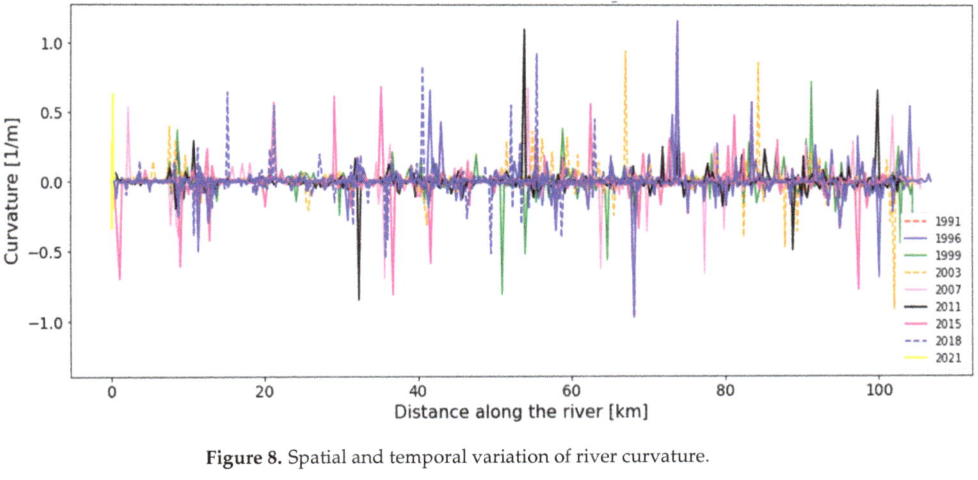

Figure 8. Spatial and temporal variation of river curvature.

4.4. Variation of River Centreline Migration

To estimate the evolution of the river, centerline migration rates were computed as described in Section 3.6. The migration rates were plotted against the distance along the river centerline and are illustrated in Figure 9. The results show three major spikes in the annual migration rates during the periods 2011 to 2015 and 2015 to 2018. This can be attributed to the changes in the main river flow path that occurred with the oxbow lake formations (during 2007 and 2011) and sediment bar formations. To have a clearer estimate of these changes, a mathematical analysis needs to be carried out including the data about riverbanks and sediment bars. The spatial variation of the annual migration rates shows that the upstream areas of the selected river segment have fewer rates of centerline migrations than the downstream segments. This can be explained considering the meandering of the high curvature bend in the river (mainly the downstream of Mahaweli River).

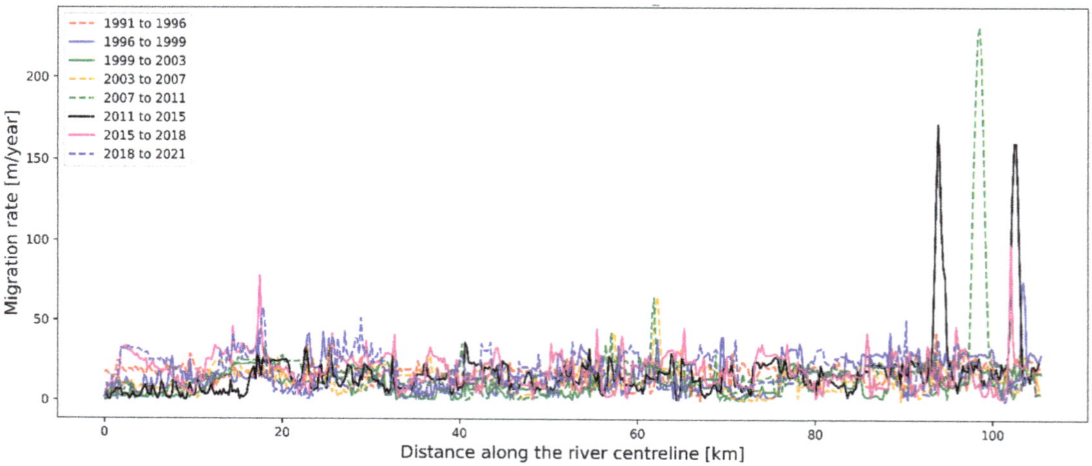

Figure 9. Spatial and temporal variation of river centerline migration rates.

When compared with the river migration rates of Deduru Oya, the Lower Mahaweli River exhibits minor values except for the three bends described above. The average centerline migration rates are within the limit 20–40 m/year.

5. Summary and Conclusions

This study focused on the longest river in Sri Lanka, whose catchment are accounts for approximately one-sixth of the country's territory. Landsat satellite images were used to identify the major planform changes over thirty years from 1991 to 2021, in Lower Mahaweli River, exhibiting the dynamic nature in its geomorphology. In the selected river segment, there were four locations where river braiding could be clearly observed. During the past three decades, there were no considerable changes in the planform of the Lower Mahaweli River, although small seasonal migrations could be seen in the upstream parts. However, the downstream segments of the river exhibited significant changes over time. With the extracted watermasks, the creation of an oxbow lake near Trincomalee was identified. The migration rates were calculated for the river centerlines; it was found that the rates were generally lower than about 30 m per annum. There were three locations where the annual migrations were more than 100 m; these may be the locations where the main flow path changed due to oxbow lake formation and sediment bar depositions. The overall results of this study highlight the need for quantitative analyses, especially in the downstream areas of the Lower Mahaweli River.

The accuracy of these geospatial analyses depends strongly on the resolution of the images. Medium resolution satellite images are a good source of data for analyzing the planform changes of a river when there are no or limited field data available. However, cloud coverage in these satellite images is a major limitation. Therefore, using satellite images that are not impeded by clouds or a lack of illumination is suggested. However, currently, such satellite data are not freely available for long-term analyses, as the images obtained with active sensors (Sentinel 1) like synthetic aperture radar (SAR) are only available for very short periods (Sentinel 1 is available only after 2014).

In the current research, the major features of the Mahaweli River planform and their temporal and spatial variations were identified. However, the methodology was designed only for a qualitative analysis as an initial study, because, to the best of the authors' knowledge, no such research has analyzed the river planform geomorphology and its variations over such a long period using remote sensing technology. The current study is therefore an introductory work for future detailed mathematical analyses.

The methodology proposed in the current study is applicable to a vast range of river planform studies, mainly for rivers which are narrower than 100 m. Nevertheless, the results of our qualitative analysis could be used to manage water resources in the downstream of the Mahaweli River. The area near the sea outlet of the Mahaweli River is interesting, as it has one of the world's most famous natural harbors.

Author Contributions: Conceptualization, U.R.; methodology, V.B. and J.T.S.; software, V.B., J.T.S. and M.B.G.; formal analysis, V.B. and J.T.S.; resources, V.B. and M.B.G.; data curation, V.B. and J.T.S.; writing—original draft preparation, V.B. and J.T.S.; writing—review and editing, U.R. and N.M.; visualization, V.B. and J.T.S.; supervision, U.R.; project administration, U.R. All authors have read and agreed to the published version of the manuscript.

Funding: This research received no external funding.

Data Availability Statement: The data presented in this research work can be obtained from the corresponding authors for research purposes.

Conflicts of Interest: The authors declare no conflict of interest.

References

1. Constantine, J.A.; Dunne, T.; Ahmed, J.; Legleiter, C.; Lazarus, E. Sediment supply as a driver of river meandering and floodplain evolution in the Amazon Basin. *Nat. Geosci.* **2014**, *7*, 899–903. [CrossRef]
2. Zanoni, L.; Gurnell, A.; Drake, N.; Surian, N. Island dynamics in a braided river from analysis of historical maps and air photographs. *River Res. Appl.* **2008**, *24*, 1141–1159. [CrossRef]
3. James, J.S.H.; Rogala, T.; Fitzpatrick, F.A. Recent Planform Changes in the Upper Mississippi River. U.S. Geological Survey. Upper Midwest Environmental Sciences Center. 2020. Available online: https://www.usgs.gov/publications/recent-planform-changes-upper-mississippi-river (accessed on 1 September 2022).
4. Rădoane, M.; Perşoiu, I.; Cristea, I.; Chiriloaei, F. River channel planform changes based on succesive cartographic data. A methodological approach. *J. Geomorphol.* **2013**, *15*, 69–88.
5. Monegaglia, F.; Zolezzi, G.; Güneralp, I.; Henshaw, A.J.; Tubino, M. Automated extraction of meandering river morphodynamics from multitemporal remotely sensed data. *Environ. Model. Softw.* **2018**, *105*, 171–186. [CrossRef]
6. Deb, M.; Das, D.; Uddin, M. Evaluation of Meandering Characteristics Using RS & GIS of Manu River. *J. Water Resour. Prot.* **2012**, *4*, 163–171. [CrossRef]
7. Dey, A.; Bhattacharya, R.K. Monitoring of River Center Line and Width—A Study on River Brahmaputra. *J. Indian Soc. Remote Sens.* **2013**, *42*, 475–482. [CrossRef]
8. Monegaglia, F. Meandering Rivers Morphodynamics—Integrating Nonlinear Modeling and Remote Sensing. 2017, p. 250. Available online: http://qmro.qmul.ac.uk/xmlui/handle/123456789/33927 (accessed on 1 September 2022).
9. Ashworth, P.J.; Best, J.L.; Roden, J.E.; Bristow, C.S.; Klaassen, G.J. Morphological evolution and dynamics of a large, sand braid-bar, Jamuna River, Bangladesh. *Sedimentology* **2000**, *47*, 533–555. [CrossRef]
10. Gilvear, D.; Winterbottom, S.; Sichingabula, H. Character of channel planform change and meander development: Luangwa River, Zambia. *Earth Surf. Process. Landf.* **2000**, *25*, 421–436. [CrossRef]
11. Mosselman, E. Morphological modelling of rivers with erodible banks. *Hydrol. Process.* **1998**, *12*, 1357–1370. [CrossRef]
12. Mosselman, E. A review of mathematical models of river planform changes. *Earth Surf. Process. Landf.* **1995**, *20*, 661–670. [CrossRef]
13. Winterbottom, S.; Gilvear, D. A GIS-based approach to mapping probabilities of river bank erosion: Regulated River Tummel, Scotland. *Regul. Rivers Res. Manag.* **2000**, *16*, 127–140. [CrossRef]

14. Boothroyd, R.; Williams, R.; Barrett, B.; Hoey, T.; Tolentino, P.; Perez, J.; Guardian, E.; David, C.; Yang, X. Detecting and Quantifying Morphological Change in Tropical Rivers Using Google Earth Engine and Image Analysis Techniques. In *River Flow 2020*; CRC Press: Boca Raton, FL, USA, 2020; pp. 1013–1021. [CrossRef]
15. Basnayaka, V.; Samarasinghe, J.T.; Gunathilake, M.B.; Muttil, N.; Hettiarachchi, D.C.; Abeynayaka, A.; Rathnayake, U. Analysis of Meandering River Morphodynamics Using Satellite Remote Sensing Data—An Application in the Lower Deduru Oya (River), Sri Lanka. *Land* **2022**, *11*, 1091. [CrossRef]
16. Hemati, M.; Hasanlou, M.; Mahdianpari, M.; Mohammadimanesh, F. A Systematic Review of Landsat Data for Change Detection Applications: 50 Years of Monitoring the Earth. *Remote Sens.* **2021**, *13*, 2869. [CrossRef]
17. Wray, R.A. Palaeochannels of the Namoi River Floodplain, New South Wales, Australia: The use of multispectral Landsat imagery to highlight a Late Quaternary change in fluvial regime. *Aust. Geogr.* **2009**, *40*, 29–49. [CrossRef]
18. Schwenk, J.; Khandelwal, A.; Fratkin, M.; Kumar, V.; Foufoula-Georgiou, E. High spatiotemporal resolution of river planform dynamics from Landsat: The RivMAP toolbox and results from the Ucayali River. *Earth Space Sci.* **2017**, *4*, 46–75. [CrossRef]
19. Nagel, G.W.; Novo, E.M.L.D.M.; Martins, V.S.; Campos-Silva, J.V.; Barbosa, C.C.F.; Bonnet, M.P. Impacts of meander migration on the Amazon riverine communities using Landsat time series and cloud computing. *Sci. Total Environ.* **2022**, *806*, 150449. [CrossRef]
20. Shelton, S.; Lin, Z. Streamflow Variability in Mahaweli River Basin of Sri Lanka during 1990–2014 and Its Possible Mechanisms. *Water* **2019**, *11*, 2485. [CrossRef]
21. Rubel, F.; Kottek, M. Observed and projected climate shifts 1901-2100 depicted by world maps of the Köppen-Geiger climate classification. *Meteorol. Z.* **2010**, *19*, 135–141. [CrossRef]
22. Withanachchi, S.S.; Köpke, S.; Withanachchi, C.R.; Pathiranage, R.; Ploeger, A. Water Resource Management in Dry Zonal Paddy Cultivation in Mahaweli River Basin, Sri Lanka: An Analysis of Spatial and Temporal Climate Change Impacts and Traditional Knowledge. *Climate* **2014**, *2*, 329–354. [CrossRef]
23. Ranasinghe, P.; Fernando, G.; Dissanayake, C.; Rupasinghe, M. Stream sediment geochemistry of the Upper Mahaweli River Basin of Sri Lanka—Geological and environmental significance. *J. Geochem. Explor.* **2008**, *99*, 1–28. [CrossRef]
24. Dissanayake, C.B.; Weerasooriya, S.V.R. The environmental chemistry of Mahaweli river, Sri Lanka. *Int. J. Environ. Stud.* **1986**, *28*, 207–223. [CrossRef]
25. Hansen, M.C.; Loveland, T.R. A review of large area monitoring of land cover change using Landsat data. *Remote Sens. Environ.* **2012**, *122*, 66–74. [CrossRef]
26. Isikdogan, F.; Bovik, A.; Passalacqua, P. RivaMap: An automated river analysis and mapping engine. *Remote Sens. Environ.* **2017**, *202*, 88–97. [CrossRef]
27. Feyisa, G.L.; Meilby, H.; Fensholt, R.; Proud, S.R. Automated Water Extraction Index: A new technique for surface water mapping using Landsat imagery. *Remote Sens. Environ.* **2014**, *140*, 23–35. [CrossRef]
28. Crist, E.P. A TM Tasseled Cap equivalent transformation for reflectance factor data. *Remote Sens. Environ.* **1985**, *17*, 301–306. [CrossRef]
29. McFeeters, S.K. The use of the Normalized Difference Water Index (NDWI) in the delineation of open water features. *Int. J. Remote Sens.* **1996**, *17*, 1425–1432. [CrossRef]
30. Xu, H. Modification of normalised difference water index (NDWI) to enhance open water features in remotely sensed imagery. *Int. J. Remote Sens.* **2006**, *27*, 3025–3033. [CrossRef]
31. Fisher, A.; Flood, N.; Danaher, T. Comparing Landsat water index methods for automated water classification in eastern Australia. *Remote Sens. Environ.* **2016**, *175*, 167–182. [CrossRef]
32. Danaher, T.; Collett, L. Development, Optimisation and Multi-Temporal Application of a Simple Landsat based Water Index. In Proceedings of the The 13th Australasian Remote Sensing and Photogrammetry Conference, Canberra, Australia, 20–24 November 2006.
33. Raheem, F.S.; Alwan, A.H. Development of a New Water Index for Landsat Operational Land Imager (OLI) Data Using Bayesian Optimization. *J. Adv. Res. Dyn. Control. Syst.* **2018**, *10*, 1–6. Available online: http://earthexplorer.usgs.gov (accessed on 1 September 2022).
34. Zhang, G.; Zheng, G.; Gao, Y.; Xiang, Y.; Lei, Y.; Li, J. Automated Water Classification in the Tibetan Plateau Using Chinese GF-1 WFV Data. *Photogramm. Eng. Remote Sens.* **2017**, *83*, 509–519. [CrossRef]
35. Xia, H.; Zhao, J.; Qin, Y.; Yang, J.; Cui, Y.; Song, H.; Ma, L.; Jin, N.; Meng, Q. Changes in Water Surface Area during 1989–2017 in the Huai River Basin using Landsat Data and Google Earth Engine. *Remote Sens.* **2019**, *11*, 1824. [CrossRef]
36. Zou, Z.; Xiao, X.; Dong, J.; Qin, Y.; Doughty, R.B.; Menarguez, M.A.; Zhang, G.; Wang, J. Divergent trends of open-surface water body area in the contiguous United States from 1984 to 2016. *Proc. Natl. Acad. Sci. USA* **2018**, *115*, 3810–3815. [CrossRef]
37. Zou, Z.; Dong, J.; Menarguez, M.A.; Xiao, X.; Qin, Y.; Doughty, R.B.; Hooker, K.V.; Hambright, K.D. Continued decrease of open surface water body area in Oklahoma during 1984–2015. *Sci. Total Environ.* **2017**, *595*, 451–460. [CrossRef] [PubMed]
38. Chen, B.; Xiao, X.; Li, X.; Pan, L.; Doughty, R.; Ma, J.; Dong, J.; Qin, Y.; Zhao, B.; Wu, Z.; et al. A mangrove forest map of China in 2015: Analysis of time series Landsat 7/8 and Sentinel-1A imagery in Google Earth Engine cloud computing platform. *ISPRS J. Photogramm. Remote Sens.* **2017**, *131*, 104–120. [CrossRef]
39. USGS. Landsat Enhanced Vegetation Index. U.S. Geological Survey. Available online: https://www.usgs.gov/landsat-missions/landsat-enhanced-vegetation-index (accessed on 3 August 2022).
40. Rohatgi, A. Webplotdigitizer: Version 4.5. 2021. Available online: https://automeris.io/WebPlotDigitizer (accessed on 1 September 2022).
41. Furbish, D.J. River-bend curvature and migration: How are they related? *Geology* **1988**, *16*, 752. [CrossRef]

land

MDPI

Article

Physical Model Test on the Interface of Loess Fill Slope

Weijia Tan [1,2], Qiangbing Huang [1,2,*] and Xing Chen [1,2]

1 Department of Geological Engineering, Chang'an University, Xi'an 710054, China
2 Key Laboratory of Western China's Mineral Resources and Geological Engineering, Ministry of Education, Chang'an University, Xi'an 710054, China
* Correspondence: qiangbinghuang123@163.com

Abstract: The interface between the filling slope and the original slope is inevitable in the process of building a city in the loess area, which will affect the deformation and stability of the filling slope. In this paper, the loess fill slope of mountain excavation and city construction project in Yan'an City, China, is taken as the research object, and, based on field investigation and sampling, the effect of Loess Fill Slope Interface (LFSI) under rainfall is revealed by physical model test. The test samples were taken from a Loess Fill in Qilipu community, Yan'an, and three layers of sensors were arranged at the left and right interfaces of the original slope and the filled slope to monitor the water content, pore water pressure and deformation and failure characteristics during the experiment. The results show that ILFS is a rainfall dominant seepage channel, and the infiltration of rainfall along the interface lags behind. In addition, the variation laws of water content and pore water pressure at the interface between fill slope and original slope under rainfall are obtained. Finally, the failure process of loess fill slope under rainfall is summarized: local mud flow failure at the toe of the slope → erosion in the middle of the slope → crack initiation on the shoulder of the slope → local slip on the slope → crack propagation on the shoulder of the slope → shallow slip on the shoulder of the slope, and the instability mechanism of loess fill slope under rainfall is further revealed. The research results can provide theoretical and experimental reference for the protection of fill slopes in loess areas.

Keywords: loess; filled slope; physical model test; interface effect; stability

Citation: Tan, W.; Huang, Q.; Chen, X. Physical Model Test on the Interface of Loess Fill Slope. *Land* **2022**, *11*, 1372. https://doi.org/10.3390/land11081372

Academic Editors: Matej Vojtek, Andrea Petroselli and Raffaele Pelorosso

Received: 5 July 2022
Accepted: 16 August 2022
Published: 22 August 2022

1. Introduction

Large-scale mountain excavation and city construction project in the Loess Plateau of China not only brings great opportunities for development, but also risks of slope disasters and safety hazards [1,2]. Loess is loose in texture and easily softened by water [3]. Rainfall is a common source of surface water, which may infiltrate along the interface between loess fill slope and original slope, affecting the stability of loess fill slope in mountain excavation and city construction project [4,5], triggering slip and instability (Figure 1). Filling body is the main part of city building project, and is the most important part of the whole project; its strength directly affects the success of the whole project. Therefore, the stability of filling slope seriously affects the safety construction and operation of the project, and it is more important to study the interface effect and instability mechanism between fill slope and original slope under rainfall [6].

For the filling slope, predecessors have done a lot of research. Day [7] studied the filled slope and concluded that the slope was unstable due to the loading of the top of the slope and the shallow slope was not cleared in time. The damaged slope can be repaired by removing the sliding soil and building a retaining wall. Cheuk et al. [8] discussed the characteristics of soil nailing in fill slope by numerical simulation. The results show that the soil nailing structure can reduce the deformation of the loose fill slope caused by rainfall infiltration, so as to maintain the stability of the slope. Zhang [9] carried out the centrifugal model test of loess high fill embankment to study the development process and distribution of embankment settlement. The results show that the settlement in the middle of the embankment is greater than that of the shoulder, and the stability of the embankment slope will be significantly reduced when the slope rate is too large or the construction speed is too fast. Duan et al. [10] took the high fill slope as an example, the deformation and stability were analyzed by finite element method and limit equilibrium method, and compared with the monitoring data in construction. It was found that the stress and displacement results calculated by finite element method were in good agreement with the field monitoring data, indicating that the combination of finite element calculation and monitoring analysis can guide the construction and reinforcement of high fill slope. Zhao et al. [11] divided the rainfall infiltration process of the newly filled slope into three stages: free infiltration, scouring infiltration and stable infiltration. During rainfall, the slope top is mainly vertical expansion and contraction deformation, while the slope surface is mainly lateral free surface displacement. Wang et al. [12] combined the relative displacement sensing technology and GSM technology to monitor the fill slope near an airport, and the monitoring results can be fed back to the monitoring station in real time, which successfully warned the collapse of the monitoring point. The interface between the original slope and the fill slope will inevitably be generated in the process of mountain excavation and city construction project, and sliding failure may occur due to the different strength properties of the soil on both sides. Through indoor physical model tests, Chang et al. [13] monitored and analyzed the hydromechanical parameters of the loess fill slope, simulated the failure mode of the loess fill slope, and proposed engineering measures to prevent and control the instability of the loess fill slope.

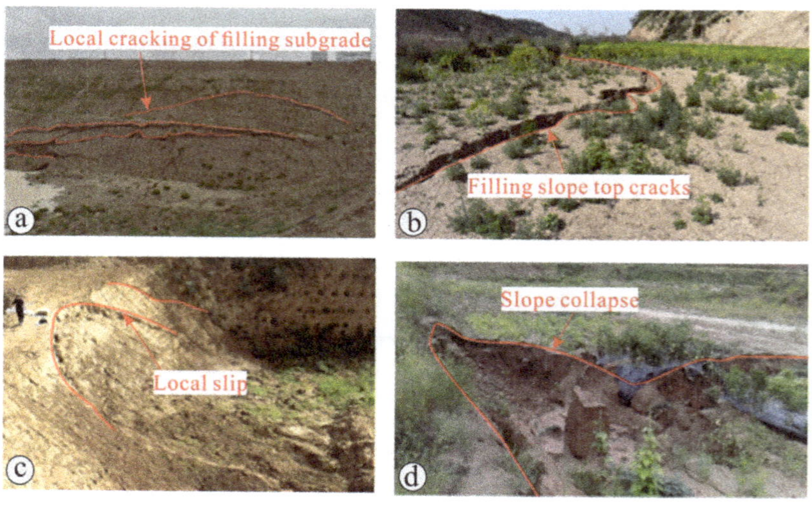

Figure 1. Loess filling slope disaster. (**a**) Local cracking of filling subgrade, (**b**) filling slope top cracks, (**c**) local slip of fill slope and (**d**) collapse of channel fill slope.

Rainfall can induce slope failure [14–16], and more attention should be paid to the effect of rainfall on loess filled slope. Through a series of laboratory slope failure tests, Tohari et al. [17] recorded the hydrological response of the model slope to the saturation process by using the volumetric soil moisture sensor, and proposed the concept of slope failure prediction method induced by rainfall. Saadatkhah et al. [18] used the instantaneous rainfall infiltration and grid-based regional slope stability analysis model, combined with spatial rainfall distribution model, found that local daily rainfall is not the only factor affecting slope stability, and long-term early rainfall may play a certain role in the formation of slope failure mechanism. Hakro [19] believed that the failure of a slope is caused by the increase of water content and pore pressure through indoor rainfall model test, and the pore pressure will increase sharply in the failure process. Gallage et al. [20] studied the influence of slope inclination on slope stability by artificial rainfall test. The results show that the slope is more prone to sudden collapse with the increase of slope angle in the process of rainfall. Model test is an important means to study the deformation and failure mechanism of slope affected by rainfall [21]. Therefore, a physical model test can be carried out on loess filled slope.

In summary, previous scholars have studied the deformation and failure process, instability mechanism, seepage field and stability of fill slope by using physical model test, field test, field monitoring, numerical simulation or a combination of multiple methods, and achieved a series of important results. However, there is a little research on the interface effect between the loess fill slope and the original slope, and the research on the variation of water content and pore pressure at the loess filling interface and the stability of the slope under rainfall conditions is not deep enough. In this paper, the variation of hydrological parameters of original slope and fill slope caused by rainfall infiltration and some understandings of the interface effect of loess fill slope are obtained by physical model test. The deformation characteristics and failure process of loess fill slope caused by rainfall are also summarized. The research results can provide reference for the deformation and stability of fill slope in mountain excavation and city construction project.

2. Experimental Design

2.1. Experimental Flume

A rigid model box with transparent organic glass on both sides is selected for physical model test, and the front end is a water tank (Figure 2a,b). The size of the model box is 3.2 m × 1.4 m × 1.5 m and the size of the water tank is 1.4 m × 0.5 m × 0.3 m.

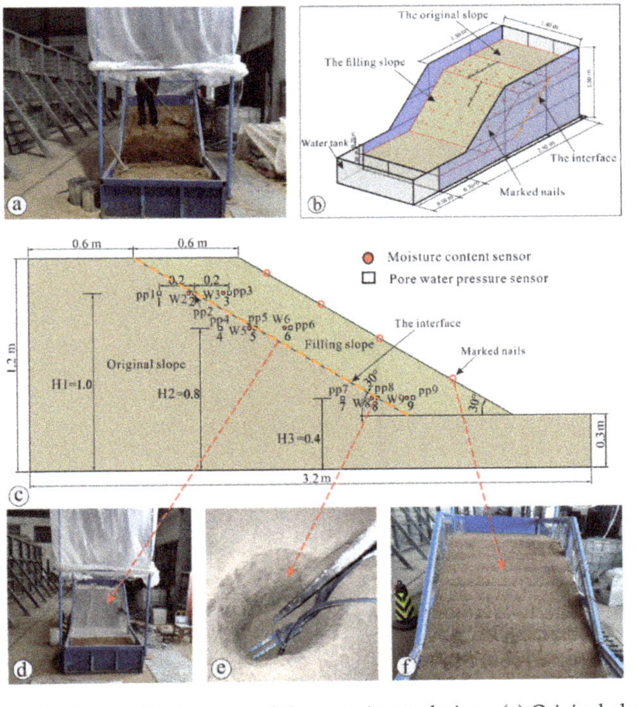

Figure 2. Schematic diagrams of the experiment design. (**a**) Original slope cutting, (**b**) model stereogram, (**c**) model side view, (**d**) slip zone model-Plastic mesh, (**e**) sensor embedding and (**f**) layout of slope tracing points-Plastic mesh.

2.2. Instrumentation

The experiment mainly includes rainfall device and sensor monitoring equipment. The rainfall device is mainly composed of water pipeline, water tank, water pump, rainfall sprinkler, portable control center and rainfall meter. In order to ensure the uniformity of rainfall, the rainfall height was designed to be 6 m according to the top area and height of the model. In the control system interface, the continuous change of rainfall intensity can be realized by adjusting the opening degree of 0–150 mm/h, and the specific rainfall intensity value can be obtained by connecting the rain gauge with the control system. The sensors are pore water pressure sensor and moisture content sensor. The range of pore water pressure sensor is −20–20 kpa, the output voltage range is 0–5 v, and the acquisition frequency can reach 1 Hz, that is, 1 s can collect a datum. The moisture content sensor can collect one datum in one minute, and the mass moisture content can be obtained by dividing the corresponding dry density.

2.3. Materials and Methods

The test samples were taken from a Loess Fill in Qilipu community, Yan'an, Shaanxi, China (Figure 3), and the retrieved soil samples were screened for 5 mm to remove large particles and impurities. According to the actual situation, the dry density of the original slope in the model test is 1.63 g/cm^3, the water content is 13.5%, and the natural gravity (the natural gravity of soil refers to the weight of soil under the condition of natural moisture content, which is equal to the ratio of the total weight of soil to the total volume of soil) is 18.53 kg/m^3. The dry density of the fill slope is 1.58 g/cm^3, the moisture content is 10%, and the natural gravity is 17.38 kg/m^3. Before the beginning of the model test, the moisture content of the packaged soil was measured by the drying method, and the moisture content was about 16.3%. Therefore, the soil needs to be turned and aired. The measured water content after airing was about 14.2%. Considering the water loss in the filling process, the original slope was filled directly with the water content. Test slope model length × wide × Height = 3.2 m × 1.4 m × 1.2 m, the filling slope height is 0.9 m, and the interface between the original slope and the filling body and the filling slope angle are taken as 30° (Figure 2).

The modeling idea is to use the dry density and moisture content to jointly control the manual compaction. Each 10 cm is divided into one layer, and the required mass of each layer is calculated. Taking into account the slope shape to be formed in the later stage, the soil that is appropriate greater than the calculated mass is weighed. The filling of the layer is carried out, and the density of the compacted soil layer is measured by the ring knife method. After meeting the requirements, the next layer of filling is carried out. Firstly, the original slope is filled on the left side, and it is placed for a week. Then, the dry density and moisture content of the filling area are controlled on the right side. The slope on the right side is filled as the filling slope. When filling the filling slope, the penetration test ring knife sample is taken to fill the filling area. The sensor part is buried using Luoyang shovel. According to the meteorological data of Yan 'an, the maximum rainfall in an hour was 62 mm (1979) and the maximum daily rainfall was 139.9 mm (1981). According to the similarity ratio of rainfall intensity, the model rainfall intensity is 12.4 mm/h. Since the uniformity of rainfall device is poor at low rainfall intensity, and considering the uniformity of rainfall time, by controlling the hourly rainfall to be 12.4 mm, 9 h–18 h per day, the rain intensity was 24.8 mm/h. Observation and data collection are carried out during and after rainfall. In the evening, the model was covered with plastic cloth to keep moisture until the slope no longer changed significantly, and the experiment lasted for eight days. Three layers of sensors were arranged on the left and right interface of the original slope and the filled slope, with elevations of 1.0 m, 0.8 m and 0.4 m, respectively (Figure 2c,e). There were six moisture measuring points and nine pore pressure measuring points in total, and the collection interval of the two sensors was 1 min, which was mainly used to analyze the interface effect on the same horizontal line.

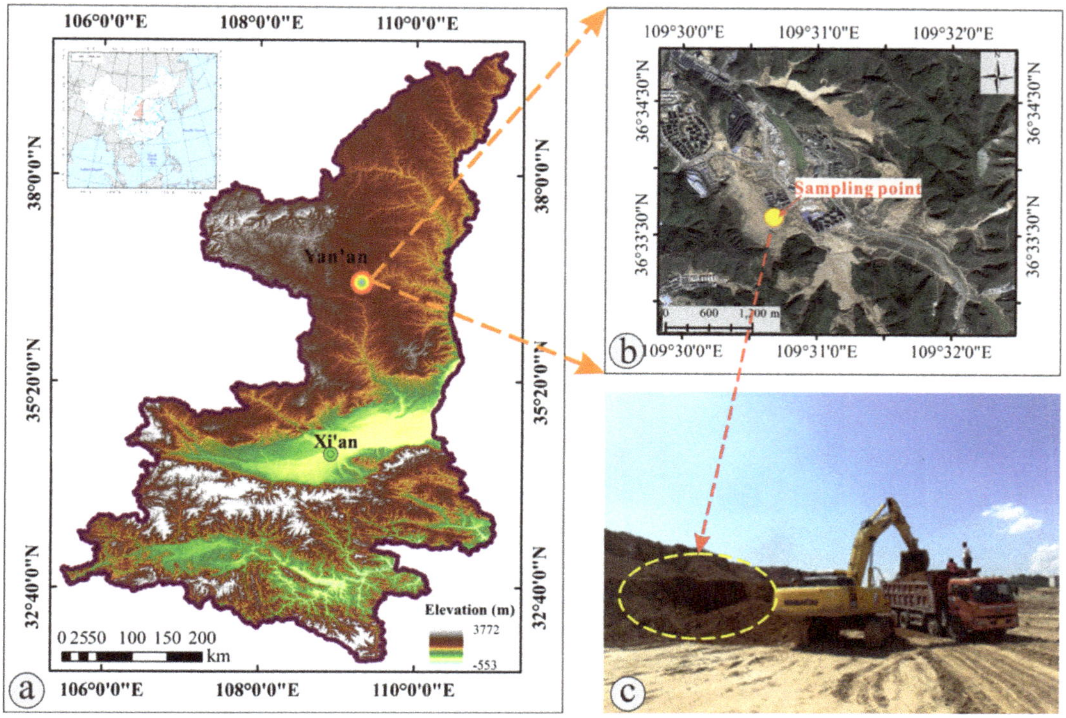

Figure 3. Location map of sampling points. (**a**) Sampling site, Shanxi Province; (**b**) loess fill in Qilipu community, Yan'an; (**c**) soil samples for test.

In order to simulate the interface effect between loess fill and original slope, plastic mesh is used as sliding zone in this test [22] (Figure 2d). After the completion of slope filling, marked nails are placed on the slope and the top of the fill slope in order to observe the deformation characteristics of the slope and the top of the fill slope and analyze the deformation and failure process of the fill slope (Figure 2f). Specifically, set for the top of the slope inserts two rows, slope inserts four rows and each row inserts five marked nails.

3. Results

3.1. Variation of Slope Water Content

3.1.1. Variation of Water Content at Top of Slope (H1 = 1.0 m)

Figure 4 shows the variation curve of water content increment with time at different measuring points at the top of the slope (H1 = 1.0 m). In the LFSI and the filling slope, the variation of water content with time under the action of rainfall is similar: the water content first increases gradually, and then gradually decreases after a period of rainfall when it reaches a certain stable value or maximum value. Before the end of the daily rainfall, the increment of volumetric water content at the interface W2 is greater than that in the filling body W3, indicating that LFSI is an easy channel for rainfall seepage. The time of water content decline at W2 measurement point was later than that at W3 after the end of rainfall, indicating that part of the water will accumulate at LFSI and continue to permeate along LFSI after rainfall, and the water at W2 at the interface is supplemented, resulting in the lag of water content decline at W2.

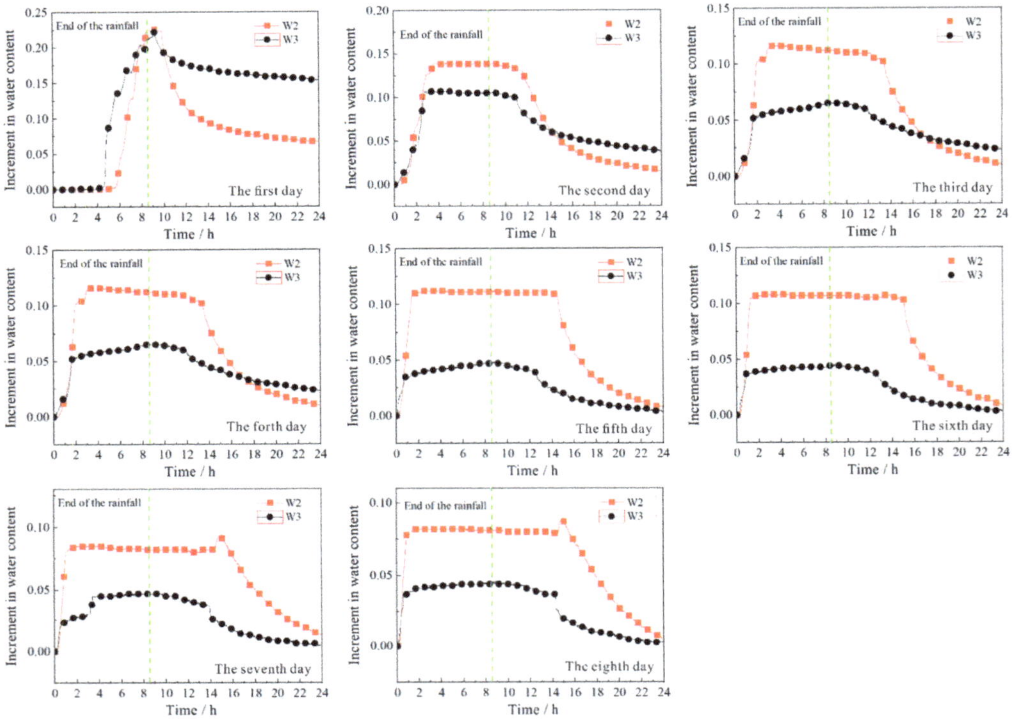

Figure 4. Variation of water content in the following days, from the first to the eighth (H1 = 1.0 m).

3.1.2. Variation of Water Content in the Middle and Upper Slope (H2 = 0.8 m)

Figure 5 shows the variation of water content in the middle and upper slope (H2 = 0.8 m) (W5 is located at LFSI and W6 is located in the filling body). It can be found that the water content in the middle and upper part of the slope increases gradually with time under the action of rainfall. When it increases to a certain stable value or maximum value, it gradually decreases after a period of rainfall, which is basically consistent with the variation trend of W2 and W3 measuring points at the top of the slope. The increment of volume moisture content before the end of rainfall shows that water content W5 is greater than W6, indicating that LFSI has infiltration advantage. The reason for the abnormality on the third day may be that the water content at W6 increased for a period after the rainfall on the second day. It is speculated that cracks occur near W6, and the rainwater is easy to gather at the measuring point, resulting in the increment of moisture content of W6. At 17 h 26 min on the third day, a shallow slope slip occurred, and the stress inside the slope was redistributed, resulting in no obvious seepage advantage at W6. Therefore, the advantageous seepage at the interface is reflected again after the fourth day. The time of water content decline in W5 after the end of rainfall was later than that in W6, indicating that water distribution at the end of rainfall was collected at the interface and continued to infiltrate along the interface, supplementing water content in W5, resulting in lagging water content decline in W5.

3.1.3. Variation of Water Content in Slope Toe (H3 = 0.4 m)

Figure 6 shows the variation of water content increment at different measuring points at the foot of slope (H3 = 0.4 m) with time (W8 is located at LFSI and W9 is located in the filling body). From the fourth day of rainfall, with the increase of rainfall time, the increment of moisture content of W8 began to be greater than that of W9, which also indicates that there is a lag phenomenon of rainfall infiltration along LFSI.

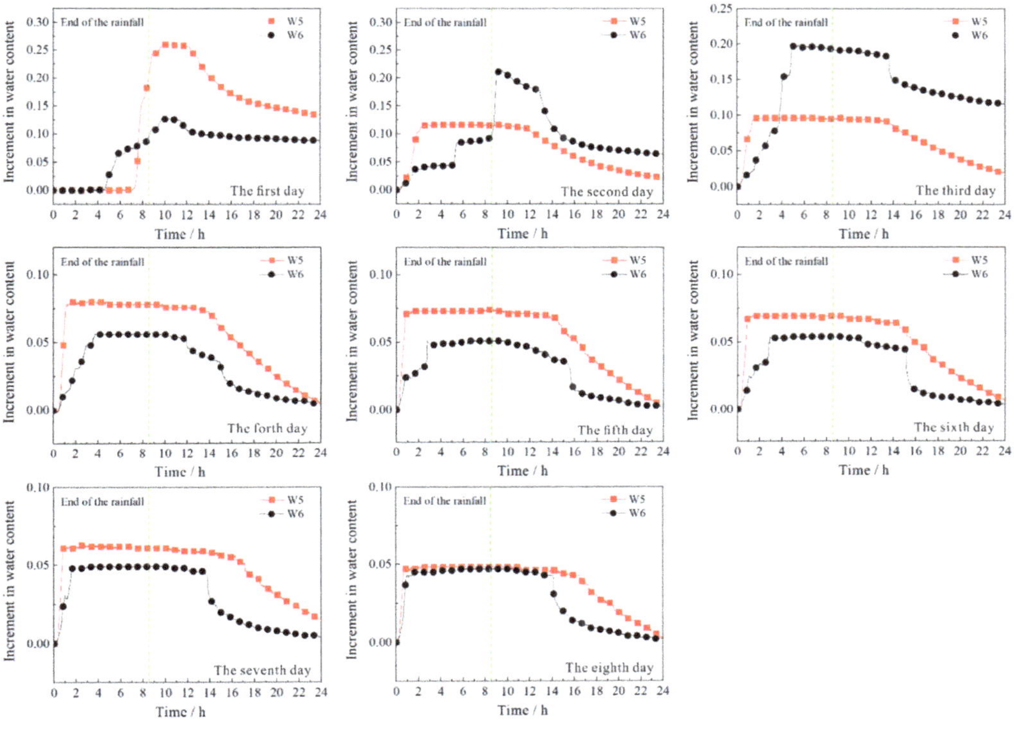

Figure 5. Variation of water content in the following days, from the first to the eighth (H2 = 0.8 m).

According to the analysis of the above test results, the essence of the "interface effect" of the loess fill slope is that the interface is a rainfall dominant seepage channel, and there is a lag phenomenon of rainfall infiltration along the interface (Figures 4–6), that is, after the end of rainfall, part of the water will continue to be collected at LFSI and penetrate along LFSI.

3.2. Variation of Pore Water Pressure of Slope

Pore water pressure increment (the pore water pressure measured at each point at each time was subtracted from the pore water pressure before the test (9 h per day)) is adopted for analysis.

3.2.1. Variation of Pore Water Pressure at Top of Slope (H1 = 1.0 m)

Figure 7 shows the curve of pore water pressure increment versus time at the top of slope (H1 = 1.0 m) (pp1 is located in the original slope, pp2 is located in the LFSI, and pp3 is located in the fill slope, where pp represents pore water pressure, the number is the corresponding measuring point position number). It can be found that the pore pressure at each measuring point increases sharply first and then decreases in the process of rainfall. The change of pore water pressure mainly comes from seepage. In the process of rainfall, rainwater infiltrates from the top of the slope and the slope surface, and seepage occurs inside the slope, which makes the pore water pressure gradually increase. As the infiltration of rainwater gradually forms a transient saturated zone on the surface of the slope body, the infiltration rate of rainwater decreases, resulting in a slow or declining increase of pore pressure. Pore water pressure decreases gradually after rainfall due to the lack of water replenishment and evaporation of slope.

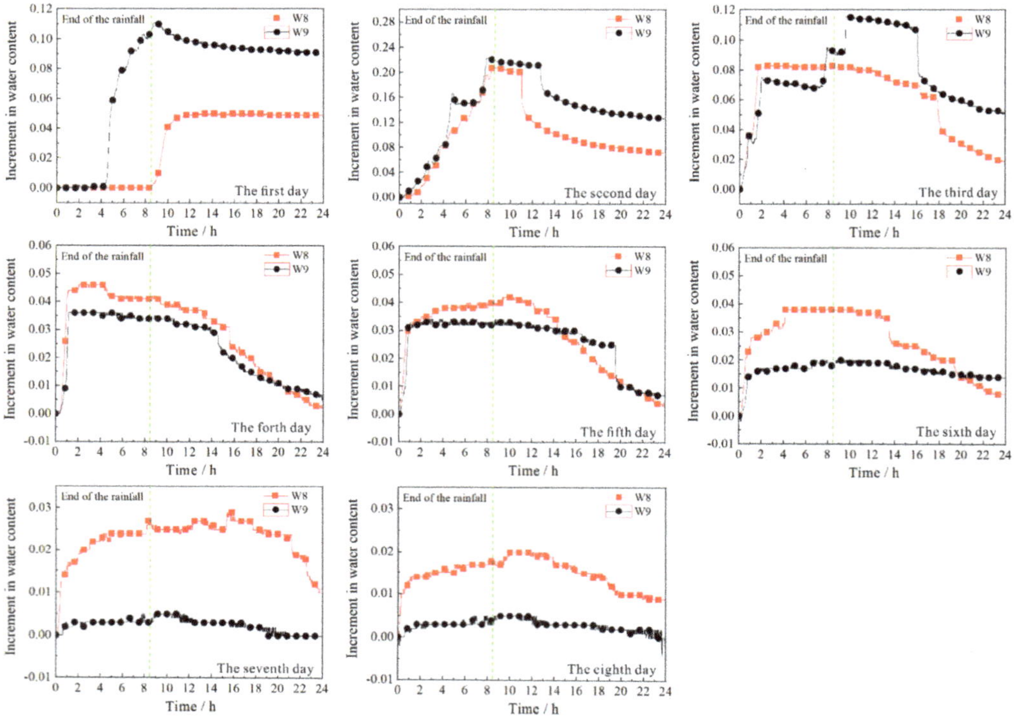

Figure 6. Variation of water content in the following days, from the first to the eighth (H3 = 0.4 m).

3.2.2. Variation of Pore Water Pressure in Middle and Upper Slope (H2 = 0.8 m)

Figure 8 shows the curve of pore water pressure increment versus time in middle and upper slope (H2 = 0.8 m) (pp5 is located in the original slope, pp6 is located in the LFSI, and pp6 is located in the fill slope). It can be found that the pore pressure at each measuring point gradually increases sharply, then increases slowly or decreases, and decreases gradually after the end of rainfall.

3.2.3. Variation of Pore Water Pressure in Slope Toe (H3 = 0.4 m)

Figure 9 shows the curve of pore water pressure increment versus time in slope toe (H3 = 0.4 m) (pp7 is located in the original slope, pp8 is located in the LFSI, and pp9 is located in the fill slope). As can be seen from Figure 9, pore water pressure at each measuring point first rises sharply, then rises slowly or decreases, and gradually decreases after the end of rainfall. The pore pressure data fluctuated significantly after the rainfall at pp9 measurement point on the fourth day, which may be caused by the mudflow destruction at the foot of slope on the third day and the rainfall on the fourth day, resulting in the development of cracks near pp9 measurement point, and the pore pressure oscillated significantly under the action of rainfall in the following days.

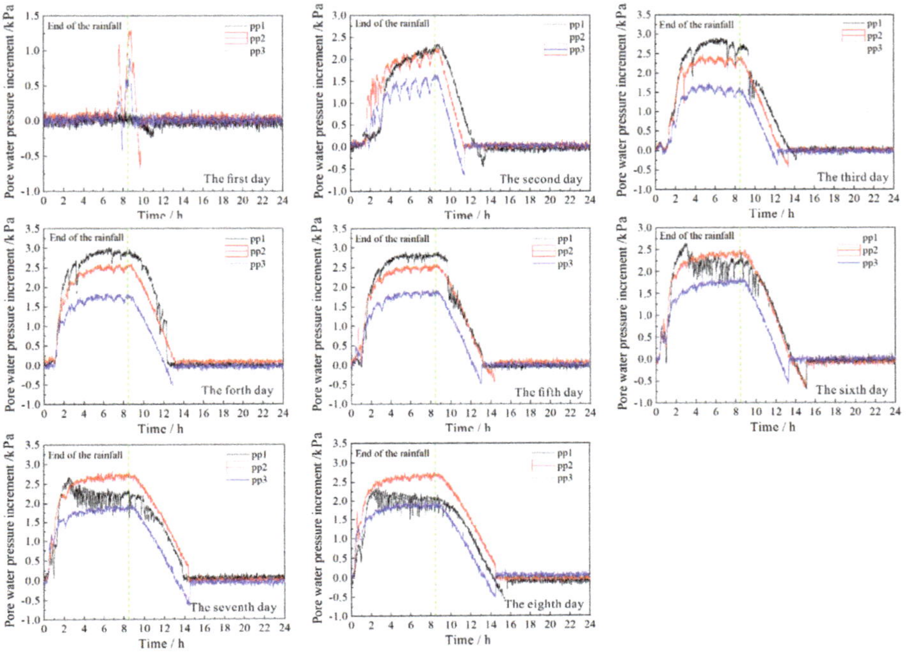

Figure 7. Variation of pore water pressure in the following days, from the first to the eighth (H1 = 1.0 m).

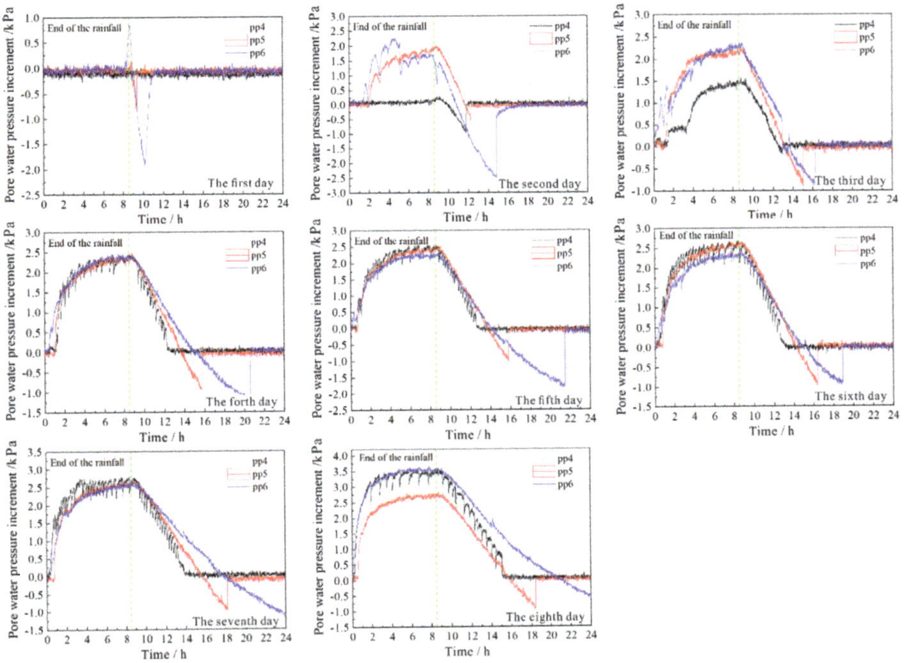

Figure 8. Variation of pore water pressure in the following days, from the first to the eighth (H2 = 0.8 m).

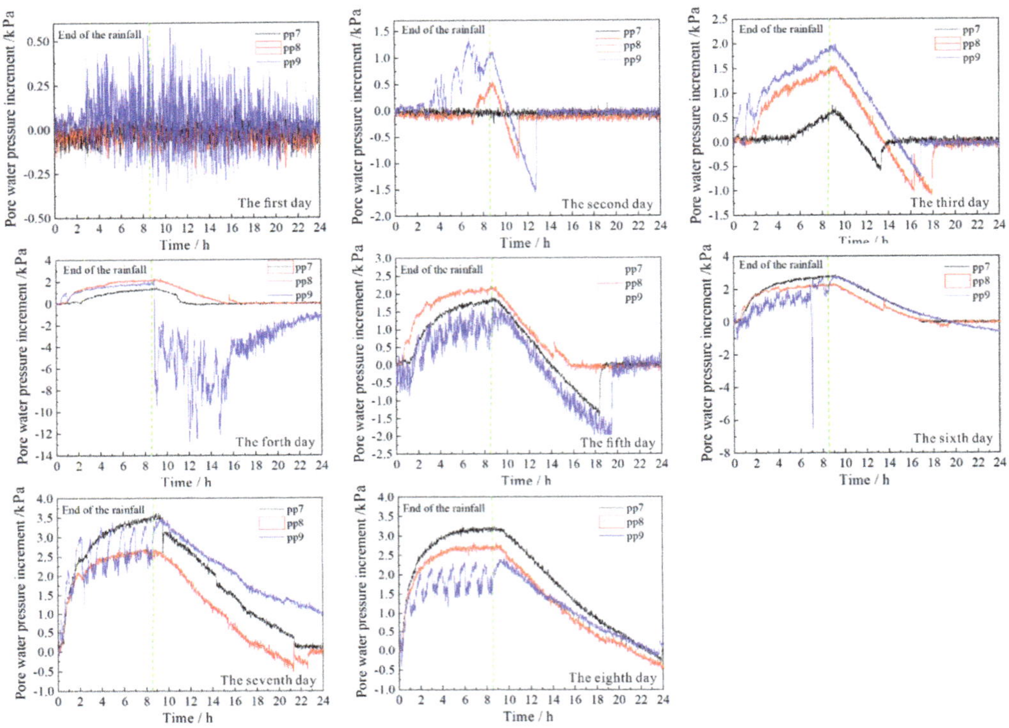

Figure 9. Variation of pore water pressure in the following days, from the first to the eighth (H3 = 0.4 m).

4. Discussion

4.1. Variation Characteristics of Water Content and Pore Pressure

Figure 10a shows the daily maximum increment change curves of water content at W2 and W3 at the top of the slope (H1 = 1.0 m). It can be seen that the maximum daily increment of water content Δw gradually decreases with the increase of rainfall days, and the cumulative increment value is 99.5% and 63.4%, respectively, which may be because with the increase of rainfall days, the soil at W2 and W3 is closer to the saturated water content [23], resulting in the decrease of the maximum increment, which is consistent with the model test conclusion of Crosta [24] and Li et al. [25]. In Figure 10b, W5 shows a similar law with the W2 and W3, and the cumulative increment value is 79.2%, while the daily maximum increment of water content of W6 in the filling slope increases first and then decreases gradually with the increase of rainfall days. It can be seen from Figure 10c that the daily maximum increment of water content at W8 and W9 increased first and then decreased gradually with the increase of rainfall days, and both reached the maximum value on the second day of rainfall, but the cumulative increment value is55.0% and 51.7%, respectively. It is speculated that this may be related to scouring damage in the middle of slope on the second day of rainfall.

Figure 11a shows the daily maximum increment change curves of pore water pressure at the top of the slope (H1 = 1.0 m). On the first to fourth day of rainfall, pp1, pp2 and pp3 near the top of the slope showed that the daily maximum pore water pressure increment $\Delta\mu$ gradually increases and the increasing speed is slower with the increase of rainfall days. Under the action of rainfall, the daily maximum pore pressure increment at LFSI (pp2) near the top of the slope is greater than that of the fill slope (pp3), which once again shows that rainfall at LFSI is easy to infiltrate. As can be seen from Figure 11b, the daily maximum pore pressure increments at pp4, pp5 and pp6 are characterized by a sharp increase at the beginning of the rainfall, then a steady increase, and finally an increase trend. The daily maximum increments at LFSI and within the filling slope are greater than those within the original slope (pp4 measuring point), while the daily maximum pore pressure increment at pp5 at LFSI is basically stable in the later stage of rainfall. It can be seen from Figure 11c that the daily maximum pore pressure increment of each measuring point pp7, pp8 and pp9 on the first to fourth day of rainfall gradually increases with the increase of rainfall days. Locally, the daily maximum pore pressure increment of pp9 at the foot of the fill slope first decreases, then increases and finally decreases, which is closely related to the process of water enrichment at the foot of the slope and water seepage after slump, and also consistent with the experimental results of Chueasamat et al. [26] and Orense et al. [27].

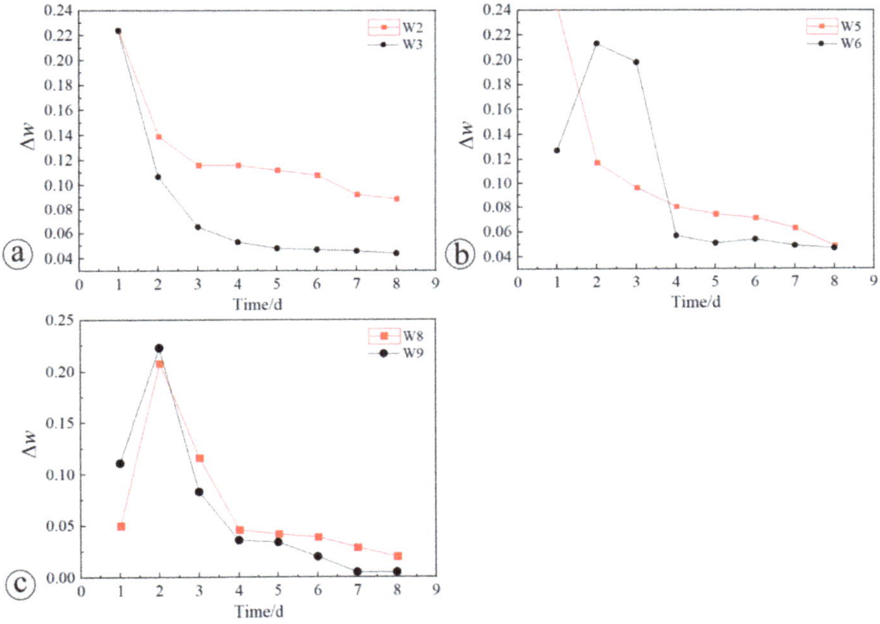

Figure 10. Daily maximum increment of moisture content at different measuring points (**a**) H1 = 1.0 m, (**b**) H2 = 0.8 m and (**c**) H3 = 0.4 m.

4.2. Analysis of Slope Deformation Characteristics and Failure Process

The increase of pore water pressure will cause slope failure during rainfall [28–30]. The foot of the filled slope is the most vulnerable place to damage under the action of rainfall [13,17,26,27], which is due to the low terrain at the toe of the slope, the rainwater is easy to collect, and the soil is soaked [17,27,31], resulting in the reduction of shear strength. At 14 h 25 min on the first day of rainfall, two mud flow failures occurred at the slope toe (Figure 12a). Figure 12b shows that there are many small-scale mud flow damages on the slope surface at 17 h 45 min after the end of rainfall on the first day, and local erosion damages occur in the middle and both sides. Figure 12c shows the picture of slope surface 18 h after the end of rainfall on the second day. The slope erosion continued to intensify on the second day of rainfall, and the erosion on the right side is the most serious. Obvious cracks appeared at 8 h 50 min before rainfall on the third day near the slope shoulder (Figure 12d), which may be related to the evaporation of water at night on the previous day. The crack L1 was about 18 cm from the slope shoulder, about 90 cm in length, and about 47 cm from the left side of the model box.

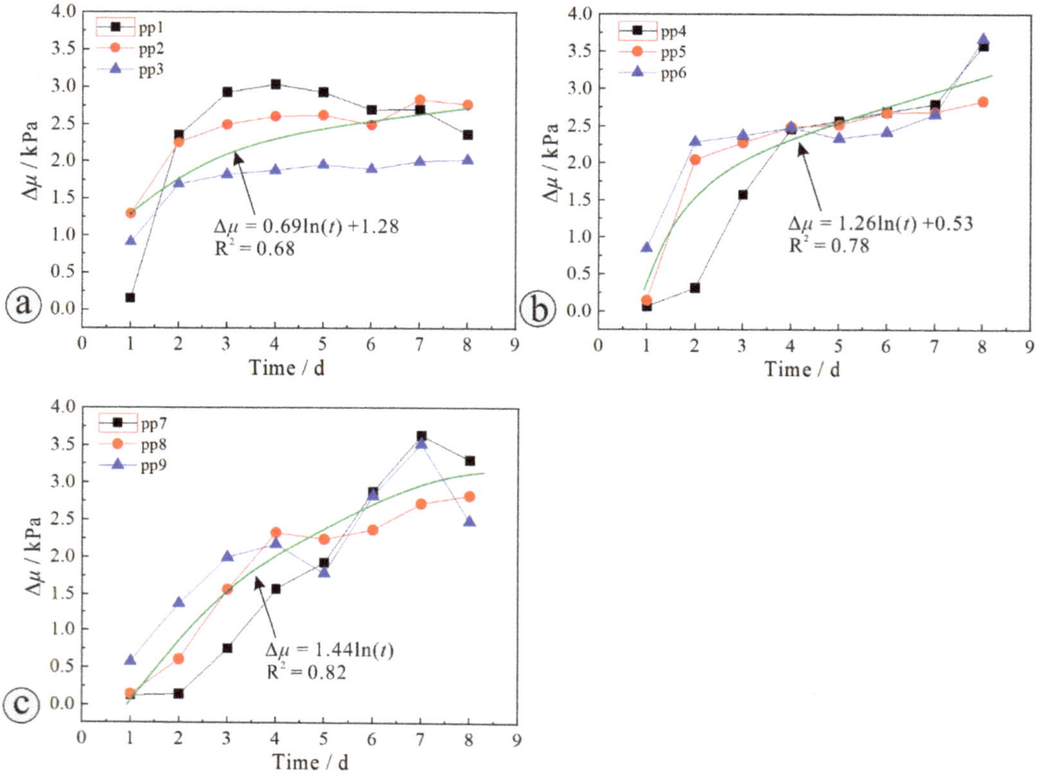

Figure 11. Daily maximum increment of pore pressure increment at different measuring points (**a**) H1 = 1.0 m, (**b**) H2 = 0.8 m and (**c**) H3 = 0.4 m.

At 17 h 26 min on the third day of rainfall, two large-scale shallow sliding failures occurred at about 0.85 m from the bottom of the slope (Figure 13a). The sliding width of the left side is about 74.2 cm and the right sliding width is about 65.8 cm, and the right sliding trailing edge was 22 cm more than the left sliding trailing edge. It can also be seen from slope monitoring landmarks 6-2 and 6-4 that there is also a small scale of sliding damage above shallow sliding. Under the action of long-term rainfall, shallow sliding failure is more likely to occur in the lower part of the slope, which is consistent with Wu et al. [32] and Kim et al. [33] model test. At 8 h 50 min before the rainfall on the fourth day, there were five cracks near the slope shoulder (Figure 13b), and four new cracks (L2, L3, L4 and L5) were generated on the slope surface and slope top near the slope shoulder. The length of L1 does not change, but the width increases; L2 and L3 are located on the slope, L2 is about 38 cm long, 22 cm away from the slope shoulder, and L3 is about 32 cm long, 35 cm away from the slope shoulder; L4 and L5 are located on the slope top, L4 is about 52 cm long, and the longest distance slope shoulder is about 14 cm and L5 is about 31 cm long, 12 cm away from the slope shoulder. The development of cracks indicates that the continuous action of rainfall produces the change of stress field in the slope [34,35], and the tensile stress is generated near the top and shoulder of the slope, which is easy to produce tensile failure [35,36].

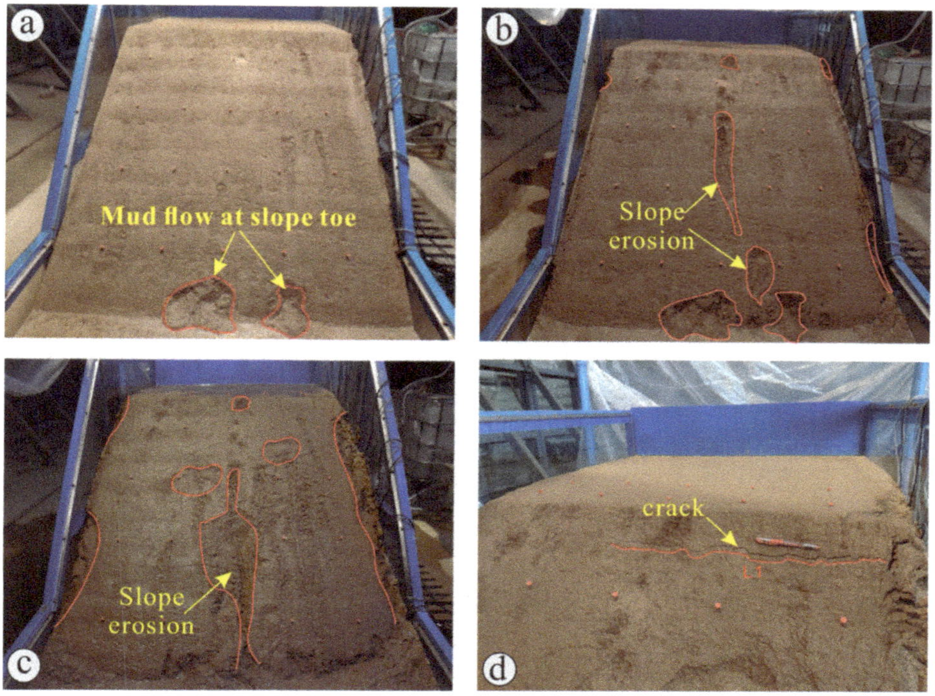

Figure 12. The deformation characteristics and failure process of slope. (**a**,**b**) Mud flow at slope toe and slope erosion on the first day, (**c**) slope erosion on the second day and (**d**) slope cracks on the third day.

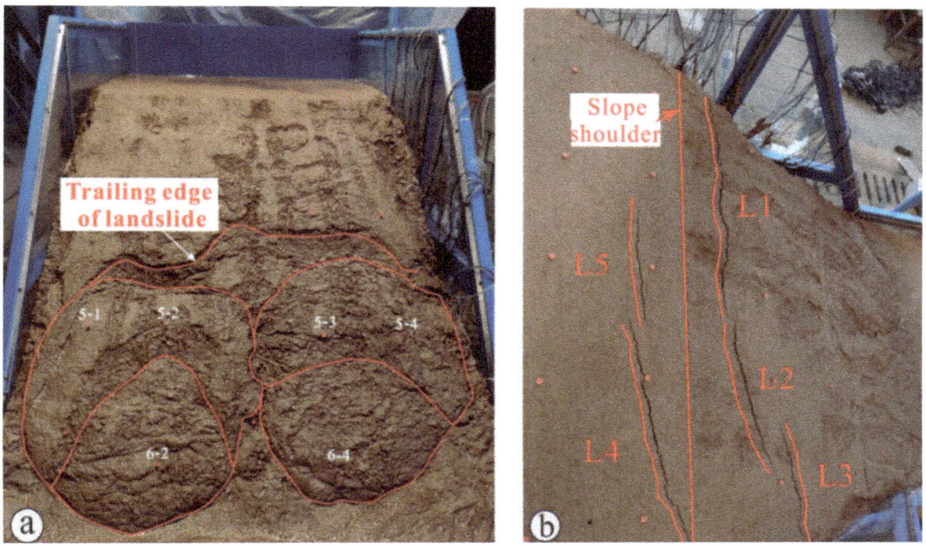

Figure 13. Large-scale shallow sliding failures. (**a**) Shallow sliding of slope toe, and (**b**) slope top and surface cracks.

Figure 14a shows the change of shoulder cracks in the first 8 h 50 min of daily rainfall from day four to day nine. Under the action of continuous rainfall, the cracks near the slope shoulder gradually increased, expanded and deepened, L4 and L5 cracks began to connect from the sixth day, and gradually developed in the later stage, but there was no significant change in the overall shape of the slope from the third day. Moreover, on the eighth day of rainfall, the slope body was still not damaged along the crack, which may be due to the blockage of the shear outlet of the sliding body on the slope shoulder by the mud flow accumulated at the slope foot (Figure 14b). Considering the gentle angle of the slope itself and no further change in the overall shape of the slope since the mud flow at the foot of the slope, the test was stopped after the rainfall on the eighth day.

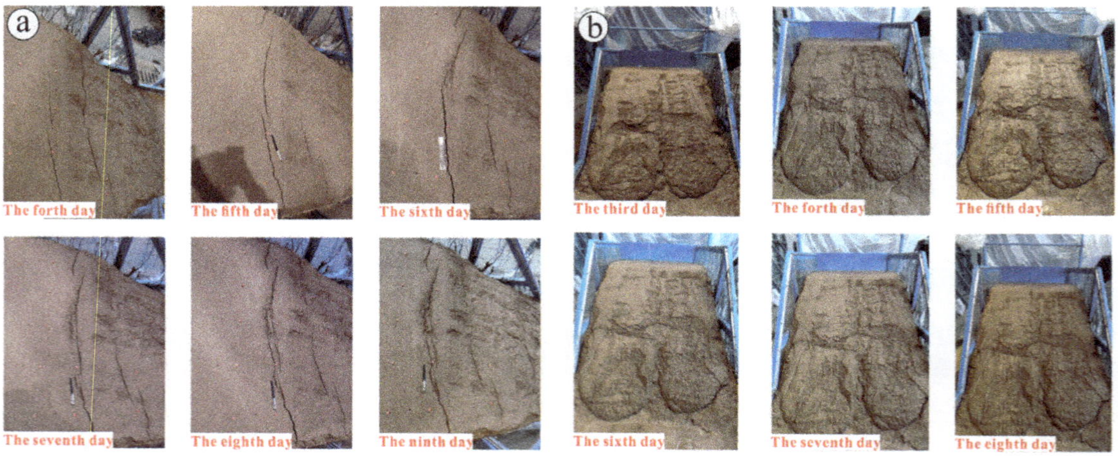

Figure 14. Deformation and failure process of fill slope under rainfall. (**a**) Crack development process near slope shoulder and (**b**) shallow sliding process.

In summary, the deformation and failure mode of loess fill slope under rainfall is shallow slip, and the failure process can be summarized as follows: local mud flow failure at the toe of the slope → erosion in the middle of the slope → crack initiation on the shoulder of the slope → local slip on the slope → crack propagation on the shoulder of the slope → shallow slip on the shoulder of the slope. Under the long-term effect of rainfall, the loess fill slope of mountain excavation and city construction project will first appear small mud flow damage at the foot of the slope, and then cracks appear at the shoulder of the slope. Rainwater infiltrates along the interface between the original slope and the filling slope and cracks, resulting in a sharp increase in the internal moisture content and pore pressure of the filling slope, which reduces the strength of the slope soil. Especially, under the action of repeated rainfall, the shallow surface soil will suffer obvious damage by dry-wet cycle [36,37], rainwater is more easily infiltrated, and the shear strength of soil will be reduced [38,39], resulting in the shallow landslide disaster of the filling slope. Therefore, it is necessary to pay attention to the treatment of slope shoulder and interface, and strengthen drainage and ecological protection measures for loess fill slope of mountain excavation and city construction project [13].

5. Conclusions

Taking the typical mountain excavation and city construction project fill slope in Yan'an as the research background, through the physical model test of filling slope considering interface effect under rainfall, the following conclusions are drawn:

(1) The essence of the "interface effect" of loess fill slope is that the interface is a dominant seepage channel of rainfall, and there is lag phenomenon of rainfall infiltration along the interface; that is, some water will continue to accumulate at the interface of loess fill slope after rainfall, and continue to infiltrate along the interface;
(2) The increase of pore water pressure will cause slope failure during rainfall, and the foot of the filled slope is the most vulnerable place to damage under the action of rainfall, which is due to the low terrain at the toe of the slope, the rainwater is easy to collect, and the soil is soaked, resulting in the reduction of shear strength. The appearance picture of slope deformation characteristics shows that shallow sliding failure is more likely to occur in the lower part of the slope under long-term rainfall;
(3) The deformation and failure mode of loess fill slope under rainfall is shallow slip, and the failure process can be summarized as follows: local mud flow failure at the toe of the slope → erosion in the middle of the slope → crack initiation on the shoulder of the slope → local slip on the slope → crack propagation on the shoulder of the slope → shallow slip on the shoulder of the slope.

The interface and its effect in the geological body is a very complex, difficult and meaningful research topic. As for the simulation of the original slope in the indoor model test of the excavated and filled slope of the City-building gully, since the original slope is formed for a long time and has its own structure in practical engineering, and it is impossible to simulate the original slope in the laboratory through simple ramming. Therefore, how to simulate the original slope needs further discussion and verification. In addition, the interface simulation is difficult in the model test. The slope is formed by tamping; therefore, the original slope and filling are simulated by plastic window gauze, and the test effect is not ideal. How to simulate the interface and reflect the interface effect in the model test needs further discussion and verification.

Author Contributions: Writing—original draft preparation, W.T.; conceptualization, Q.H.; writing—review and editing, Q.H. and X.C.; visualization, W.T. All authors have read and agreed to the published version of the manuscript.

Funding: This work was supported by The National key research and development program of China (NO.2021YFC3000404), The State Key Research Development Program of China (Grant No. 2017YFD0800501) and Major projects of the National Natural Science Foundation of China (41790443) are gratefully acknowledged.

Institutional Review Board Statement: Not applicable.

Informed Consent Statement: Not applicable.

Data Availability Statement: Not applicable.

Conflicts of Interest: The authors declare no conflict of interest.

References

1. Peng, J.; Fan, Z.; Wu, D.; Huang, Q.; Wang, Q.; Zhuang, J.; Che, W. Landslides triggered by excavation in the loess plateau of China: A case study of Middle Pleistocene loess slopes. *J. Asian Earth Sci.* **2019**, *171*, 246–258. [CrossRef]
2. Li, P.; Qian, H.; Wu, J. Environment: Accelerate research on land creation. *Nat. News* **2014**, *510*, 29. [CrossRef] [PubMed]
3. Derbyshire, E.; Dijkstra, T.; Smalley, I.; Li, Y. Failure mechanisms in loess and the effects of moisture content changes on remoulded strength. *Quat. Int.* **1994**, *24*, 5–15. [CrossRef]
4. Wang, X.; Lian, B.; Liu, K.; Luo, L. Trigger mechanism of loess-mudstone landslides inferred from ring shear tests and numerical simulation. *J. Mt. Sci.* **2021**, *18*, 2412–2426. [CrossRef]
5. Zhang, S.; Zhao, L.; Delgado-Tellez, R.; Bao, H. A physics-based probabilistic forecasting model for rainfall-induced shallow landslides at regional scale. *Nat. Hazards Earth Syst. Sci.* **2018**, *18*, 969–982. [CrossRef]
6. Solgi, A.; Naghdi, R.; Zenner, E.K.; Hemmati, V.; Behjou, F.K.; Masumian, A. Evaluating the Effectiveness of Mulching for Reducing Soil Erosion in Cut Slope and Fill Slope of Forest Roads in Hyrcanian Forests. *Croat. J. For. Eng.* **2021**, *42*, 259–268. [CrossRef]
7. Day, R.W. Fill-Slope Failure and Repair. *J. Perform. Constr. Facil.* **1992**, *6*, 161–168. [CrossRef]
8. Cheuk, C.; Ng, C.; Sun, H. Numerical experiments of soil nails in loose fill slopes subjected to rainfall infiltration effects. *Comput. Geotech.* **2005**, *32*, 290–303. [CrossRef]

9. Zhang, W.; Tang, L. Settlement and deformation laws of high loess-filled embankment based on centrifugal model test. *Appl. Mech. Mater.* **2011**, *90*, 222–229. [CrossRef]
10. Duan, L.; He, G.; Yi, J.; Li, Y. Numerical simulation and monitoring analysis for high-fill slope. *Adv. Mater. Res.* **2013**, *639*, 274–278. [CrossRef]
11. Zhao, J.; Wang, J.; Gu, T. Study on deformation characteristic and mechanism of expansive soil fill slope under precipitation-evaporation cycle. *J. Nat. Disasters* **2013**, *22*, 116–124.
12. Wang, H.; Tuo, X.; Zhang, G.; Peng, F. Panzhihua airport landslide (3 October 2009) and an emergency monitoring and warning system based on the internet of things. *J. Mt. Sci.* **2013**, *10*, 873–884. [CrossRef]
13. Chang, Z.; Huang, F.; Huang, J.; Jiang, S.; Zhou, C.; Zhu, L. Experimental study of the failure mode and mechanism of loess fill slopes induced by rainfall. *Eng. Geol.* **2021**, *280*, 105941. [CrossRef]
14. Askarinejad, A.; Akca, D.; Springman, S. Precursors of instability in a natural slope due to rainfall: A full-scale experiment. *Landslides* **2018**, *15*, 1745–1759. [CrossRef]
15. Rosi, A.; Peternel, T.; Jemec-Aufli, M.; Komac, M.; Segoni, S.; Casagli, N. Rainfall thresholds for rainfall-induced landslides in Slovenia. *Landslides* **2016**, *13*, 1571–1577. [CrossRef]
16. Lee, L.; Kassim, A.; Gofar, N. Performances of two instrumented laboratory models for the study of rainfall infltration into unsaturated soils. *Eng. Geol.* **2011**, *117*, 78–89. [CrossRef]
17. Tohari, A.; Nishigaki, M.; Komatsu, M. Laboratory Rainfall-Induced Slope Failure with Moisture Content Measurement. *J. Geotech. Geoenviron. Eng.* **2007**, *133*, 575–587. [CrossRef]
18. Saadatkhah, N.; Kassim, A.; Lee, L.M.; Rahnamarad, J. Spatiotemporal regional modeling of rainfall-induced slope failure in Hulu Kelang, Malaysia. *Env. Earth Sci.* **2015**, *73*, 8425–8441. [CrossRef]
19. Hakro, M.; Harahap, I. Laboratory experiments on rainfall-induced flowslide from pore pressure and moisture content measurements. *Nat. Hazards Earth Syst. Sci. Discuss.* **2015**, *3*, 1575–1613.
20. Gallage, C.; Abeykoon, T.; Uchimura, T. Instrumented model slopes to investigate the effects of slope inclination on rainfall-induced landslides. *Soils Found.* **2021**, *61*, 160–174. [CrossRef]
21. Hu, X.; He, C.; Zhou, C.; Xu, C.; Zhang, H.; Wang, Q.; Wu, S. Model test and numerical analysis on the deformation and stability of a landslide subjected to reservoir filling. *Geofluids* **2019**, *2019*, 5924580. [CrossRef]
22. Liu, H.; Men, Y.; Li, X.; Yan, J.; Zhang, T. Study on slip material in landslide model tests. *J. Catastrophol.* **2011**, *26*, 10–13. (In Chinese)
23. Tu, X.; Kwong, A.; Dai, F.; Tham, L.; Min, H. Field monitoring of rainfall infiltration in a loess slope and analysis of failure mechanism of rainfall-induced landslides. *Eng. Geol.* **2009**, *105*, 134–150. [CrossRef]
24. Crosta, G. Regionalization of rainfall thresholds: An aid to landslide hazard evaluation. *Environ. Geol.* **1998**, *35*, 131–145. [CrossRef]
25. Li, A.; Tham, L.; Yue, Z.; Lee, C.; Law, K. Comparison of field and laboratory soil-water characteristic curves. *J. Geotech. Environ. Eng.* **2005**, *131*, 1176–1179. [CrossRef]
26. Chueasamat, A.; Hori, T.; Saito, H.; Sato, T.; Kohgo, Y. Experimental tests of slope failure due to rainfalls using 1g physical slope models. *Soils Found.* **2018**, *58*, 290–305. [CrossRef]
27. Orense, R.; Shimoma, S.; Maeda, K.; Towhata, I. Instrumented model slope failure due to water seepage. *J. Nat. Disaster Sci.* **2004**, *26*, 15–26. [CrossRef]
28. Li, W.; Lee, L.; Cai, H.; Li, H.; Dai, F.; Wang, M. Combined roles of saturated permeability and rainfall characteristics on surficial failure of homogeneous soil slope. *Eng. Geol.* **2013**, *153*, 105–113. [CrossRef]
29. Dyer, M. Performance of flood embankments in England and Wales. Proceedings of the 490 Institution of Civil Engineers. *Water Manag.* **2004**, *157*, 177–186.
30. Huang, C.; Ju, Y.; Hwu, L.; Lee, J. Internal soil moisture and piezometric responses to rainfall-induced shallow slope failures. *J. Hydrol.* **2009**, *370*, 39–51.
31. Hakro, M.; Harahap, I.; Memon, I. Model experiment on rainfall-induced slope failures with moisture content measurements. *Res. J. Appl. Sci. Eng. Technol.* **2016**, *13*, 122–130. [CrossRef]
32. Wu, L.; Zhang, L.; Zhou, Y.; Xu, Q.; Yu, B.; Liu, G.; Bai, L. Theoretical analysis and model test for rainfall induced shallow landslides in the red-bed area of Sichuan. *Bull. Eng. Geol. Environ.* **2018**, *77*, 1343–1353. [CrossRef]
33. Min, S.; Yuichi, O.; Taro, U.; Jin, K.; Young, S. Effect of seepage on shallow landslides in consideration of changes in topography: Case study including an experimental sandy slope with artificial rainfall—Sciencedirect. *CATENA* **2018**, *161*, 50–62.
34. Qi, S.; Vanapalli, S. Simulating hydraulic and mechanical responses of unsaturated expansive soil slope to rainfall: Case study. *Int. J. Geomech.* **2018**, *18*, 05018002.1–05018002.17. [CrossRef]
35. Pu, X.; Wang, L.; Wang, P.; Chai, S. Study of shaking table test of seismic subsidence loess landslides induced by the coupling effect of earthquakes and rainfall. *Nat. Hazards* **2020**, *103*, 923–945. [CrossRef]
36. Zhang, J.; Luo, Y.; Zhou, Z.; Victor, C.; Duan, M. Research on the rainfall-induced regional slope failures along the Yangtze river of Anhui, China. *Landslides* **2021**, *18*, 1801–1821. [CrossRef]
37. Yang, L.; Kaisheng, C.; Mengfei, L.; Yingchao, W. Study on failure of red clay slopes with different gradients under dry and wet cycles. *Bull. Eng. Geol. Environ.* **2020**, *79*, 4609–4624. [CrossRef]

38. Lian, B.; Peng, J.; Zhan, H.; Huang, Q.; Wang, X.; Hu, S. Formation mechanism analysis of irrigation-induced retrogressive loess landslides. *CATENA* **2020**, *195*, 104441. [CrossRef]
39. Wang, X.; Wang, J.; Zhan, H.; Li, P.; Hu, S. Moisture content effect on the creep behavior of loess for the catastrophic Baqiao landslide. *CATENA* **2020**, *187*, 104371. [CrossRef]

land

MDPI

Article

Analysis of Meandering River Morphodynamics Using Satellite Remote Sensing Data—An Application in the Lower Deduru Oya (River), Sri Lanka

Vindhya Basnayaka [1], Jayanga T. Samarasinghe [2], Miyuru B. Gunathilake [3,4], Nitin Muttil [5,6,*], Dileepa C. Hettiarachchi [7], Amila Abeynayaka [8] and Upaka Rathnayake [1]

1 Department of Civil Engineering, Faculty of Engineering, Sri Lanka Institute of Information Technology, Malabe 10115, Sri Lanka; vindhya.b@sliit.lk (V.B.); upaka.r@sliit.lk (U.R.)
2 Department of Earth Environmental and Resource Sciences, University of Texas, El Paso, TX 79968, USA; jthambanged@miners.utep.edu
3 Environment and Natural Resources, Norwegian Institute of Bioeconomy and Research, 1433 Ås, Norway; miyuru.gunathilake@nibio.no
4 Water, Energy and Environmental Engineering Research Unit, Faculty of Technology, University of Oulu, 8000 Oulu, Finland
5 Institute for Sustainable Industries & Liveable Cities, Victoria University, P.O. Box 14428, Melbourne, VIC 8001, Australia
6 College of Engineering and Science, Victoria University, Melbourne, VIC 8001, Australia
7 Department of Irrigation, Office of Director of Irrigation, Airport Road, Anuradhapura 50000, Sri Lanka; dileepa1@gmail.com
8 Institute for Global Environmental Strategies (IGES), 2108-11 Kamiyamaguchi, Hayama 240-0115, Japan; abeynayaka@iges.or.jp
* Correspondence: nitin.muttil@vu.edu.au

Citation: Basnayaka, V.; Samarasinghe, J.T.; Gunathilake, M.B.; Muttil, N.; Hettiarachchi, D.C.; Abeynayaka, A.; Rathnayake, U. Analysis of Meandering River Morphodynamics Using Satellite Remote Sensing Data—An Application in the Lower Deduru Oya (River), Sri Lanka. *Land* **2022**, *11*, 1091. https://doi.org/10.3390/land11071091

Academic Editors: Matej Vojtek, Andrea Petroselli and Raffaele Pelorosso

Received: 2 July 2022
Accepted: 15 July 2022
Published: 16 July 2022

Abstract: River meandering and anabranching have become major problems in many large rivers that carry significant amounts of sediment worldwide. The morphodynamics of these rivers are complex due to the temporal variation of flows. However, the availability of remote sensing data and geographic information systems (GISs) provides the opportunity to analyze the morphological changes in river systems both quantitatively and qualitatively. The present study investigated the temporal changes in the river morphology of the Deduru Oya (river) in Sri Lanka, which is a meandering river. The study covered a period of 32 years (1989 to 2021), using Landsat satellite data and the QGIS platform. Cloud-free Landsat 5 and Landsat 8 satellite images were extracted and processed to extract the river mask. The centerline of the river was generated using the extracted river mask, with the support of semi-automated digitizing software (WebPlotDigitizer). Freely available QGIS was used to investigate the temporal variation of river migration. The results of the study demonstrated that, over the past three decades, both the bend curvatures and the river migration rates of the meandering bends have generally increased with time. In addition, it was found that a higher number of meandering bends could be observed in the lower (most downstream) and the middle parts of the selected river segment. The current analysis indicates that the Deduru Oya has undergone considerable changes in its curvature and migration rates.

Keywords: river meandering; river morphology; centerline migration; remote sensing; satellite images

1. Introduction

River morphodynamics are an intriguing subject and much research has been carried out to study the dynamics of the fluvial processes. Meandering is one important river process that has attracted the attention of researchers worldwide. The process is associated with outer bank erosion and deposition at the inner bank. The velocity and the shear stress distribution asymmetry in curved river segments are the reasons for these erosions and accretion [1]. Estimating and predicting the meandering features of a certain river over time have great importance in engineering and geological applications, such as management of infrastructure, erosion, and agricultural land. Meandering has become a prevailing issue for many major rivers in the world that carry a significant amount of sediment load, including the Amazon River and Yangtze River, and many other alluvial riverine environments. The process is highly dynamic in nature and needs to be assessed with a close examination over a considerable period that is sufficient to capture the dynamics of the river morphology. According to Constantine et al. [2], these large meandering rivers with high sediment transport rates can sometimes have migrations of several meters in a year.

The studies on river meandering in the literature can be categorized into three types based on the method they follow in extracting the meandering parameters. Some studies have adopted analytical methods to analyze the process [3], whereas others have used field measurements [4,5]. Furthermore, some studies have used numerical models to extract the meandering features of rivers [6]. However, with the recent development of remote sensing technologies, the data available for analysis have significantly increased. These remotely sensed data are widely used in many disciplines and for different applications, including geospatial analysis, agricultural management, morphological analysis, etc. Hence, in recent years, several algorithms and programs have been developed to analyze these remotely sensed data [7]. The ability to capture real-time environments using remote sensing technologies within a short time and improved computing capacities have increased the use of remotely sensed images in the estimation of river morphodynamics [8].

Many Sri Lankan rivers are facing major issues due to river meandering, such as the erosion of agricultural land, reduction of flood plain areas, and effects on infrastructure, such as bridge failures, etc. The erosion of outer banks can result in the exposure of the bridge abutments and subsequent failures. Therefore, understanding and predicting the meandering morphodynamics of a certain river have great importance in river management. The Kelani River, the Kalu River, the Deduru Oya, the Gin River, and the Mahaweli River are a few examples of rivers where a significantly meandering process can be observed [9–11]. The meandering process can clearly be witnessed downstream of the rivers. However, research efforts to identify the meandering problem are still lacking for many major rivers in Sri Lanka. Several studies have discussed the meandering behavior of Sri Lankan rivers; however, most were focused on various issues related to the meandering process, such as water quality [9–13]. None tried to analyze these rivers in Sri Lanka by using Landsat-based remote sensing data to represent their meandering behavior. Therefore, the search for sustainable solutions in a country like Sri Lanka, which has a radial river system, is challenging. Thus, the research presented herein aimed to provide information about these morphodynamical features for one of the major rivers in Sri Lanka, the Deduru Oya. This is the first attempt to use Landsat remote sensing data in Sri Lankan rivers to capture the meandering behavior, and the findings will be helpful in designing and managing the riverine environment of the Deduru Oya.

Remote Sensing Applications in the Fluvial Context

Remotely sensed satellite data are used in many applications, including geological studies, agricultural management, coastal management, spatial planning, land-use management, river management, disaster risk analysis, and GIS mapping [14–18]. Among these applications, the raster-based digital elevation models are useful. The raster-based digital elevation models (DEMs) take satellite images as an input to derive elevations. These DEMs are used in many studies as sources of topographic data because they are easy to access and available over the whole world. Satellite-derived meteorological data (rainfall data, temperature data, evapotranspiration estimates, etc.) are used in data-scarce regions in particular, and these satellite technologies are becoming a good source of data because of their free availability and easy access [19,20]. The recent development of these technologies has significantly increased the availability of satellite images and the tools and software used for analysis. Therefore, in the monitoring and modeling of fluvial processes, remote sensing data are frequently used to estimate river floods and to detect the fluvial planform changes and the controlling processes [21].

The morphological changes in the Bislak and Cagayan Rivers in the Philippines were studied by Boothroyd et al. [22] for a period of over 30 years using Landsat satellite images and the Google Earth Engine (GEE) cloud platform. To estimate the river planform changes, riverbank erosion, and accretion, Rowland et al. [23] introduced an algorithm called Spatially Continuous Riverbank Erosion and Accretion Management (SCREAM), which uses satellite remote sensing data to extract river properties. RivaMap is another algorithm that has been developed to analyze and map river features from remotely sensed satellite images [24]. Indian satellite remote sensing (IRS) data were used by Boruah et al. [25] to map the geomorphology and physical habitats of the Brahmaputhra River close to the Himalayas, where the river is braided and has instabilities.

Combining the available knowledge from this literature, the meandering behavior of the Deduru Oya in Sri Lanka is qualitatively presented in this research.

2. Study Area

Sri Lanka is a water-rich country that has 103 major river basins. The Deduru Oya is one of the major rivers in Sri Lanka, originating in the Western part of the central highlands (in Matale and Kandy districts) and reaching the Indian Ocean on the west coast of Sri Lanka near the Chilaw urban area. Nearly 97% of the river basin lies in the northwestern province and only 3% belongs to the central province [26]. The basin is located in the intermediate agro-climatic zone. Based on the terrain features, the catchment of the Deduru Oya is categorized into two classes: uplands and lowlands. The mainstream of the river has a length of 115 km and an annual discharge of 1608 million cubic meters (MCM) into the sea [27,28]. The average annual rainfall received by the catchment is 1609 mm and the maximum amounts are recorded in October and November [28].

The digital elevation model (DEM) shows that the lower Deduru Oya river basin has almost flat terrain compared to the upland area, which could be a reason for the frequently observed river meandering. Furthermore, the downstream area of the river is facing the problem of erosion and could undergo the possible exposure of bridge foundations at the outer banks and accretion of sediment at the inner banks (see the area in the red square in Figure 1a). In addition, much damage to the infrastructure due to eroded banks can be easily observed (see Figure 1b). Furthermore, the meandering bend and its sand deposition can be clearly seen in Figure 1c. These were observed by the authors during their field visits to the area.

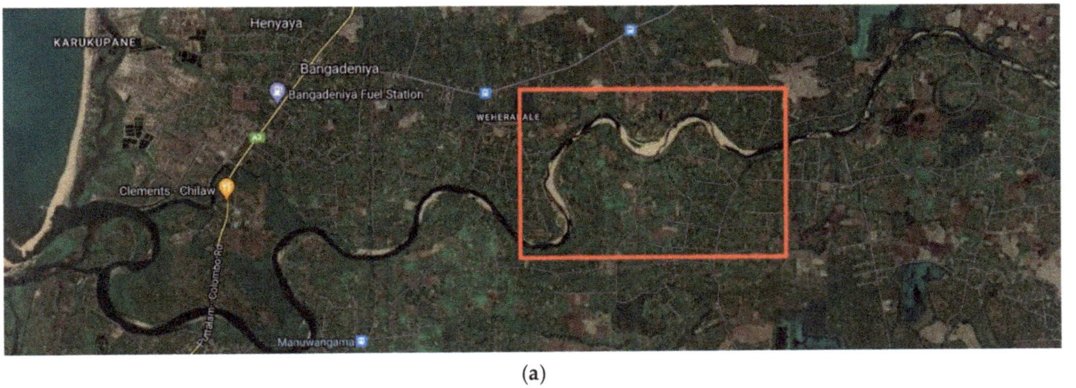

(a)

Figure 1. *Cont.*

(**b**) (**c**)

Figure 1. River meandering behavior: (**a**) meandering behavior and sediment deposition along the Deduru Oya (source: Google maps); (**b**) eroded riverbanks and damaged houses (photo credit—D.C.H); (**c**) meandering bend (photo credit—D.C.H).

The study area reached from downstream of the Deduru Oya reservoir dam to the river mouth (sea outfall), which is about a 70 km long stretch along the river. The Deduru Oya river basin and the selected river segment are presented in Figure 2.

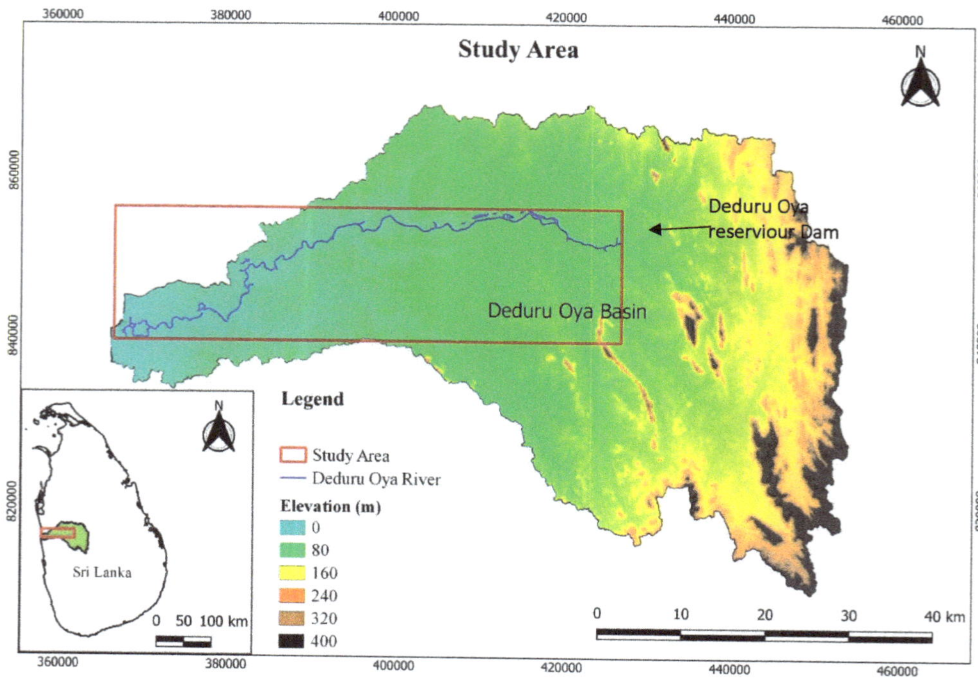

Figure 2. Map of the Deduru Oya river basin and the river segment selected for the analysis.

3. Data and Methods

The meandering morphodynamics of the river were estimated through a spatiotemporal analysis of the river planform using freely available remotely sensed satellite data. Many geospatial analysis tools can calculate pixel-based indices using different band combinations from the remotely sensed images. The mapping of different features, such as water, vegetation, and soil properties, was carried out using combinations of these pixel-based indices.

The remotely sensed satellite image data were used to identify the river mask, planform, and centerline in this study. Through a temporal analysis performed for a selected period, the meandering dynamics of the Deduru Oya were estimated using remotely sensed satellite data, semi-automated digitizing software, and a QGIS mapping tool.

3.1. Landsat Satellite Data Acquisition

Four Landsat products are available: Landsat 1–5 MSS, Landsat 4–5 TM, Landsat 7, and Landsat 8–9. In the current study, we used the Landsat 4–5 Thematic Mapper (TM), Landsat 7 Enhanced Thematic Mapper (ETM+), and Landsat 8 Operational Land Imager (OLI) and Thermal Infrared Sensors (TIRS) products to calculate the meandering morphodynamics. The temporal availability, resolutions, and number of bands for each Landsat data type are given in Table 1.

Table 1. Landsat data availability (https://earthexplorer.usgs.gov/, accessed on 2 May 2022).

Landsat Data	Resolution	Available Period	Number of Bands
Landsat 1–5 MSS	60 m	1972–2012	04
Landsat 4–5 TM	30 m	1982–2012	07
Landsat 7 ETM+	30 m	1999 to present	08
Landsat 8 OLI and TIRS	30 m	April 2013 to present	11
Landsat 9 OLI	30 m	February 2022 to present	11

The data were extracted from the USGS Earthexplorer website (https://earthexplorer.usgs.gov/, accessed on 2 May 2022) and covered 32 years from 1989 to 2021. These Landsat images were filtered based on the cloud cover criterion. The maximum cloud cover threshold for the land area was taken to be 20%, thereby removing the images where the river was significantly covered by clouds. Further, there were Landsat images with transverse no-data stripes in the images taken after 2003. These images were also excluded when selecting the satellite images for our analysis. Additionally, poorly georeferenced images among the available Landsat images were removed. The selected Landsat data and the dates of acquisition of the images are listed in Table 2.

Table 2. Landsat data used for the present analysis.

Year	Landsat Data Type	Extracted Date
1989	Landsat 5 TM	2 December 1989
1994	Landsat 5 TM	11 September 1994
2001	Landsat 5 TM	14 September 2001
2005	Landsat 5 TM	17 March 2005
2008	Landsat 5 TM	4 November 2008
2021	Landsat 8 OLI	23 October 2020

These Landsat images can be downloaded as tiles and each tile contains 5000 × 5000 30 m pixels obtained on a given day. Before the analysis, the images were cropped to our study area to reduce the image processing time. As mentioned in Section 3.1, this area was the lower Deduru Oya, downstream of the Deduru Oya reservoir. Figure 3 shows two Landsat images of the downstream Deduru Oya taken in 2008 and 2010.

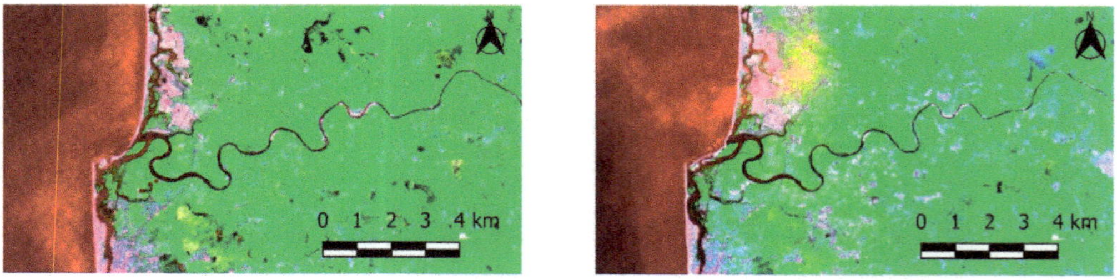

(**a**) (**b**)

Figure 3. Illustrative maps of meandering features of the Deduru Oya using Landsat data: (**a**) for 2008—Landsat false color composite bands on LT05_L1TP_142055_20081229_20200828_02_T1_B (1, 4, 7); (**b**) for 2010—Landsat false color composite bands on LT05_L1TP_142055_20100218_20200824_02_T1_B (1, 4, 7).

3.2. Extraction of Water Mask

One important aspect of the Landsat data is the availability of multispectral bands, and these bands can be combined to calculate indices that represent the relative abundances of different features, such as water, vegetation, and built-up areas [29]. Due to this capability, Landsat images are commonly used when calculating land use/land cover and vegetation properties, when identifying water bodies, and for flood mapping. Considering the above properties, it was decided to use the Landsat images for the extraction of the river water masks.

The first step of the process was to extract the river mask using the different Landsat band collections from the Landsat images. Landsat satellite data types, including Landsat 4–5 TM, Landsat 7 ETM+ and Landsat 8 OLI, and Landsat 8 TIRS, were used to extract the river mask. The extraction was performed based on three indices calculated using the Landsat image bands.

1. *NDVI*—normalized difference vegetation index
2. *MNDWI*—modified normalized difference water index
3. *EVI*—enhanced vegetation Index

MNDWI is frequently used when identifying water bodies because of its ability to suppress the noise from built-up land. However, misclassifications can still happen when using only the *MNDWI*, mainly due to mixed distributions of water and vegetation, especially in wetland areas [30]. Therefore, it has been suggested in the literature [30–33] that *MNDWI* should be used together with *NDVI* and *EVI* to avoid these classification errors. The calculated values for the *NDVI* represent the percentage of vegetation in each pixel (30 m × 30 m pixels) and the *MNDWI* indicates water and non-water pixels. The *NDVI*, *MNDWI*, and *EVI* are defined by Equations (1)–(3). In these equations, ρ_{NIR} is a near-infrared band and ρ_{SWIR1} is a mid-infrared band; for example, in Landsat 5 TM images, band 4 represents the infrared band and band 5 represents the shortwave infrared band, and in the Landsat 8 images, band 5 represents the infrared band and band 6 represents the shortwave infrared band.

$$NDVI = \frac{\rho_{NIR} - \rho_{Red}}{\rho_{NIR} + \rho_{Red}} \tag{1}$$

$$MNDWI = \frac{\rho_{Green} - \rho_{SWIR1}}{\rho_{Green} + \rho_{SWIR1}} \tag{2}$$

$$EVI = \frac{\rho_{NIR} - \rho_{Red}}{\rho_{NIR} + 6\rho_{Red} - 7.5\rho_{Blue}} \tag{3}$$

Considering the above facts and the previous work by Xia et al. [30], it was decided to use the criteria *MNDWI* > *NDVI* and *MNDWI* > *EVI* to extract the water pixels, with the water signals being stronger than the vegetation signals. Further, to avoid the interference of wetland vegetation, the condition $EVI < 0.1$ was used. The water mask extraction process was automated using a simple python script and GIS environment. The resulting raster images were then used to extract the river centerline.

3.3. River Centreline Delineation

River planform extraction was performed with the extracted water mask. The centerline of the river had to be identified using the river mask created in the previous step. Several methods have been proposed in the literature to identify the centerline of a certain river. One method is to identify the shortest flow path in a multi-branching channel. The main channel might not always be the shortest flow path, especially in cases where there are chute channels. However, the main channel has a higher discharge compared to the chute channels.

A semi-automated digitizing software package called WebPlotDigitizer version 4.5 [34] was used to extract the river centerline coordinates from the water mask images. After importing the 2D water mask image to the WebPlotDigitizer, it was calibrated using the axis coordinates of a few selected known points. The software allows the user to perform the digitization manually or using automatic digitization algorithms. There are several automatic extraction algorithms available in the software for extracting a large number of data points within a short time. In our study, we used the averaging window method as the automatic extraction algorithm. Then, these automatically extracted points were adjusted manually to fit them to the actual river by comparing the background image of the water mask and the points generated by the automatic extraction algorithm. Finally, the extracted points were saved as comma-separated value (.csv) files and these data were used to calculate the geometric parameters for the channel. A brief overview of the overall methodology followed in the study is presented in Figure 4.

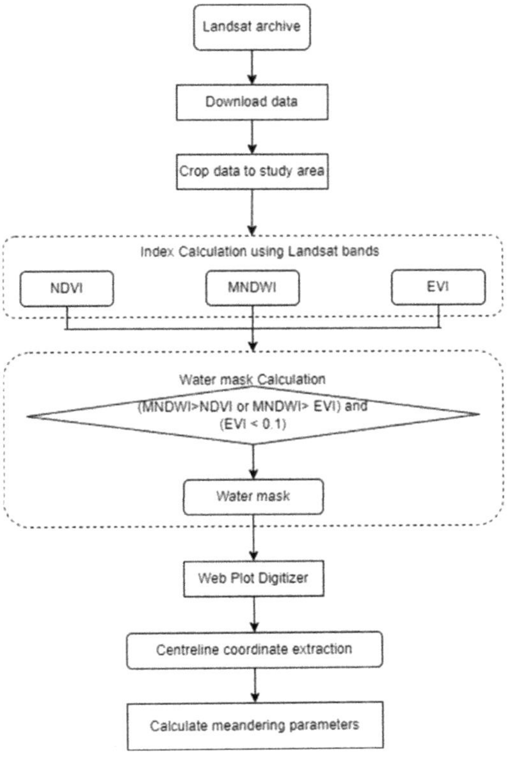

Figure 4. Overview of the methodology of the study.

3.4. Estimating Planform Geometry

For a certain point (x_i, y_i), the centerline arc length (S_i), the inflection angle (θ_i), and the curvature (c_i) were calculated using Equations (4)–(6).

$$S_i = \sqrt{\left(x_i - x_{(i-1)}\right)^2 + \left(y_i - y_{(i-1)}\right)^2} \tag{4}$$

$$\theta_i = tan^{-1}\left(\frac{y_i - y_{(i-1)}}{x_i - x_{(i-1)}}\right) \tag{5}$$

$$c_i = \frac{\theta_{(i+1)} - \theta_{(i-1)}}{S_{(i+1)} - S_{(i-1)}} \tag{6}$$

The zero-crossing points of the spatial variation of the centerline curvature (c_i) indicate the inflection points or the separation of individual meander bends [8]. Therefore, the variation of the centerline curvature was plotted against the river length to identify the individual meandering bends of the Deduru Oya. After separating the bends, the geometry of meandering bends was estimated based on two parameters. Meandering length (M_L) was defined as the axial length of an individual meander along the direction of flow, and the sinuosity as the ratio between the actual channel length and the direct axial length of the river [35].

3.5. Estimation of Centreline Migration

The evolution of a planform can be analyzed by estimating the centerline migration rates. This was carried out by calculating the rates of migration in the river centerline between two consecutive years. This was undertaken with the use of the Geometric Attribute tools in the QGIS software package. First, transects were created to each river centerline with constant distances. Then, the intersecting points on the river centerlines were ranked based on the distance along the river, and distances were calculated between the transects based on their ranks. Finally, the annual river migration rates were estimated by calculating the distance of the migration during the selected time interval and dividing the migrated distance by the time taken for the migration.

4. Results and Discussion

4.1. River Planform

The river masks extracted based on the three indices described in Section 3.3 are presented in Figure 5 (data *only shown for 1989, 2005, and 2021*). The results indicated that the river masks in some images did not have continuous connections (*see the dashed red circles*). This can be attributed to the misclassifications that occurred due to the low resolution of the satellite images and the possible cloud cover in some images. The selected river segment of the Deduru Oya has a river width of about 50 m upstream and 200 m near the outlet. However, the resolution of Landsat 5 and Landsat 8 images is 30 m. Therefore, the representation of the river was difficult for the narrow sections. This was one of the major limitations of our study of the morphology of the Deduru Oya. Further, this limitation could also have reduced the accuracy of the delineated river planform. Misclassifications due to the mixture of water, sediment, and vegetation can also cause erroneous water mask classifications.

In addition, the blue patch shown in Figure 5c (*see green dashed rectangle*) represents the recently constructed Deduru Oya reservoir, which is mainly used for irrigation purposes. However, the study took place downstream of this reservoir.

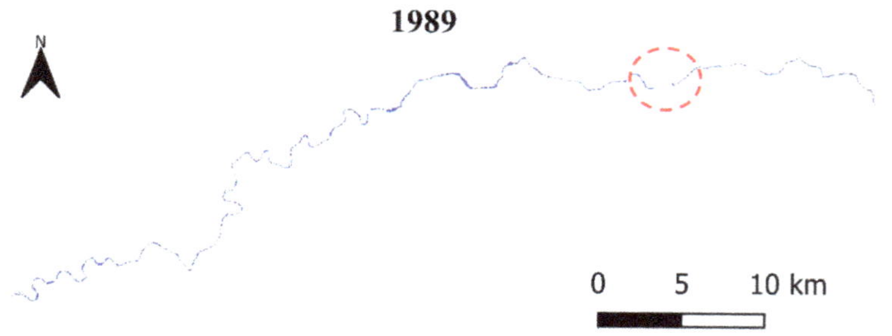

(**a**)

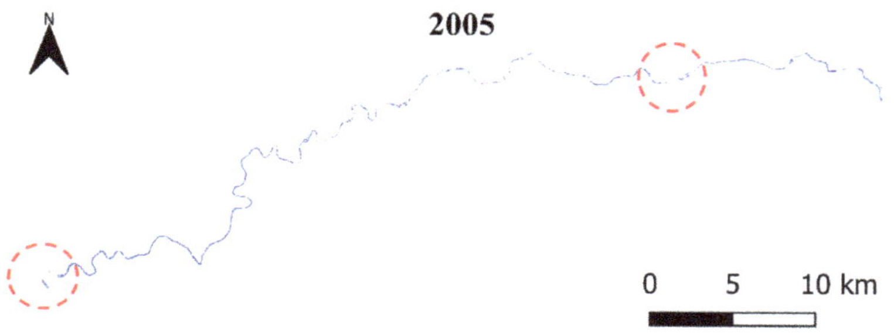

(**b**)

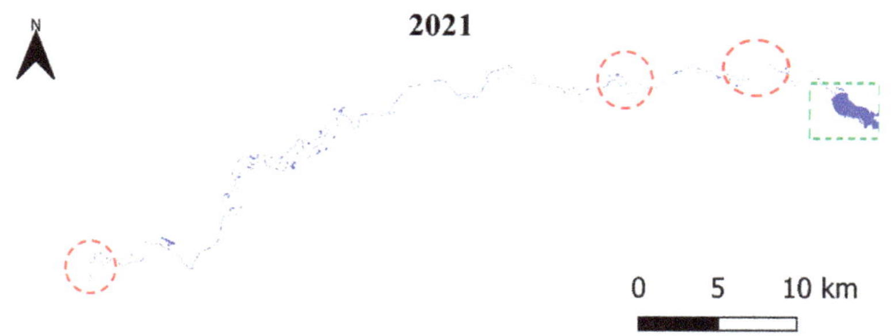

(**c**)

Figure 5. Extracted water masks for: (**a**) 1989; (**b**) 2005; (**c**) 2021.

4.2. River Centerline Variation

The centerlines were generated for the selected six years with the use of a digitization tool. The extracted centerlines for each year are presented in Figure 6. Significant variations can be clearly seen at the bends of the Deduru Oya. This is clear evidence of river meandering in the Deduru Oya. The changes are significant and, therefore, the riverbanks and the surrounding vicinity at the bends are highly vulnerable to erosion (sedimentation).

The spatial distributions for the river curvature for the six analyzed years are shown in Figure 7. The figure clearly shows the possible meander bends and their locations along the river centerline. In addition, the values for the curvature give an idea of the meandering sinuosity. When analyzing the temporal changes in the river curvature, the highest degree of curvature in the bends was observed in the years 2021 and 2005, while the lowest bend curvature values were observed for 1989. These curvature estimates were used to separate the individual bends.

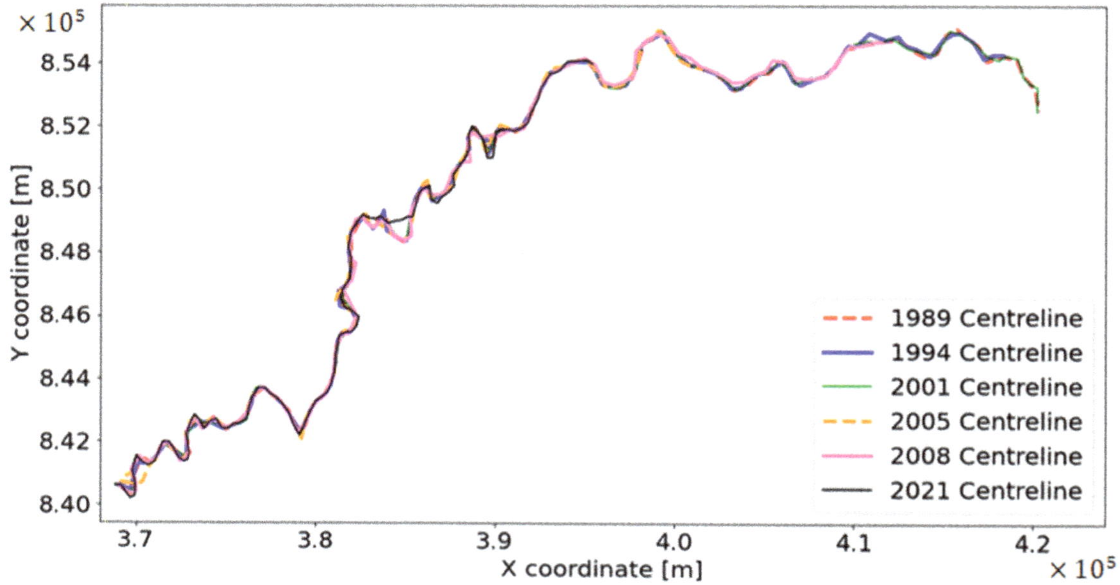

Figure 6. Temporal variation of river centerline.

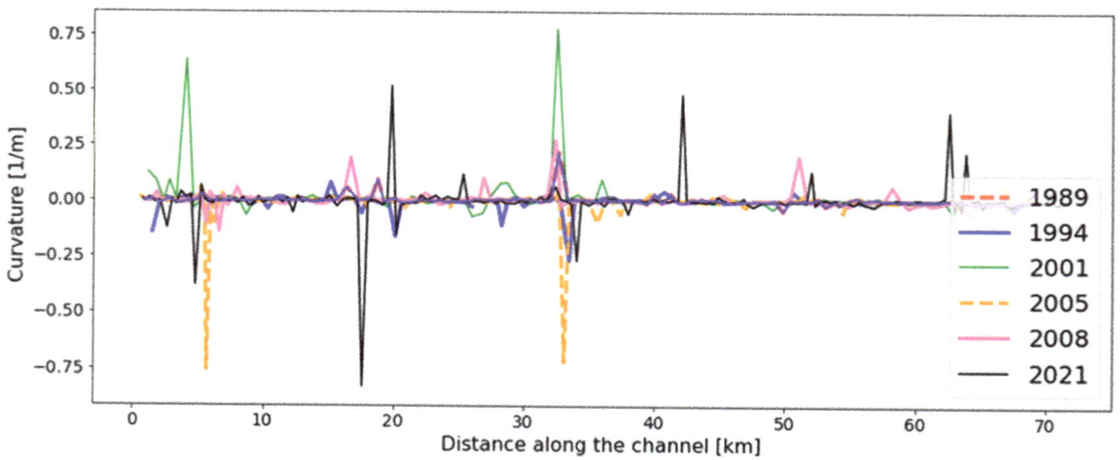

Figure 7. Spatiotemporal variation of curvature along the river centerline.

4.3. River Centerline Migration

The rate of total migration along the channel was plotted and the graph is presented in Figure 8. This was used to identify the variations in the migration rates on the spatial and temporal scales.

According to the estimated annual centerline migration rates, the minimum migration amounts were observed during the period from 1989 to 1994 and the maximum amount of total river migration occurred during the period from 2005 to 2008. These migration rates are significant and can also be seen in Figure 2. The damage resulting from these migrations is thus disastrous.

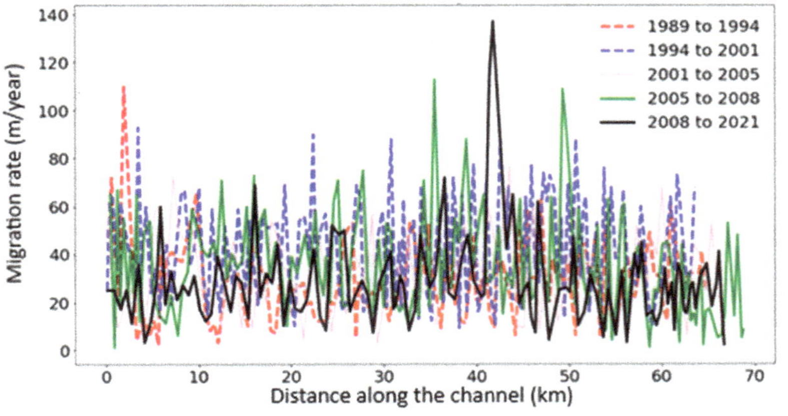

Figure 8. Centerline migration rates for the Deduru Oya.

The meandering bends identified in the study region of the Deduru Oya are shown in Figure 9 in a coordinate system. These bends were further analyzed, and the meandering length and sinuosity of each bend were calculated using the Geometric Attribute tools in QGIS.

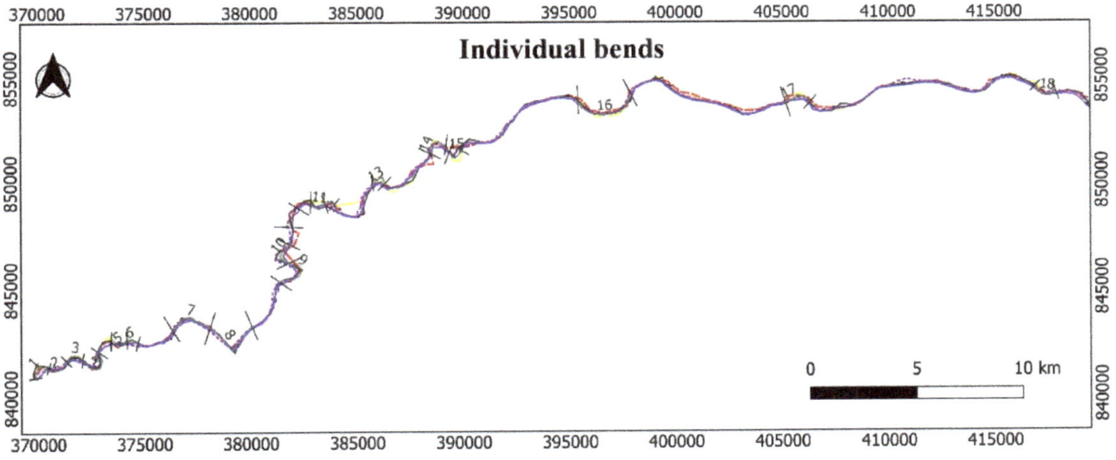

Figure 9. Separation of individual meander bends along the Deduru Oya.

Different types of meandering bend migration are described in the literature [35,36] depending on the direction and nature of the migrations. The concept of an inflection point was used to separate the individual meander bends and qualitatively study the types of bends. These bends are shown in Figure 10. Deb et al. [35] and Lagasse [36] classified meander bend migrations according to the patterns of the centreline migrations of rivers.

The bends detected in Figure 9 were compared with these different modes of meandering loop development described in the literature and classified by visually inspecting the changes in the river channel during the selected period. These results are shown in Table 3.

The extension bend migration type was the most frequently found bend migration type in the study area (see Table 3). Therefore, the curvature is increasing in the Deduru Oya. In addition, the maximum meander lengths were observed to be at the fourth, seventh, eighth, and ninth bends (see Figure 9). These bends were located downstream of the study area. In addition, the nine bend, which was classified as a translation, exhibited the maximum sinuosity. Therefore, this bend is moving along the river. Interestingly, the 16th bend showcased the migration type involving a neck cutoff by a chute. Therefore, the meandering behavior of the bend is becoming reduced over time and straightening the river at the 16th bend. Furthermore, the lowest meander length was observed at the 12th bend.

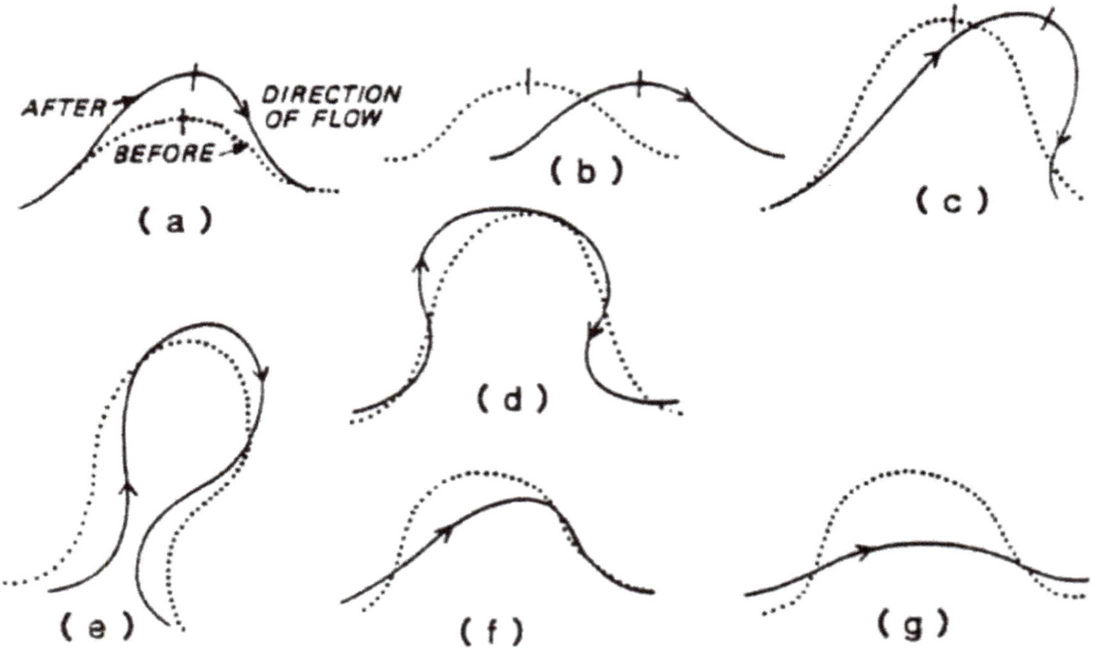

Figure 10. Different types of bend migrations: (**a**) extension; (**b**) translation; (**c**) rotation; (**d**) conversion to a compound loop; (**e**) neck cutoff by closure; (**f**) diagonal cutoff by chute; (**g**) neck cutoff by chute (extracted from Lagasse [36]).

Table 3. Classification of meandering bend migrations in the lower Deduru Oya.

Bend ID	Location Coordinates		Meander Length (m)	Sinuosity	Type of Bend Migration
	Starting Point	Endpoint			
1	(369,891, 840,926)	(370,469, 841,322)	1147	1.67	Diagonal cutoff by chute
2	(370,469, 841,322)	(371,392, 841,643)	1074	1.12	Rotation
3	(371,392, 841,643)	(372,085, 841,713)	970	1.26	Extension
4	(372,085, 841,713)	(372,744, 842,194)	1747	2.14	Extension
5	(373,439, 842,829)	(374,175, 842,607)	958	1.28	Extension
6	(374,175, 842,607)	(374,658, 842,408)	602	1.14	Conversion to compound loop
7	(376,310, 843,127)	(378,050, 843,167)	2137	1.22	Extension
8	(378,050, 843,167)	(380,069, 843,259)	2937	1.48	Extension
9	(381,376, 845,347)	(381,803, 846,241)	2052	2.16	Translation
10	(381,803, 846,241)	(381,770, 847,127)	1239	1.42	Conversion to compound loop
11	(382,838, 849,184)	(383,484, 849,056)	793	1.13	Extension
12	(383,484, 849,056)	(383,897, 848,928)	581	1.39	Extension
13	(385,793, 849,733)	(386,464, 850,102)	948	1.74	Translation
14	(388,671, 851,246)	(389,186, 851,590)	1304	1.75	Rotation
15	(389,427, 851,590)	(389,864, 851,679)	850	1.41	Extension
16	(395,413, 853,972)	(397,947, 854,106)	3136	1.23	Neck cutoff by chute
17	(405,221, 853,671)	(406,338, 853,766)	1361	1.18	Extension
18	(417,035, 854,669)	(417,950, 854,127)	1277	1.17	Translation

The identification of the bend migration type is important for any future riverbank conservation and protection services. The planners can introduce necessary steps suitable for each river bend.

5. Summary and Conclusions

Studying river migration patterns is an important factor when designing critical infrastructure, such as bridges, roads, etc. In addition, it is highly important for managing river flows. The results of this study will therefore be useful for planning river management activities (conservation) and designing infrastructure close to the river.

The current study was conducted to develop a simple method to investigate the river meandering features of a narrow river using freely available remote sensing satellite data and GIS tools. The method was applied to a meandering river in Sri Lanka (the Deduru Oya) to investigate the change in the active channel during the past three decades from 1989 to 2021. This is the first such study undertaken for the river basins in Sri Lanka and, more importantly, for the Deduru Oya (as per the authors' knowledge). The method developed used Landsat satellite image data to identify the river planform and this was then digitized to calculate the river geometry, the changes over time, and the meandering features of the Deduru Oya.

The current analysis indicated that the highest number of meandering bends could observed in the lower (most downstream) and the middle parts of the selected river segment. It was also observed that the Deduru Oya has undergone considerable change in its curvature and migration rates. According to the results obtained, the curvatures of the critical meandering bends have increased over time between 1989 and 2021. Similarly, the annual meandering rates were observed to be higher in recent years compared to the past. The meandering bends identified were classified according to the bend types defined by Lagasse [36]. This information can be effectively used by the Irrigation Department of Sri Lanka to restore and conserve the riverbanks and their surroundings. This is very important as the Deduru Oya has been identified as one of the rivers in Sri Lanka that has been most intensely mined for construction sand. Therefore, the riverbanks are often mined legally and illegally. The riverine ecosystem always affects the meandering process and vice versa. For example, the existence of vegetation close to the river banks slows down meander bend migration, and the river meandering can also turn the vegetated areas into wetlands, completely changing the ecosystem features. Thus, the findings of this research are very useful for the management of the river ecosystem. In addition, the Deduru Oya floods annually. Knowledge about the meandering patterns of the Deduru Oya river will help to reduce flood-related erosion risks in the riverine environment. Therefore, the results can be used for flood risk analysis of the Deduru Oya river. Furthermore, the mathematical representations of these meandering bends could also be investigated in a future study.

The usefulness of Landsat satellite data for studying river channel morphodynamics is widely acknowledged in the literature [2,8,22,35–39]. However, the accuracy of the geomorphological applications of these satellite data is highly dependent on the resolution of the images and the river width. Analyzing the narrow channels (i.e., river width < 100 m) using medium-resolution images (Landsat, Sentinel, etc.) can reduce the accuracy of the results [21]. The width of the Deduru Oya changes from about 50 m in the upstream areas to about 200 m near the sea outfall. Hence, there are limitations regarding feature extraction and classification when using these freely available Landsat images, which can cause misclassifications and reduce the accuracy of the extracted river representation. Therefore, it is recommended to use satellite data that have higher resolutions in future studies. Another limitation of the current method is that it is only able to estimate the centerline changes over time and cannot be used to identify sediment bars or to determine the dynamics of sediment in rivers. Therefore, the method could be developed to calculate the sediment bar dynamics in the Deduru Oya in future research.

Author Contributions: Conceptualization, U.R.; methodology, V.B. and J.T.S.; software, V.B., J.T.S. and M.B.G.; formal analysis, V.B. and J.T.S.; resources, M.B.G. and D.C.H.; data curation, V.B. and J.T.S.; writing—original draft preparation, V.B. and J.T.S.; writing—review and editing, U.R., N.M. and A.A.; visualization, V.B. and J.T.S.; supervision, U.R.; project administration, U.R. and N.M.; funding acquisition, N.M. All authors have read and agreed to the published version of the manuscript.

Funding: This research received no external funding.

Institutional Review Board Statement: Not applicable.

Informed Consent Statement: Not applicable.

Data Availability Statement: The data presented in this research work can be obtained from the corresponding authors for research purposes.

Conflicts of Interest: The authors declare no conflict of interest.

References

1. Sylvester, Z.; Durkin, P.; Covault, J.A.; Sharman, G.R. High curvatures drive river meandering. *Geology* **2019**, *47*, e486. [CrossRef]
2. Constantine, J.A.; Dunne, T.; Ahmed, J.; Legleiter, C.; Lazarus, E.D. Sediment supply as a driver of river meandering and floodplain evolution in the Amazon Basin. *Nat. Geosci.* **2014**, *7*, 899–903. [CrossRef]
3. Ikeda, S.; Parker, G.; Sawai, K. Bend theory of river meanders. Part 1. Linear development. *J. Fluid Mech.* **1981**, *112*, 363–377. [CrossRef]
4. Hooke, J.M.; Yorke, L. Channel bar dynamics on multi-decadal timescales in an active meandering river. *Earth Surf. Process. Landf.* **2011**, *36*, 1910–1928. [CrossRef]
5. Seminara, G. Meanders. *J. Fluid Mech.* **2006**, *554*, 271–297. [CrossRef]
6. Asahi, K.; Shimizu, Y.; Nelson, J.; Parker, G. Numerical simulation of river meandering with self-evolving banks. *J. Geophys. Res. Earth Surf.* **2013**, *118*, 2208–2229. [CrossRef]
7. Monegaglia, F. Meandering Rivers Morphodynamics–Integrating Nonlinear Modeling and Remote Sensing. Ph.D. Thesis, University of London, London, UK, 2017; p. 250. Available online: http://qmro.qmul.ac.uk/xmlui/handle/123456789/33927 (accessed on 15 May 2022).
8. Monegaglia, F.; Zolezzi, G.; Güneralp, I.; Henshaw, A.J.; Tubino, M. Automated extraction of meandering river morphodynamics from multitemporal remotely sensed data. *Environ. Model. Softw.* **2018**, *105*, 171–186. [CrossRef]
9. Surasinghe, T.; Kariyawasam, R.; Sudasinghe, H.; Karunarathna, S. Challenges in Biodiversity Conservation in a Highly Modified Tropical River Basin in Sri Lanka. *Water* **2019**, *12*, 26. [CrossRef]
10. Liyanage, G.; Illango, A.; Manage, P. Prevalence and Quantitative Analysis of Antibiotic Resistance Genes (ARGs) in Surface and Groundwater in Meandering Part of the Kelani River Basin in Sri Lanka. *Water Air Soil Pollut.* **2021**, *232*, 351. [CrossRef]
11. Jayapadma, J.; Souma, K.; Ishidaira, H.; Magome, J.; Wickramaarachchi, T. The Effect of Incorporation of Embankment Information for Flood Simulation of the Gin River, Sri Lanka. *J. Disaster Res.* **2022**, *17*, 475–486. [CrossRef]
12. Mahagamage, M.; Pathirage, M.; Manage, P. Contamination Status of *Salmonella* spp., *Shigella* spp. and *Campylobacter* spp. in Surface and Groundwater of the Kelani River Basin 2020, Sri Lanka. *Water* **2020**, *12*, 2187. [CrossRef]
13. Jayasuriya, A. A new forest vegetation type in Sri Lanka: Dry Canal-associated Evergreen Forest. *Ceylon J. Sci.* **2019**, *48*, 375–381. [CrossRef]
14. Huang, Y.; Chen, Z.; Yu, T.; Huang, X.; Gu, X. Agricultural remote sensing big data: Management and applications. *J. Integr. Agric.* **2018**, *17*, 1915–1931. [CrossRef]
15. Sheffield, J.; Wood, E.; Pan, M.; Beck, H.; Coccia, G.; Serrat-Capdevila, A.; Verbist, K. Satellite Remote Sensing for Water Resources Management: Potential for Supporting Sustainable Development in Data-Poor Regions. *Water Resour. Res.* **2018**, *54*, 9724–9758. [CrossRef]
16. Zhao, M.; Zhou, Y.; Li, X.; Cao, W.; He, C.; Yu, B.; Li, X.; Elvidge, C.D.; Cheng, W.; Zhou, C. Applications of Satellite Remote Sensing of Nighttime Light Observations: Advances, Challenges, and Perspectives. *Remote Sens.* **2019**, *11*, 1971. [CrossRef]
17. Fathololoumi, S.; Vaezi, A.; Alavipanah, S.; Ghorbani, A.; Saurette, D.; Biswas, A. Improved digital soil mapping with multitemporal remotely sensed satellite data fusion: A case study in Iran. *Sci. Total Environ.* **2020**, *721*, 137703. [CrossRef] [PubMed]
18. Alvarez-Vanhard, E.; Corpetti, T.; Houet, T. UAV & satellite synergies for optical remote sensing applications: A literature review. *Sci. Remote Sens.* **2021**, *3*, 100019. [CrossRef]
19. Perera, H.; Fernando, S.; Gunathilake, M.; Sirisena, T.; Rathnayake, U. Evaluation of Satellite Rainfall Products over the Mahaweli River Basin in Sri Lanka. *Adv. Meteorol.* **2022**, *2022*, 1926854. [CrossRef]
20. Makumbura, R.; Samarasinghe, J.; Rathnayake, U. Multidecadal Land Use Patterns and Land Surface Temperature Variation in Sri Lanka. *Appl. Environ. Soil Sci.* **2022**, *2022*, 2796637. [CrossRef]
21. Gilvear, D.; Bryant, R. Analysis of remotely sensed data for fluvial geomorphology and river science. In *Tools in Fluvial Geomorphology*; John Wiley & Sons, Ltd.: Hoboken, NJ, USA, 2016; pp. 103–132. [CrossRef]

22. Boothroyd, R.J.; Williams, R.D.; Hoey, T.B.; Barrett, B.; Prasojo, O.A. Applications of Google Earth Engine in fluvial geomorphology for detecting river channel change. *Wiley Interdiscip. Rev. Water* **2021**, *8*, e21496. [CrossRef]
23. Rowland, J.C.; Shelef, E.; Pope, P.A.; Muss, J.; Gangodagamage, C.; Brumby, S.P.; Wilson, C.J. A morphology independent methodology for quantifying planview river change and characteristics from remotely sensed imagery. *Remote Sens. Environ.* **2016**, *184*, 212–228. [CrossRef]
24. Isikdogan, F.; Bovik, A.; Passalacqua, P. RivaMap: An automated river analysis and mapping engine. *Remote Sens. Environ.* **2017**, *202*, 88–97. [CrossRef]
25. Boruah, S.; Gilvear, D.; Hunter, P.; Sharma, N. Quantifying channel planform and physical habitat dynamics on a large braided river using satellite data—the Brahmaputra, India. *River Res. Appl.* **2008**, *24*, 650–660. [CrossRef]
26. Wickramaarachchi, T.N. Preliminary assessment of surface water resources-a study from Deduru Oya Basin of Sri Lanka. In *APHW Proceeding*; Suntec International Convvention and Exhibition Center: Singapore, 2004.
27. Jayasena, H.; Chandrajith, R.; Gangadhara, K. Water Management in Ancient Tank Cascade Systems (TCS) in Sri Lanka: Evidence for Systematic Tank Distribution. *J. Geol. Soc. Sri Lanka* **2011**, *14*, 29–34.
28. PGSomaratne, K.J.; Perera, L.R.; Ariyaratne, B.R.; Bandaragoda, D.J.; Makin, I.W. *Developing Effective Institutions for Water Resources Management: A Case Study in the Deduru Oya Basin, Sri Lanka*; Working Paper 58; IWMI: Colombo, Sri Lanka, 2003; p. 104. Available online: http://www.iwmi.cgiar.org/publications/iwmi-working-papers/iwmi-working-paper-58/ (accessed on 30 May 2022).
29. Boothroyd, R.; Nones, M.; Guerrero, M. Deriving Planform Morphology and Vegetation Coverage from Remote Sensing to Support River Management Applications. *Front. Environ. Sci.* **2021**, *9*, 657354. [CrossRef]
30. Xia, H.; Zhao, J.; Qin, Y.; Yang, J.; Cui, Y.; Song, H.; Ma, L.; Jin, N.; Meng, Q. Changes in water surface area during 1989–2017 in the Huai River Basin using Landsat data and Google earth engine. *Remote Sens.* **2019**, *11*, 1824. [CrossRef]
31. Chen, B.; Xiao, X.; Li, X.; Pan, L.; Doughty, R.; Ma, J.; Dong, J.; Qin, Y.; Zhao, B.; Wu, Z.; et al. A mangrove forest map of China in 2015: Analysis of time series Landsat 7/8 and Sentinel-1A imagery in Google Earth Engine cloud computing platform. *ISPRS J. Photogramm. Remote Sens.* **2017**, *131*, 104–120. [CrossRef]
32. Zou, Z.; Dong, J.; Menarguez, M.A.; Xiao, X.; Qin, Y.; Doughty, R.B.; Hooker, K.V.; David Hambright, K. Continued decrease of open surface water body area in Oklahoma during 1984–2015. *Sci. Total Environ.* **2017**, *595*, 451–460. [CrossRef]
33. Zou, Z.; Xiao, X.; Dong, J.; Qin, Y.; Doughty, R.B.; Menarguez, M.A.; Zhang, G.; Wang, J. Divergent trends of open-surface water body area in the contiguous United States from 1984 to 2016. *Proc. Natl. Acad. Sci. USA* **2018**, *115*, 3810–3815. [CrossRef]
34. Rohatgi, A. Webplotdigitizer: Version 4.5. 2021. Available online: https://automeris.io/WebPlotDigitizer (accessed on 15 June 2022).
35. Deb, M.; Das, D.; Uddin, M. Evaluation of Meandering Characteristics Using RS & GIS of Manu River. *J. Water Resour. Prot.* **2012**, *04*, 163–171. [CrossRef]
36. Lagasse, P.F. *Handbook for Predicting Stream Meander Migration*; National Cooperative Highway Research Program; Transportation Research Board, National Research Council: Washington, DC, USA, 2004.
37. Dey, A.; Bhattacharya, R.K. Monitoring of River Center Line and Width-A Study on River Brahmaputra. *J. Indian Soc. Remote Sens.* **2014**, *42*, 475–482. [CrossRef]
38. Hamilton, S.K.; Kellndorfer, J.; Lehner, B.; Tobler, M. Remote sensing of floodplain geomorphology as a surrogate for biodiversity in a tropical river system (Madre de Dios, Peru). *Geomorphology* **2007**, *89*, 23–38. [CrossRef]
39. Wang, D.; Ma, Y.; Liu, X.; Huang, H.Q.; Huang, L.; Deng, C. Meandering-anabranching river channel change in response to flow-sediment regulation: Data analysis and model validation. *J. Hydrol.* **2019**, *579*, 124209. [CrossRef]

land

MDPI

Article

A Heuristic Method to Evaluate the Effect of Soil Tillage on Slope Stability: A Pilot Case in Central Italy

Evelina Volpe [1], Stefano Luigi Gariano [1,*], Francesca Ardizzone [1], Federica Fiorucci [1] and Diana Salciarini [2]

1 CNR IRPI, via della Madonna Alta 126, 06128 Perugia, Italy; evelina.volpe@irpi.cnr.it (E.V.); francesca.ardizzone@irpi.cnr.it (F.A.); federica.fiorucci@irpi.cnr.it (F.F.)

2 Department of Civil and Environmental Engineering, University of Perugia, Via G. Duranti 93, 06125 Perugia, Italy; diana.salciarini@unipg.it

* Correspondence: stefano.luigi.gariano@irpi.cnr.it; Tel.: +39-075-7014-424

Abstract: Among the various predisposing factors of rainfall-induced shallow landslides, land use is constantly evolving, being linked to human activities. Between different land uses, improper agricultural practices can have a negative impact on slope stability. Indeed, unsustainable soil tillage can modify the mechanical properties of the soils, leading to a possible increase of the instability phenomena. However, the effects of soil tillage on slope stability are poorly investigated. To address this topic, the PG_TRIGRS model (a probabilistic, geostatistic-based extension of TRIGRS) was applied to a cultivated, landslide-prone area in central Italy, thoroughly studied and periodically monitored through systematic image analysis and field surveys. A heuristic approach was adopted to quantitatively evaluate the effect of soil tillage on the mechanical properties of the soil: after a first run of the model with unbiased parameters, the slope stability analysis was carried out assuming several percentages of reduction of the effective soil cohesion to mimic an increasing impact of soil tillage on the strength conditions. Then, a comparison between observed landslides and the spatial distribution of the probability of failure derived from the application of PG_TRIGRS was carried out. A back analysis with contingency matrix and skill scores was adopted to search for the best compromise between correct and incorrect model outcomes. The results show that soil tillage caused a 20 to 30% reduction in soil cohesion in the analyzed area.

Keywords: land use; landslide modelling; shallow landslides; soil cohesion; soil tillage

Citation: Volpe, E.; Gariano, S.L.; Ardizzone, F.; Fiorucci, F.; Salciarini, D. A Heuristic Method to Evaluate the Effect of Soil Tillage on Slope Stability: A Pilot Case in Central Italy. *Land* **2022**, *11*, 912. https://doi.org/10.3390/land11060912

Academic Editors: Matej Vojtek, Andrea Petroselli and Raffaele Pelorosso

Received: 29 April 2022
Accepted: 12 June 2022
Published: 15 June 2022

1. Introduction

Shallow landslides induced by rainfall are very common phenomena that occur in hilly and mountainous areas, causing loss of human life and environmental and economic damage [1,2]. The main triggering factors of shallow landslides are represented by intense or prolonged rainfall [3–5], while the main predisposing factors are represented by lithology, morphology [6,7], and soil conditions—such as land cover and land use [8,9]. In particular, land use is constantly evolving, and its changes affect landslide occurrence [10–14]. Indeed, changes in vegetation cover have an impact on the landscape diversity [15] and relevant effects on the hydrological processes and mechanical structure of the soil, with either positive or negative consequences for slope stability [16–19]. Moreover, agricultural practices, on one hand, contribute positively to landscape and on the other hand, can have a negative impact on the slope stability. As an example, unsustainable agricultural practices characterized by heavy mechanization, such as soil tillage, can generate an excessive pressure on the soil, making the soil more susceptible to instability and degradation phenomena [20].

Several scientific contributions, among others, show the positive effects of vegetation cover on slope stability [21–25]: from a hydrological point of view, the vegetation dissipates most of the kinetic energy of the raindrops, weakening the erosion action, with a degree of interception depending on the density of the leaves and the size of the plant [26]; from a geo-mechanical point of view, the most important effect is represented by the mechanical reinforcement exerted by the roots [22,27–30], consisting of an increase in the shear strength, included in the Mohr–Coulomb failure criterion as additional soil cohesion [31,32]. In particular, the main contribution of roots to soil strength is related to the presence of small roots in the most superficial soil layers, which increase the compound matrix strength. Such an effect has been largely known as root reinforcement [33]. Another action involves large roots which intersect the shear surface and mobilize a soil–root friction force instead of the entire tensile strength [14]. Moreover, root reinforcement depends on the density of roots in soils and root diameters. As an example, grasses provide significant reinforcement to the shallower layers of soil, while woody roots of trees and shrubs provide reinforcement over a greater depth of soil through a combination of both fine, fibrous roots and coarser, woody roots [34].

Conversely, the effects of agricultural practices, and in particular of soil tillage, on slope stability conditions are poorly investigated. In this case, some studies show that the modification of soil's mechanical properties can be related to the tools used to till the land. In particular, some studies investigated how the soil characteristics vary in response to practices with the aim of assessing the efficiency of agricultural machinery, e.g., [35]. As an example, tillage with a rotary paraplow can be considered as a conservation technique leaving the soil state unchanged [36]. Other works, e.g., [37], show that the effect of tillage tools on the soil depends on the nature of the soil (fine-grained or coarse-grained soil). In the case of paddy soils, different laboratory experiments were conducted to determine the effect of tillage operations on the soil's physical, rheological and mechanical properties [38,39]. A negative impact of the tillage operation was found in soil bulk density, while a change in the rheological behavior of paddy soil according to the variation of the moisture content was observed [38]. Some authors have performed field-scale analyses aimed at evaluating the effect of different tillage techniques (conventional or conservation) on soil erosion, also in comparison with non-tilled cases [40–42]. Overall, considerable increases in soil erosion, runoff, and sediment loss are observed in the cases with conventional tillage as compared with the cases with soil saving technologies (conservation tillage or no-tillage) [41,42]. On the other hand, no significant differences in hydraulic conductivity were observed [40]. Recently, Straffelini and co-authors proposed a physical modeling approach to assess runoff and soil erosion in vineyards under different soil managements and observed that continuous tillage aggravated soil erosion as compared to reference tillage, single tillage, and nectariferous [43].

To our knowledge, no articles regarding particularly the impact of soil tillage on slope stability and shallow landslide occurrence in hilly environments are currently present in the literature. Indeed, most studies have focused on steep, often terraced landscapes [see, e.g., [20] for further references] or on erosion hazards, e.g., [41,42]. Quantitative measurements of the link between tillage operations and soil mechanical characteristics are not straightforward. However, there is a general consensus that tillage, plowing, and leveling generate a modification of the mechanical properties of the soils involved, leading to a possible increase in the propensity to slope failures of an area. In particular, soil tillage could lead to a decrease of the soil cohesion (e.g., due to soil disaggregation) and an increase of the friction angle (e.g., due to soil compaction). Quantitative data on this topic are rarely available: a table reporting quantitative changes in the mechanical parameters was found only in Albiero et al. (2014) [44].

In this paper, we analyze the decrease of the soil cohesion due to soil tillage, through a back-analysis approach and with the application of a probabilistic, physically-based model for the triggering of rainfall-induced landslides [45] in an agricultural environment. The choice of a probabilistic model is motivated to take into account the natural changes of the physical and mechanical properties of soils and rocks, which are characterized by high variability in space both in horizontal and vertical dimensions [46]. In probabilistic approaches, the safety level of the slope is given by the probability of failure (PoF), i.e., the probability associated with a value of factor of safety ≤ 1. Probabilistic approaches can provide a high level of reliability when a detailed description of the study area is available in terms of slope topography and physical, mechanical, and hydraulic soil properties. For landslide prediction, probabilistic approaches, which assume input data as random variables defined through their probability density functions, are more suitable than deterministic approaches, which assume the input data without uncertainty [47].

The model is applied to the Collazzone area, a cultivated area located in central Italy, characterized by a high susceptibility to landslides. The area has been the subject of several studies, e.g., [48–50] and is periodically monitored through systematic image analysis and on-site surveys [51,52]. This allowed a preliminary quantitative assessment of the effect of the crops on the stability conditions of the area. Through a back-analysis approach and with the support of sensitivity indices, a quantitative evaluation on the effect induced by soil tillage on the mechanical properties of the soil has been provided. The contribution is divided as follows: after the introduction, an overview of the theoretical aspects of the landslide model and the method used in this study are illustrated in Section 2. The description of the study area and the database available for the mechanical soil characterization are part of this paragraph. In Section 3, after defining the geotechnical and hydrological assumptions considered for the reliability analysis, the results of the model are shown and discussed. The conclusions and future research developments represent the final section of the paper (Section 4).

2. Materials and Methods

2.1. Study Area and Data

The Collazzone area extends for 80 km^2 in the Perugia province, Umbria region, central Italy (Figure 1A). The geology of the area has been investigated several times [53,54]: it consists of the alternation of recent fluvial deposits along the valley bottoms, continental gravel sand and clay, travertine deposits, sandstone and marl in various percentages and thinly layered limestone. The digital elevation model (with a 20 m resolution) reveals that the territory is mainly hilly with elevations ranging between 145 and 634 m a.s.l.; the slopes that stand on the area have a gradient varying between 5° and 50°, with the highest values in the northern part of the area (Figure 1B).

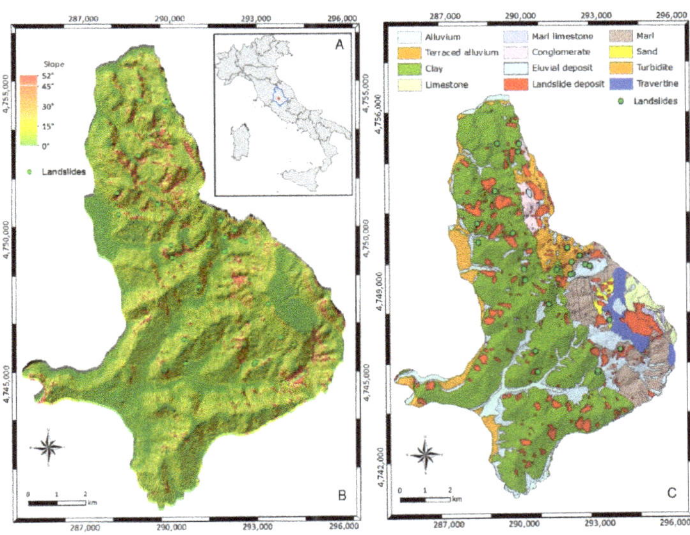

Figure 1. (**A**) Localization of the study area (in red) within the Umbria region (blue borders); (**B**) slope distribution in the area; (**C**) geotechnical classification of soil types according to Fanelli et al., (2016) [55]; the green circles represent the landslides.

For the purposes of this work, defining the mechanical parameters needed in input by the model: the geotechnical classification of the Perugia province proposed by Fanelli et al., (2016) [55] was adopted, which includes an estimation of the physical and mechanical properties of the soil types constituting the near-surface cover. According to this classification, which reports the geotechnical parameters and a general description of the stratigraphy, the central part of the study area is covered mainly by clays, while the rocks (e.g., marl and travertine) are distributed in the south-eastern part of the study area (Figure 1C).

The area studied represents a high percentage of cultivated land (Figure 2) and the soil is inventoried as arable land. According to the last Agriculture Report, published on the Umbria region website (https://www.regione.umbria.it/agricoltura/statistica, last accessed on 28 April 2022), about 75% of the total agricultural area is used. According to the reference soil groups of the international soil classification [56] and to the map of the soils of the Umbria region [57], most of the area can be classified as a *calcaric cambisol.*

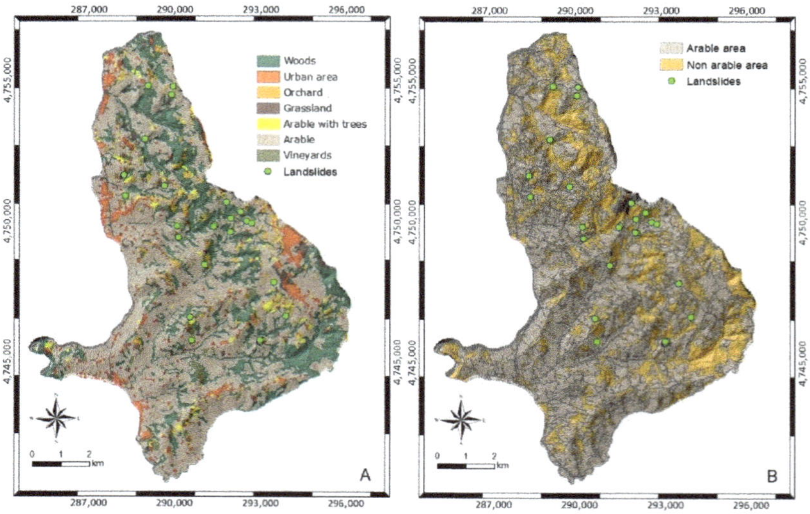

Figure 2. Maps showing (**A**) the distribution of land use and (**B**) a picture of the cultivated and uncultivated areas in the study area.

Despite the presence of vegetation, the area presents a high landslide susceptibility, typically, the slope failures are triggered chiefly by meteorological events, including intense and prolonged rainfall and rapid snow melting [53]. Using aerial and satellite images, Fiorucci and co-authors [51] estimated the landslide mobilization rates in the area in the period 2004–2005 and speculated that the remarkably high yearly rate of landslide mobilization observed in the area in the analyzed period might be due to the agricultural and land use practices. Considering the interesting case study for the period 1941–2005, a multi-temporal landslide inventory map analyzing different sets of aerial photographs and field surveys is available for the area [53]; even now, numerous annual surveys continue to be made in the area and an ongoing mapping has also been made using remote sensing product such as Lidar, monoscopic and stereoscopic satellite images [51,52,58], and field surveys carried out after intense or prolonged rainfall, when images were not available. The last survey was carried out by CNR-IRPI on 20 December 2020, following the copious rain that affected the area in the first ten days of the month. In particular, on 8 December from 03:00 A.M. to 15:00 P.M. (local time), 50 mm of cumulative rainfall fell (Figure 3), corresponding to a mean intensity of 4 mm/h (0.07 mm/min) and a peak intensity of 9.2 mm/h (0.16 mm/min).

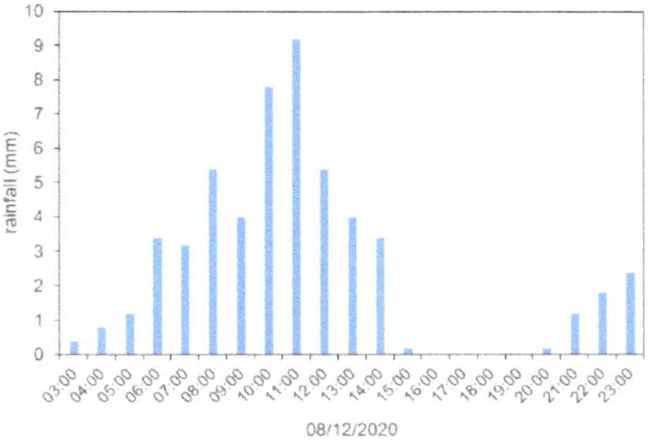

Figure 3. Bar chart showing the hourly rainfall measured by the rain gauge located in Collazzone on 8 December 2020. Data provided by the administration of the Umbria region.

The landslide information was obtained through a reconnaissance survey of the area (20 December 2020) driving and walking along main, secondary, and farm roads [59]. The investigators stopped at viewing points to check slopes where single or multiple landslides were identified and took photos of each landslide or group of landslides using a camera provided with GPS, and prepared a rapid (raw) mapping of the landslide. In the laboratory, the geolocated photographs were used to improve the location of the individual landslides and to characterize the type and the size. Landslides identified in the field and in the photographs were mapped on Google Earth. The main weakness of the reconnaissance inventory is the completeness, since from viewing points some slopes are not entirely visible, and an undetermined number of landslides may not have been identified and mapped.

During this last field survey, an event inventory was defined, including 26 shallow landslides. For the aim of this work, the landslides located in areas classified as rocks according to Fanelli et al. (2016) [55] and in uncultivated areas were excluded from the analysis; thus only 19 landslides (shown in Figure 2) were considered for modelling and analysis. As can be seen from Figures 1 and 2, these 19 landslides are located in the central portion of the area; the soils involved (Figure 1C) are clays, characterized by a cohesive resistance. Figure 4 shows an example of a shallow landslide (soil slide) mapped during the field survey.

Figure 4. Example of shallow landslide (soil slide) in the study area. Photo taken on 20 December 2020, by CNR-IRPI.

2.2. Method

In order to evaluate the effect of soil tillage on the stability conditions of the area, the method applied in this research includes the comparison between landslides observed in situ and the PoF distribution deriving from the application of the PG_TRIGRS model. PG_TRIGRS (probabilistic, geostatistic-based, transient rainfall infiltration and grid-based slope stability model) is a probabilistic extension [45] of the deterministic version implemented in the original TRIGRS code, developed by Baum et al., (2008) [60], which treats each cell of the study area, subdivided into a GIS grid, as an infinite slope. Referring to deterministic slope stability analysis, TRIGRS allows the coefficient of failure F_S along the slip surface to be evaluated with respect to translational sliding as:

$$F_s(Z,t) = \frac{\tau_f}{\tau_m} = \frac{\tan \phi\prime}{\tan \alpha} + \frac{c' - \gamma_w \psi(Z,t) \tan \phi\prime}{\gamma Z \sin \alpha \cos \alpha}, \quad (1)$$

where τ_m is the mobilized shear stress, τ_f is the available shear strength, c' and ϕ' are the effective cohesion and friction angle of the soil, respectively, $\psi = u/\gamma_w$ is pressure head, u is the pore water pressure, γ_w is the water unit weight, α is the slope angle and γ is the unit weight of the soil. In transient flow conditions, the factor of safety varies with Z and t, due to the evolution with time and space of the pressure head ψ generated by the rainfall infiltration process [60].

In the deterministic analysis, F_s depends on 12 parameters, briefly represented as:

$$F_s \ = f(\alpha, h, d_w, \gamma_s, c', \varphi\prime, E_{ed}, k_s, \theta_s, \theta_r, a_{\alpha,}, I_{LT}), \quad (2)$$

in which, in addition to the quantities identified above, θ_s and θ_r are the saturated and residual volumetric water content, E_{ed} the soil stiffness, d_w the initial pre-storm water table depth, a_α and I_{LT} the pre-storm infiltration rate parameters; h the thickness of the soil cover and k_s the hydraulic conductivity.

The parameters present in Equation (2) are deterministic quantities, considered exact values without uncertainty, but soils and rocks are described by parameters characterized by high variability in space both in horizontal and vertical dimension [61]. For instance, mechanical properties show their uncertainty not only from site to site and within a given stratigraphy, but also within homogeneous covers, as a consequence of natural deposition processes [62]. When the randomness of the quantities is considered, the stability conditions are expressed by the PoF and the input quantities are random variables. This is the approach followed by PG_TRIGRS which considers random variables c', ϕ' and k_s; the PoF is evaluated cell-by-cell, through the probabilistic point estimate method approach. The code, validated over different areas of the Umbria Region characterized by different landslide susceptibility levels [46,63,64], has been used to evaluate the PoF distribution linked to the rainfall event described in Section 2.1. In particular, different analyses were carried out considering different scenarios of reducing the mechanical properties of the soil linked to soil tillage.

Considering the spatial distributions of slopes and soils (Figure 1), it could be observed that the landslides were not localized in areas with higher slopes. Conversely, the landslides were mostly localized in human-modified clay soils in the central part of the study area, where the land is more easily cultivated than in the eastern part of the area, where fine-grained soils alternate with rocks.

These findings highlight the impact of agricultural practices on the soil's mechanical characteristics. In fact, arable land often requires conventional tillage that is able to alter the intrinsic physical properties of the soil, such as soil structure, porosity, pore-size distribution, aggregation, particle size distribution, water retention capacity and permeability but above all the effective cohesion of soil [65].

Moreover, agricultural practices induce changes on the slopes. Indeed, in the rainiest periods there are freshly tilled soils without vegetation, while in the driest periods they are covered with vegetation. To this purpose, two images of the study area gathered in

April 2018 and October 2019 and are shown in Figure 5. Moving from autumn to spring, the seasonal transformations undergone by the slopes can be seen; these changes are very likely associated with agricultural practices that are supposed to reduce soil strength.

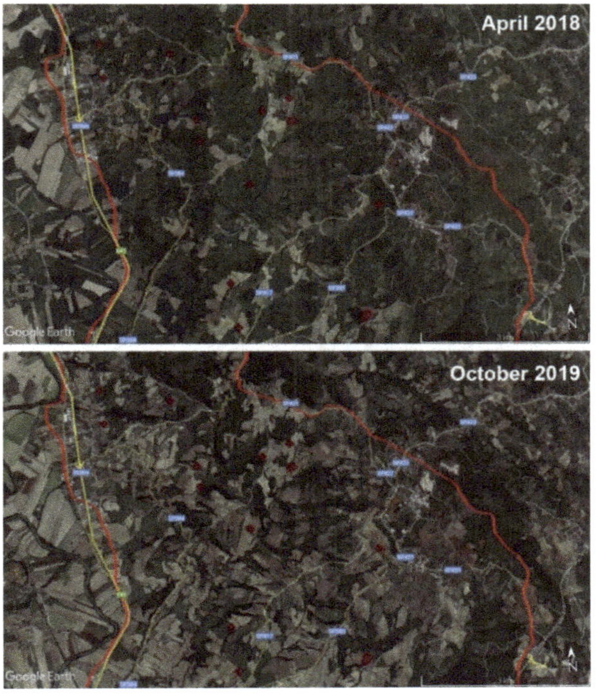

Figure 5. Two aerial images of the central part of the study area (red dots represent shallow landslides). In October, the area is devoid of any vegetation (many brown areas); in April, several zones are fully vegetated (green). Background images from Google Earth.

In the absence of quantitative information and specific studies on this topic, to take into account the impact of soil tillage in the analysis of the stability conditions of the area, a heuristic approach was here adopted. In particular, the slope analysis was carried out assuming a decrease of the effective cohesion c' equal to 10, 20, 30, 40 and 50% in all arable areas (cf. Figure 2B) to mimic an increasing impact of soil tillage on the geomechanical conditions. In all model runs, the same triggering rainfall event was considered as input. In this way, only the effect of soil tillage on the stability conditions was evaluated.

3. Results and Discussion

3.1. Statistical Analysis

As mentioned, the random variables considered in the PG_TRIGRS approach are: (i) the effective cohesion c'; (ii) the effective friction angle φ' and (iii) the saturated hydraulic conductivity k_s. The random variability is described, for each variable, by theoretical probability density function (pdf); the pdf definition was evaluated on the basis of in situ measurements available for different soils of the Perugia Province [55]. The correlation coefficient (ρ) between c' and φ' is assumed to be equal to -0.5 [66–68], while the correlation between k_s and the soil strength parameters is assumed equal to 0. Major details of the stochastic characterization of the mechanical properties of the soil are presented in the work of Fanelli et al., (2016) [55] and Salciarini et al. [45,69]; for further information, the stochastic soil characterization is summarized in Table 1.

Table 1. The table summarizes the stochastic parameters (mean value and coefficient of variation, CoV) used to characterize the soils in the study area (according to Fanelli et al., 2016 [55]). In the reliability analyses, rocks such as limestone, marl limestone, conglomerate, marl and travertine are not considered.

	c' (kPa)	φ' (°)	Ks (m/s)	CoV (c′)	CoV (φ′)	CoV (ks)
Alluvium	3	28	5×10^{-8}	0.25	0.05	0.9
Terraced alluvium	6	29	1×10^{-7}	0.25	0.05	0.9
Clay	15	26	5×10^{-11}	0.25	0.05	0.9
Eluvial deposit	7	29	1×10^{-9}	0.25	0.05	0.9
Landslide deposit	0	31	5×10^{-8}	0.40	0.09	2.4
Sand	0	30	5×10^{-5}	0.40	0.09	2.4
Turbidite	5	30	5×10^{-7}	0.25	0.05	0.9

In the absence of detailed information or recorded data, the initial pre-storm water table depth (d_w) was set equal to 50% of h, and the steady pre-storm infiltration rate (I_{LT}) was assumed to be negligible. The soil covers were considered completely saturated ($S_r = 1$) and the soil thickness did not exceed 2 m. The stability analyses were conducted under saturated conditions, as conservative assumptions, therefore evaluations about the estimated soil moisture or the groundwater level change were not considered in the study.

3.2. Modelling Results

Starting from the soil characterization shown in Table 1, the model was applied to the study area considering the rainfall event described in Section 2.1 (Figure 3). The results show that the area is mostly stable (Figure 6). The stability conditions vary considering the critical rainfall event, but the PoF values are negligible in most part of the area. A small area with PoF between 0.3 and 0.5 is located in the northwestern part, which is characterized by high slopes and Turbiditic soils (with low values of c′).

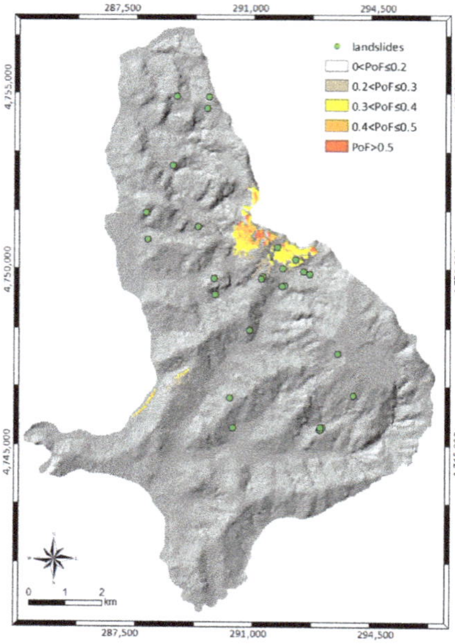

Figure 6. The image shows the PoF distribution considering natural conditions for the soil cover and the rainfall event described in Figure 3. For the soil characterization, the mechanical properties used are reported in Table 1.

Observing the landslide distribution, we noticed that the movements occurred on the portions of the study area characterized by modest slopes; the soil mainly involved is clay (according to Fanelli et al., 2016 [55]). This evidence suggested that agricultural techniques such as soil tillage could significantly reduce the effective cohesion of the soil covers. As previously mentioned, a reduction in the average value of cohesion was considered for the arable soils in the study area. Figure 7 shows the PoF spatial distributions, evaluated at the end of the considered rainfall event, assuming reductions of c' in arable areas respectively equal to 20 (Figure 7A), 30 (Figure 7B), 40 (Figure 7C) and 50% (Figure 7D).

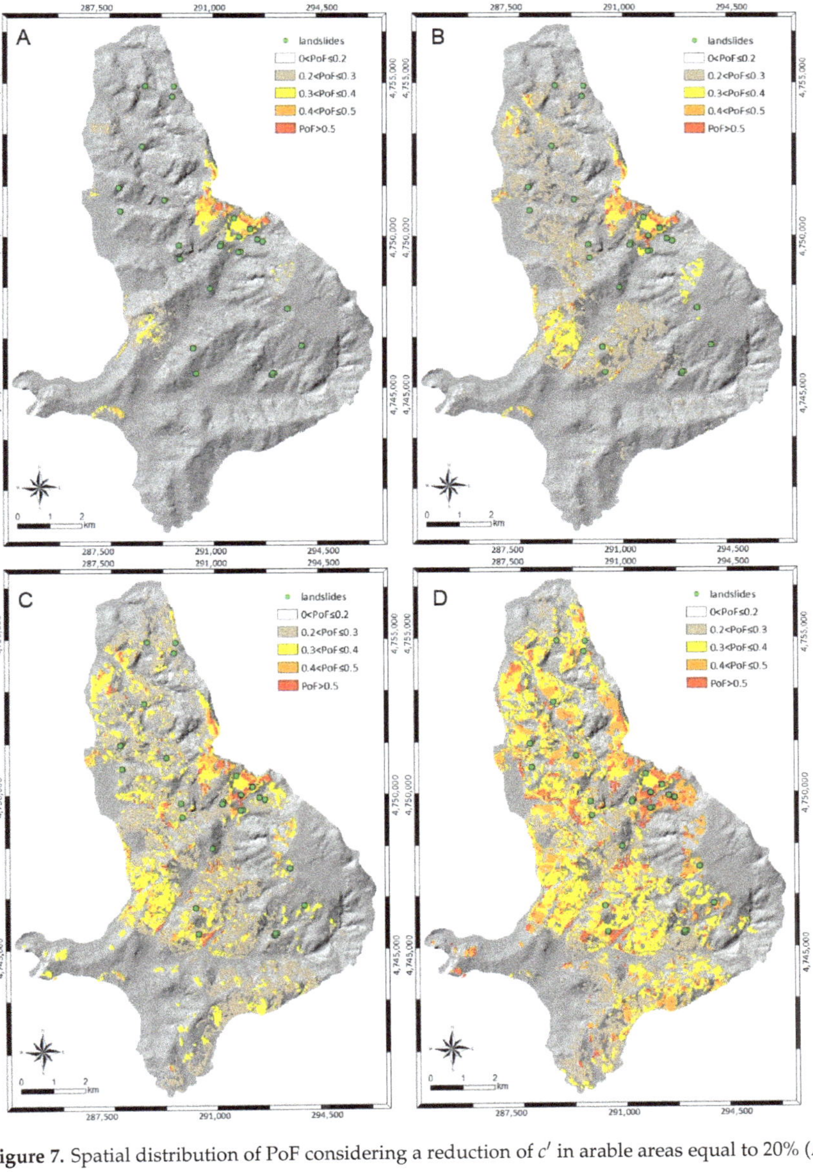

Figure 7. Spatial distribution of PoF considering a reduction of c' in arable areas equal to 20% (**A**), 30% (**B**), 40% (**C**), and 50% (**D**).

The 20% reduction for c' amplified the portions of the area originally identified by PG_TRIGRS and shown in Figure 5; however, the central sector of the area, where shallow landslides were localized, was still affected by a negligible PoF. The central sector is characterized by gentle slopes, ranging from a minimum of 3° to a maximum of 18°, therefore, the triggering causes of the observed landslides were likely due to the changes in the mechanical characteristics of the soil.

As expected, higher probabilities of failure are observed considering a 30% of reduction for the effective cohesion of the soils (Figure 7B). This decrease of c' produces, in the central part of the area, a large sector characterized by PoF values between 20 and 30%; and most landslides are correctly predicted by the model. The additional decrease of 40% for c' provides only an increase in extension of the critical areas already identified, without significantly improving the model prediction. Finally, considering a reduction of 50% for c', a large part of the area becomes unstable.

Figure 8 summarizes the quantitative results obtained from the analyses. Each row represents the distribution of the cells in the various PoF classes, while for each column a different reduction of c' has been considered. As the reduction of c' increases, as can be expected, more pixels move towards classes with higher PoF. The pixels decrease, with reference to first class (white), is minimum passing from a reduction of about 10% to a reduction of 20% of c', while it becomes important considering the c' reduction of 30 to 50%. Pixels belonging to the orange class ($0.2 < \text{PoF} \leq 0.3$) also increase passing from the reduction of c' of 20 to 30%.

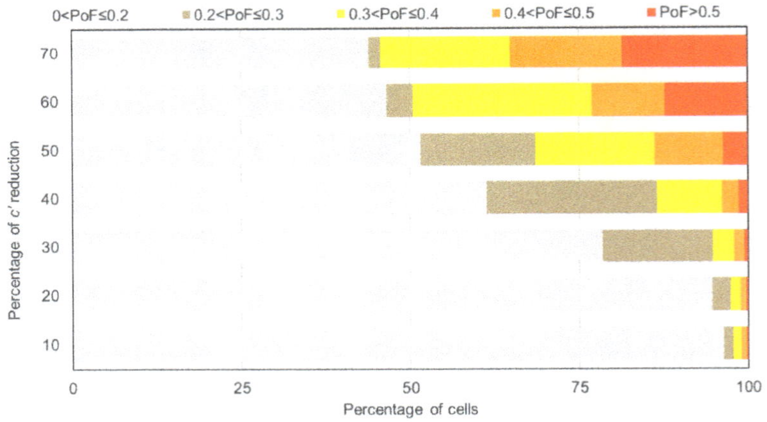

Figure 8. Pixel distribution in the five classes of PoF, according to the different reductions of c'.

Relating to the spatial distribution of PoF with the landslides observed in situ, the c' reduction corresponding to the absence of landslides in the first class ($0 < \text{PoF} \leq 0.2$) is 50%. In other words, considering a reduction of c' equal to 50%, the pixels were the landslides that have been measured and all were located in the second class (PoF > 20%). The model identified accurately the pixels characterized by landslide occurrence but incorrectly classified the other stable pixels.

3.3. Reliabaility Analysis

In order to evaluate the reliability of the proposed method, we used the standard contingencies and metrics adopted for model evaluation. For each case of c' reduction, we calculate the number of cells with: landslides correctly hindcasted by the model (true positives, TP); landslides mapped but not hindcasted by the model (false negatives, FN); positive outcome from the model and absence of mapped landslides (false positives, FP); no landslides mapped and negative outcome from the model (true negatives, TN). Given that the outcome of the model is probabilistic and not deterministic, we set three increasing thresholds values for PoF to define a positive and a negative outcome. In particular, we identified the values of 0.2, 0.3, 0.5 as the threshold PoF value for calculating the contingencies. As an example, in the case of the PoF threshold fixed at 0.2, we considered a positive outcome of the model to be a cell with a PoF > 0.2 and a negative outcome a cell with a PoF ≤ 2; in the case of 0.5 threshold, a positive (negative) outcome was a cell with a PoF equal or higher (lower) than 0.5. Therefore, for each case of c' reduction and for each PoF threshold value, we calculated the four contingencies and some skill scores: the true positive rate, TPR = TP/(TP + FN); the false positive rate, FPR = FP/(FP + TN); the Hansen–Kuipers skill score, HK = TP − FN; the predictive power, Pp = TP/(TP + FP). Detailed results are reported in Table 2. As an example, with reference to the lowest Pof threshold (0.2), the best HK value is obtained considering a 40% reduction of c'.

Table 2. Values of the calculated skill scores for the five cases of c' reduction and the three PoF threshold values. For each case of c' reduction, the average value for each skill score is also reported. The best average values for each skill score are in bold.

C' Reduction	PoF Threshold	Skill Scores			
		FPR (%)	Pp (%)	TPR (%)	HK (%)
10%	0.2	3.6	5.2	11.1	7.5
	0.3	62.5	8.4	100.0	37.5
	0.5	9.1	0.0	0.0	−9.1
	Mean	25.1	4.5	37.0	12.0
20%	0.2	5.3	3.6	11.1	5.8
	0.3	49.9	7.1	100.0	50.1
	0.5	9.9	0.0	0.0	−9.9
	Mean	21.7	3.6	37.0	**15.3**
30%	0.2	21.6	2.6	33.3	11.8
	0.3	24.1	3.6	33.3	9.2
	0.5	11.4	10.6	25.0	13.6
	Mean	**19.0**	**5.6**	30.6	11.5
40%	0.2	38.6	2.8	63.0	24.3
	0.3	55.5	0.8	23.5	−32.0
	0.5	9.8	4.8	20.0	10.2
	Mean	34.6	2.8	35.5	0.9
50%	0.2	48.4	2.5	70.4	22.0
	0.3	64.8	2.0	52.6	−12.2
	0.5	11.6	10.3	37.5	25.9
	Mean	41.6	4.9	**53.5**	11.9

Finally, we calculated the mean values of the skill scores for each case of c' reduction, averaging the values obtained for the three PoF thresholds values (Table 2). The average TPR closest to 1 (i.e., the optimal value) was observed for a c' reduction equal to 50%; however, this condition also generated many FP, thus, other considerations were needed. Indeed, if we considered the average values of FPR and HK, the best model performance was obtained considering a 20 and 30% reduction of c', respectively. Overall, the best compromise between correct and incorrect model outcomes was obtained considering a 20% reduction of c' between 20% (best values for HK) and 30% (best value for FPR and Pp).

4. Conclusions

The paper presents a study focusing on the effect of soil tillage due to agricultural activity on slope stability conditions in a hilly environment. To this aim, a back analysis has been performed using the PG_TRIGRS model on the study area of Collazzone, central Italy. The code has already been used for the prediction of shallow landslides in the Umbria

Region, but it has not been tested on areas where landslides, triggered by a specific rainfall event, have been directly observed. The proposed probabilistic approach which assumes input data as random variables defined through their probability density functions, can provide a higher level of reliability than deterministic approaches, which assume the input data without uncertainty.

For this preliminary study, we decided to perform the stability analyses under saturated conditions and we have not considered the randomness of the initial water table depth d_w in order to have only three random variables in input (c', φ', and k_s). This choice allowed us to reduce the computational times and to better identify the effect of soil tillage on the soil's mechanical properties. However, future upgrades of the method will include the piezometric variability, the unsaturated conditions, and the soil thickness analysis. The study area selected for this purpose is of particular scientific interest because it is characterized by intense agricultural activity that greatly increases its susceptibility to landslides. The latter appears to be related to agricultural management systems which modify the physical and mechanical properties of the soil. During the year the ground is tilled and denuded in the rainiest period, while in the summer and spring months the slopes appear covered by vegetation. The continuous change in soil conditions alters the mechanical characteristics of the soil in a non-negligible way. The use of unsustainable agricultural machinery should be limited to avoid loss of cohesion in the cover soil.

From the results obtained in this study, with reference to the area analyzed, soil tillage due to agricultural activity caused a reduction in cohesion of between 20 and 30%; this estimation agrees with the results obtained in studies aimed at evaluating the effects produced by specific tillage on soil's mechanical properties [44]. With reference to the results, the slope stability model is able to predict surface landslides but it classifies, as a precaution, the areas not affected by landslides. In any case, in order to safeguard the Collazzone area from shallow sliding phenomena, it is necessary to practice planned and appropriately selected agricultural techniques.

The method presented in this work is quantitative and reproducible, thus can be applied in other areas with similar environmental contexts, also for comparison with in situ and laboratory tests to refine and optimize the evaluation of the effect of soil tillage on slope stability.

Author Contributions: Conceptualization, E.V. and D.S.; methodology, E.V., S.L.G., F.A., F.F. and D.S.; analysis, E.V. and S.L.G.; data curation, F.A., F.F. and E.V.; writing—original draft preparation, E.V. and S.L.G.; writing—review and editing, S.L.G., F.A., F.F. and D.S. All authors have read and agreed to the published version of the manuscript.

Funding: This research received no external funding.

Institutional Review Board Statement: Not applicable.

Informed Consent Statement: Not applicable.

Data Availability Statement: Not applicable.

Acknowledgments: Rainfall provided by the administration of the Umbria region. The authors would like to thank two anonymous reviewers for their comments and suggestion, which were useful in improving the manuscript.

Conflicts of Interest: The authors declare no conflict of interest.

References

1. Froude, M.J.; Petley, D.N. Global Fatal Landslide Occurrence from 2004 to 2016. *Nat. Hazards Earth Syst. Sci.* **2018**, *18*, 2161–2181. [CrossRef]
2. Haque, U.; da Silva, P.F.; Devoli, G.; Pilz, J.; Zhao, B.; Khaloua, A.; Wilopo, W.; Andersen, P.; Lu, P.; Lee, J.; et al. The Human Cost of Global Warming: Deadly Landslides and Their Triggers (1995–2014). *Sci. Total Environ.* **2019**, *682*, 673–684. [CrossRef]
3. Sidle, R.C.; Ochiai, H. *Landslides: Processes, Prediction, and Land Use;* Water Resources Monograph; American Geophysical Union: Washington, DC, USA, 2006; Volume 18, ISBN 978-0-87590-322-4.

4. Knevels, R.; Brenning, A.; Gingrich, S.; Heiss, G.; Lechner, T.; Leopold, P.; Plutzar, C.; Proske, H.; Petschko, H. Towards the Use of Land Use Legacies in Landslide Modeling: Current Challenges and Future Perspectives in an Austrian Case Study. *Land* **2021**, *10*, 954. [CrossRef]
5. Liu, J.; Wu, Z.; Zhang, H. Analysis of Changes in Landslide Susceptibility According to Land Use over 38 Years in Lixian County, China. *Sustainability* **2021**, *13*, 10858. [CrossRef]
6. Montgomery, D.R.; Dietrich, W.E. A Physically Based Model for the Topographic Control on Shallow Landsliding. *Water Resour. Res.* **1994**, *30*, 1153–1171. [CrossRef]
7. Van Westen, C.J. The Modelling of Landslide Hazards Using GIS. *Surv. Geophys.* **2000**, *21*, 241–255. [CrossRef]
8. Persichillo, M.G.; Bordoni, M.; Meisina, C.; Bartelletti, C.; Barsanti, M.; Giannecchini, R.; D'Amato Avanzi, G.; Galanti, Y.; Cevasco, A.; Brandolini, P.; et al. Shallow Landslides Susceptibility Assessment in Different Environments. *Geomat. Nat. Hazards Risk* **2017**, *8*, 748–771. [CrossRef]
9. Brandolini, P.; Pepe, G.; Capolongo, D.; Cappadonia, C.; Cevasco, A.; Conoscenti, C.; Marsico, A.; Vergari, F.; Del Monte, M. Hillslope Degradation in Representative Italian Areas: Just Soil Erosion Risk or Opportunity for Development? *Land Degrad. Dev.* **2018**, *29*, 3050–3068. [CrossRef]
10. Glade, T. Landslide Occurrence as a Response to Land Use Change: A Review of Evidence from New Zealand. *CATENA* **2003**, *51*, 297–314. [CrossRef]
11. Reichenbach, P.; Busca, C.; Mondini, A.C.; Rossi, M. The Influence of Land Use Change on Landslide Susceptibility Zonation: The Briga Catchment Test Site (Messina, Italy). *Environ. Manag.* **2014**, *54*, 1372–1384. [CrossRef]
12. Pisano, L.; Zumpano, V.; Malek, Ž.; Rosskopf, C.M.; Parise, M. Variations in the Susceptibility to Landslides, as a Consequence of Land Cover Changes: A Look to the Past, and Another towards the Future. *Sci. Total Environ.* **2017**, *601–602*, 1147–1159. [CrossRef]
13. Gariano, S.L.; Petrucci, O.; Rianna, G.; Santini, M.; Guzzetti, F. Impacts of Past and Future Land Changes on Landslides in Southern Italy. *Reg. Environ. Chang.* **2018**, *18*, 437–449. [CrossRef]
14. Persichillo, M.G.; Bordoni, M.; Meisina, C. The Role of Land Use Changes in the Distribution of Shallow Landslides. *Sci. Total Environ.* **2017**, *574*, 924–937. [CrossRef]
15. Falcucci, A.; Maiorano, L.; Boitani, L. Changes in Land-Use/Land-Cover Patterns in Italy and Their Implications for Biodiversity Conservation. *Landsc. Ecol.* **2007**, *22*, 617–631. [CrossRef]
16. Greenway, D.R. Vegetation and Slope Stability. In *Slope Stability*; Wiley: Chichester, UK, 1987; pp. 187–230.
17. Schmidt, K.M.; Roering, J.J.; Stock, J.D.; Dietrich, W.E.; Montgomery, D.R.; Schaub, T. The Variability of Root Cohesion as an Influence on Shallow Landslide Susceptibility in the Oregon Coast Range. *Can. Geotech. J.* **2001**, *38*, 995–1024. [CrossRef]
18. Schwarz, M.; Preti, F.; Giadrossich, F.; Lehmann, P.; Or, D. Quantifying the Role of Vegetation in Slope Stability: A Case Study in Tuscany (Italy). *Ecol. Eng.* **2010**, *36*, 285–291. [CrossRef]
19. Wu, T.H. Root Reinforcement of Soil: Review of Analytical Models, Test Results, and Applications to De sign. *Can. Geotech. J.* **2013**, *50*, 259–274. [CrossRef]
20. Tarolli, P.; Straffelini, E. Agriculture in Hilly and Mountainous Landscapes: Threats, Monitoring and Sustainable Management. *Geogr. Sustain.* **2020**, *1*, 70–76. [CrossRef]
21. Endo, T. Effect of Tree Roots upon the Shear Strength of Soil. *Jpn. Agric. Res. Q.* **1980**, *12*, 12–115.
22. Wu, T.H.; McKinnell III, W.P.; Swanston, D.N. Strength of Tree Roots and Landslides on Prince of Wales Island, Alaska. *Can. Geotech. J.* **1979**, *16*, 19–33. [CrossRef]
23. Wu, T.H.; Beal, P.E.; Lan, C. In-Situ Shear Test of Soil-Root Systems. *J. Geotech. Eng.* **1988**, *114*, 1376–1394. [CrossRef]
24. Bischetti, G.B.; Chiaradia, E.A.; Epis, T.; Morlotti, E. Root Cohesion of Forest Species in the Italian Alps. *Plant Soil* **2009**, *324*, 71–89. [CrossRef]
25. Salciarini, D.; Volpe, E.; Di Pietro, L.; Cattoni, E. A Case-Study of Sustainable Countermeasures against Shallow Landslides in Central Italy. *Geosciences* **2020**, *10*, 130. [CrossRef]
26. Salciarini, D.; Morbidelli, R.; Cattoni, E.; Volpe, E. Physical and Numerical Modelling of the Response of Slopes under Different Rainfalls, Inclinations and Vegetation Conditions. *Ital. Geotech. J.* **2022**, *2*, 7–21.
27. Waldron, L.J. The Shear Resistance of Root-Permeated Homogeneous and Stratified Soil. *Soil Sci. Soc. Am. J.* **1977**, *41*, 843–849. [CrossRef]
28. Operstein, V.; Frydman, S. The Influence of Vegetation on Soil Strength. *Proc. Inst. Civ. Eng.-Ground Improv.* **2000**, *4*, 81–89. [CrossRef]
29. Stokes, A.; Atger, C.; Bengough, A.G.; Fourcaud, T.; Sidle, R.C. Desirable Plant Root Traits for Protecting Natural and Engineered Slopes against Landslides. *Plant Soil* **2009**, *324*, 1–30. [CrossRef]
30. Löbmann, M.T.; Geitner, C.; Wellstein, C.; Zerbe, S. The Influence of Herbaceous Vegetation on Slope Stability—A Review. *Earth-Sci. Rev.* **2020**, *209*, 103328. [CrossRef]
31. Cazzuffi, D.; Corneo, A.; Crippa, E. Slope Stabilisation by Perennial "Gramineae" in Southern Italy: Plant Growth and Temporal Performance. *Geotech. Geol. Eng.* **2006**, *24*, 429–447. [CrossRef]
32. Norris, J.; Greenwod, J.R. Assessing the Role of Vegetation on Soil Slopes in Urban Areas. In Proceedings of the 10th IAEG International Congress, Nottingham, UK, 6 September 2006.

33. Bordoni, M.; Cislaghi, A.; Vercesi, A.; Bischetti, G.B.; Meisina, C. Effects of Plant Roots on Soil Shear Strength and Shallow Landslide Proneness in an Area of Northern Italian Apennines. *Bull. Eng. Geol. Environ.* **2020**, *79*, 3361–3381. [CrossRef]
34. Masi, E.B.; Segoni, S.; Tofani, V. Root Reinforcement in Slope Stability Models: A Review. *Geosciences* **2021**, *11*, 212. [CrossRef]
35. Horn, R. Time Dependence of Soil Mechanical Properties and Pore Functions for Arable Soils. *Soil Sci. Soc. Am. J.* **2004**, *68*, 1131–1137. [CrossRef]
36. Albiero, D.; Maciel, A.J.S.; Gamero, C.A.; Lanças, K.P.; Mion, R.L.; Viliotti, C.A.; Monteiro, L.A. Dimensional Analysis of Soil Properties after Treatment with the Rotary Paraplow, a New Conservationist Tillage Tool. *Span. J. Agric. Res.* **2011**, *9*, 693. [CrossRef]
37. Bullock, M.S.; Nelson, S.D.; Kemper, W.D. Soil Cohesion as Affected by Freezing, Water Content, Time and Tillage. *Soil Sci. Soc. Am. J.* **1988**, *52*, 770–776. [CrossRef]
38. Mari, I.A.; Changying, J.; Leghari, N.; Chandio, F.A.; Arslan, C.; Hassan, M. Impact of Tillage Operation on Soil Physical, Mechanical and Rhelogical Properties of Paddy Soil. *Bulg. J. Agric. Sci.* **2017**, *21*, 940–946.
39. Jiang, Q.; Cao, M.; Wang, Y.; Wang, J.; He, Z. Estimation of Soil Shear Strength Indicators Using Soil Physical Properties of Paddy Soils in the Plastic State. *Appl. Sci.* **2021**, *11*, 5609. [CrossRef]
40. Castellini, M.; Ventrella, D. Impact of Conventional and Minimum Tillage on Soil Hydraulic Conductivity in Typical Cropping System in Southern Italy. *Soil Tillage Res.* **2012**, *124*, 47–56. [CrossRef]
41. Biddoccu, M.; Ferraris, S.; Opsi, F.; Cavallo, E. Long-Term Monitoring of Soil Management Effects on Runoff and Soil Erosion in Sloping Vineyards in Alto Monferrato (North–West Italy). *Soil Tillage Res.* **2016**, *155*, 176–189. [CrossRef]
42. Carretta, L.; Tarolli, P.; Cardinali, A.; Nasta, P.; Romano, N.; Masin, R. Evaluation of Runoff and Soil Erosion under Conventional Tillage and No-till Management: A Case Study in Northeast Italy. *CATENA* **2021**, *197*, 104972. [CrossRef]
43. Straffelini, E.; Pijl, A.; Otto, S.; Marchesini, E.; Pitacco, A.; Tarolli, P. A High-Resolution Physical Modelling Approach to Assess Runoff and Soil Erosion in Vineyards under Different Soil Managements. *Soil Tillage Res.* **2022**, *222*, 105418. [CrossRef]
44. Albiero, D.; da Silva Maciel, A.J.; Monteiro, L.D.A.; Lanças, K.P.; Gamero, C.; Melo, R.P.; Araújo, M.C.D. Effects in Mechanical Properties and Structure of the Soil after Tillage with Rotary Paraplow. *Afr. J. Agric. Res.* **2014**, *9*, 1855–1866. [CrossRef]
45. Salciarini, D.; Tamagnini, C.; Ronchi, F.; Volpe, E.; Fanelli, G. An Approach for Large-Scale Soil Characterization for the Application of Non-Structural Landslide Risk Mitigation. *Riv. Ital. Geotec.* **2017**, *3*, 7–21. [CrossRef]
46. Salciarini, D.; Volpe, E.; Cattoni, E. Probabilistic vs. Deterministic Approach in Landslide Triggering Prediction at Large-Scale. In *Proceedings of the Geotechnical Research for Land Protection and Development*; Calvetti, F., Cotecchia, F., Galli, A., Jommi, C., Eds.; Springer International Publishing: Cham, Switzerland, 2020; pp. 62–70.
47. Volpe, E.; Ciabatta, L.; Salciarini, D.; Camici, S.; Cattoni, E.; Brocca, L. The Impact of Probability Density Functions Assessment on Model Performance for Slope Stability Analysis. *Geosciences* **2021**, *11*, 322. [CrossRef]
48. Guzzetti, F.; Reichenbach, P.; Cardinali, M.; Galli, M.; Ardizzone, F. Probabilistic Landslide Hazard Assessment: An Example in the Collazzone Area, Central Italy. In Proceedings of the 3rd Probabilistic Workshop, Technical Systems and Natural Hazards, Vienna, Austria, 24–25 November 2005; Volume 7, pp. 173–182.
49. Raia, S.; Alvioli, M.; Rossi, M.; Baum, R.L.; Godt, J.W.; Guzzetti, F. Improving Predictive Power of Physically Based Rainfall-Induced Shallow Landslide Models: A Probabilistic Approach. *Geosci. Model Dev.* **2014**, *7*, 495–514. [CrossRef]
50. Lombardo, L.; Opitz, T.; Ardizzone, F.; Guzzetti, F.; Huser, R. Space-Time Landslide Predictive Modelling. *Earth-Sci. Rev.* **2020**, *209*, 103318. [CrossRef]
51. Fiorucci, F.; Cardinali, M.; Carlà, R.; Rossi, M.; Mondini, A.C.; Santurri, L.; Ardizzone, F.; Guzzetti, F. Seasonal Landslide Mapping and Estimation of Landslide Mobilization Rates Using Aerial and Satellite Images. *Geomorphology* **2011**, *129*, 59–70. [CrossRef]
52. Ardizzone, F.; Cardinali, M.; Galli, M.; Guzzetti, F.; Reichenbach, P. Identification and Mapping of Recent Rainfall-Induced Landslides Using Elevation Data Collected by Airborne Lidar. *Nat. Hazards Earth Syst. Sci.* **2007**, *7*, 637–650. [CrossRef]
53. Guzzetti, F.; Galli, M.; Reichenbach, P.; Ardizzone, F.; Cardinali, M. Landslide Hazard Assessment in the Collazzone Area, Umbria, Central Italy. *Nat. Hazards Earth Syst. Sci.* **2006**, *6*, 115–131. [CrossRef]
54. Guzzetti, F.; Reichenbach, P.; Ardizzone, F.; Cardinali, M.; Galli, M. Landslide Hazard Assessment, Vulnerability Estimation and Risk Evaluation: An Example from the Collazzone Area (Central Umbria, Ialy). *Geogr. Fis. E Din. Quat.* **2009**, *32*, 183–192.
55. Fanelli, G.; Salciarini, D.; Tamagnini, C. Reliable Soil Property Maps over Large Areas: A Case Study in Central Italy. *Environ. Eng. Geosci.* **2016**, *22*, 37–52. [CrossRef]
56. IUSS Working Group WRB. *World Reference Base for Soil Resources 2014, Update 2015. International Soil Classification System for Naming Soils and Creating Legends for Soil Maps*; World Soil Resources Reports; F.A.O.: Rome, Italy, 2015; ISBN 978-92-5-108370-3.
57. Regione Umbria, Assessorato Agricoltura e Foreste Carta dei Suoli Della Regione Dell'Umbria. 2010. Available online: http://www.umbriageo.regione.umbria.it/ (accessed on 8 June 2022).
58. Ardizzone, F.; Fiorucci, F.; Santangelo, M.; Cardinali, M.; Mondini, A.C.; Rossi, M.; Reichenbach, P.; Guzzetti, F. Very-High Resolution Stereoscopic Satellite Images for Landslide Mapping. In *Landslide Science and Practice: Volume 1: Landslide Inventory and Susceptibility and Hazard Zoning*; Margottini, C., Canuti, P., Sassa, K., Eds.; Springer: Berlin/Heidelberg, Germany, 2013; pp. 95–101, ISBN 978-3-642-31325-7.
59. Cardinali, M.; Galli, M.; Guzzetti, F.; Ardizzone, F.; Reichenbach, P.; Bartoccini, P. Rainfall Induced Landslides in December 2004 in South-Western Umbria, Central Italy: Types, Extent, Damage and Risk Assessment. *Nat. Hazards Earth Syst. Sci.* **2006**, *6*, 237–260. [CrossRef]

60. Baum, R.L.; Savage, W.Z.; Godt, J.W. *TRIGRS—A Fortran Program for Transient Rainfall Infiltration and Grid-Based Regional Slope-Stability Analysis, Version 2.0*; USGS Open-File Report 2008-1159; US Geological Survey: Reston, VA, USA, 2008; p. 75.
61. Fenton, G.A.; Griffiths, D.V. *Risk Assessment in Geotechnical Engineering*; John Wiley & Sons, Inc.: Hoboken, NJ, USA, 2008; ISBN 978-0-470-28470-4.
62. Lacasse, S.; Nadim, F. *Uncertainties in Characterising Soil Properties*; ASCE: Reston, VA, USA, 1996; pp. 49–75.
63. Salciarini, D.; Volpe, E.; Kelley, S.A.; Brocca, L.; Camici, S.; Fanelli, G.; Tamagnini, C. Modeling the Effects Induced by the Expected Climatic Trends on Landslide Activity at Large Scale. *Procedia Eng.* **2016**, *158*, 541–545. [CrossRef]
64. Salciarini, D.; Brocca, L.; Camici, S.; Ciabatta, L.; Volpe, E.; Massini, R.; Tamagnini, C. Physically Based Approach for Rainfall-Induced Landslide Projections in a Changing Climate. *Proc. Inst. Civ. Eng.-Geotech. Eng.* **2019**, *172*, 481–495. [CrossRef]
65. Nacinovic, M.G.G.; Mahler, C.F.; Avelar, A. de S. Soil Erosion as a Function of Different Agricultural Land Use in Rio de Janeiro. *Soil Tillage Res.* **2014**, *144*, 164–173. [CrossRef]
66. Lumb, P. Safety Factors and the Probability Distribution of Soil Strength. *Can. Geotech. J.* **1970**, *7*, 225–242. [CrossRef]
67. Wolff, T.F. *Probabilistic Slope Stability in Theory and Practice*; ASCE: Reston, VA, USA, 1996; pp. 419–433.
68. Li, H.-Z.; Low, B.K. Reliability Analysis of Circular Tunnel under Hydrostatic Stress Field. *Comput. Geotech.* **2010**, *37*, 50–58. [CrossRef]
69. Salciarini, D.; Fanelli, G.; Tamagnini, C. A Probabilistic Model for Rainfall—Induced Shallow Landslide Prediction at the Regional Scale. *Landslides* **2017**, *14*, 1731–1746. [CrossRef]

land

MDPI

Article

Land Use in Flood-Prone Areas and Its Significance for Flood Risk Management—A Case Study of Alpine Regions in Austria

Lena Junger [1,*], Severin Hohensinner [2], Karin Schroll [3], Klaus Wagner [3] and Walter Seher [1]

1 Institute of Spatial Planning, Environmental Planning and Land Rearrangement, Department of Landscape, Spatial and Infrastructure Sciences, University of Natural Resources and Life Sciences, 1190 Vienna, Austria; walter.seher@boku.ac.at
2 Institute of Hydrobiology and Aquatic Ecosystem Management, Department of Water, Atmosphere and Environment, University of Natural Resources and Life Sciences, 1180 Vienna, Austria; severin.hohensinner@boku.ac.at
3 Federal Institute of Agricultural Economics, Rural and Mountain Research, 1030 Vienna, Austria; karin.schroll@bab.gv.at (K.S.); klaus.wagner@bab.gv.at (K.W.)
* Correspondence: lena.junger@boku.ac.at; Tel.: +43-1-47654-85513

Abstract: Increasing flood damage has led to a rising importance of land use in flood risk management policies, commonly referred to as the spatial turn in flood risk management. This includes policies aiming at making space for rivers, which, in practice, lead to an increasing demand for land. Although research has been conducted on the variety of policies, the resulting land use conflicts in flood-prone areas have not been paid much attention to. This paper therefore analyses the current land use and its changes in Alpine flood-prone areas in Austria. The results show that space for rivers has been decreasing due to human activities (e.g., river straightening and channel narrowing) since the middle of the 19th century, and settlements have been expanding into flood-prone areas. Furthermore, the share of valuable agricultural land (which is important for food production) located in flood hazard zones is higher in more mountainous areas. Given the limited space for permanent settlement in Alpine regions, these land use changes exert pressure on the availability of land suitable for flood risk management. Therefore, making space for rivers as part of flood risk management policies faces considerable restrictions in Alpine areas.

Keywords: flood risk; land use; settlement development; agricultural land use; flood-prone areas; Alpine regions

Citation: Junger, L.; Hohensinner, S.; Schroll, K.; Wagner, K.; Seher, W. Land Use in Flood-Prone Areas and Its Significance for Flood Risk Management—A Case Study of Alpine Regions in Austria. *Land* **2022**, *11*, 392. https://doi.org/10.3390/land11030392

Academic Editors: Matej Vojtek, Andrea Petroselli and Raffaele Pelorosso

Received: 17 February 2022
Accepted: 4 March 2022
Published: 7 March 2022

1. Introduction

Flooding is one of the most damaging natural hazards worldwide, with flood risks ever increasing [1]. This can, among other things, be linked to climate change, which is likely to further intensify flooding. Detailed evaluations of climate change impacts on flooding at local level, however, are still inconclusive [2,3]. A second important driver of increasing flood damage is socio-economic development, including an increase in wealth, population development, and settlement expansion in flood-prone areas [4–6].

In order to reduce flood risks, a shift from controlling rivers and hazard prevention to an integrated approach of flood risk management can be observed [7–10]. Integrated flood risk management includes structural and non-structural measures, moving away from a solely engineering task to the inclusion of different disciplines [11,12]. With this paradigm shift, spatial planning has become a crucial part of flood risk management, and related policies are seen as an essential means to prevent flood damage [11]. Scholarly literature uses the term 'spatial turn' to describe the increasing relevance of (mainly undeveloped) land for flood risk management [13,14]. According to van Ruiten and Hartmann [14], three aspects can be regarded as valid indicators for this spatial turn in flood risk management: "[...] *the policy of more space for the river, an integrated approach to the issue, and an approach beyond structural measures"*. To operationalize the spatial turn, Löschner [13] goes into more detail, including, among other aspects, *"policy efforts to widen rivers and improve the connectedness between rivers and floodplains"*.

Research has been conducted on the implementation of policies addressing 'making space for rivers' [15]. This includes research on the 'Room for River Directive' in the Netherlands [16–18], analysis of the practical implementation of giving space to rivers in Germany [19], and studying the application of the 'Room for River Directive' in Canada [20]. Furthermore, in the UK, there is research on the effectiveness of national planning policy in achieving compromises between space for rivers and space for people. This underlines the *"conflict between land and water"* [21].

This paper builds upon the aspect of 'making space for rivers', which, in practice, means an increasing demand of land both for river restoration and flood risk management. In consequence, this demand of land for rivers can lead to land use conflicts [21]. Although the implementation of related policies has been widely explored, there is research demand on the quality and extent of land use conflicts in flood-prone areas, particularly in areas with limited space for permanent settlement. This research will address this gap by looking at land cover changes and the current land use in flood-prone areas in Austrian Alpine regions. Considering the aspects of transformation of water bodies, settlement development, and agricultural land, this paper will answer following research questions. How did water bodies and settlements in flood-prone areas change between 1826–1857 and 2016 in Austrian Alpine regions? How are flood-prone areas in Alpine regions currently used (focusing on settlements and agriculture)?

A historic view on the human transformation of Alpine water bodies shows the decreasing space for rivers. As part of flood risk management, structural measures have been applied, including river channelization. Research on the channelization of Alpine rivers shows that, in comparison to other European countries, rivers in Austria have been altered more intensely [22]. These regulations left rivers with limited land availability [23]. Thus, river regulations are the preceding development, making it necessary nowadays to consciously implement policies in order to return space to rivers. However, it is not simply narrowing riverbeds that affect runoff. The use of flood-prone areas for settlements has an additional impact on discharge due to an increase in sealed surfaces water that cannot infiltrate into the ground, impacting the flood event [23]. Besides the relevance of widening rivers as a part of flood risk management, the ecological benefits must not be ignored [24]. The use of floodplains for agriculture as well as settlements often has a negative impact on the aquatic ecosystems [25,26].

Furthermore, a view on settlement development in flood-prone areas will underline the increasing land-take and will show the land use pressure in flood-prone areas. Research has been conducted on the exposure of settlements to flooding on different levels, showing a general increase in built-up land in hazard zones [5,27–29]. Our research will add to this broad spectrum of research on flood exposure by analyzing settlement development in areas up to a one in a 300-year flooding on a cross-regional level.

Besides settlement development, this paper presents an analysis of agricultural land in flood-prone areas, with a focus on land valuable for regional food production. A considerable share of areas in flood-prone areas is used for agricultural purposes [30]. The potential for mutual impacts is accordingly great, which makes it important for inclusion in this research.

The paper aims at showing land use dynamics in flood-prone areas and the resulting consequences for flood risk management, including the limitations of giving space to rivers.

2. Materials and Methods

2.1. Study Area

The research presented in this paper focuses on Alpine regions in Austria, and even though they are not homogeneous, they share similar spatial challenges. First of all, Alpine areas are prone to multiple natural hazards, including snow avalanches, landslides, and floods; floods, however, cause the most economic damage [31]. Furthermore, the space for permanent settlement is limited due to topography. According to the Alpine Convention, 65% of Austrian territory is part of the Alps. However, only 35.7% of the area for permanent settlement is situated within the Alps. This results in only 21.18% of the Austrian Alps being suitable for permanent settlement [32]. In addition to the limited space, the on-going conversion of (mainly) farmland to housing, commercial and industrial areas, and traffic and recreation infrastructure increases the pressure on open-space land uses. For Austria, this so-called land-take amounted to approximately 12 hectares per day in 2020 [33]. The increasing land-take in combination with limited space for permanent settlement particularly exerts pressure on existing and available land resources for flood runoff and flood retention [34].

The research is based on GIS analysis of spatial data on the land use of flood-prone areas, including settlements, water bodies, and agricultural land. For water bodies and settlements, historical as well as current data exists for the catchment area of the rivers Rhine (Vorarlberg), Salzach (Salzburg), and Drava (East Tyrol and Carinthia) (see Figure 1). The agricultural land use was analyzed based on current data, which was available for the whole of Austria.

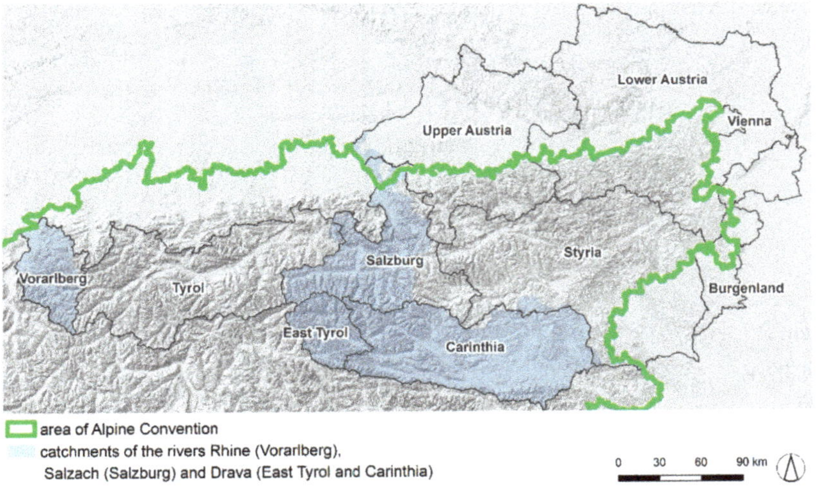

Figure 1. Study area (source of borders: Federal Office of Metrology and Surveying (BEV), Alpine Convention; source of digital terrain model: GELAENDE—basemap.at, accessed on 28 February 2022).

2.2. Data

The historical settlement data and the data on water bodies are based on the Franziscean Cadastre (1826–1859), which was digitalized by Hohensinner et al. [35]. The use of historical data can lead to some inaccuracies; the Franciscean cadastre shows variations in accuracy between intensely used valley areas and more remote locations higher up [35]. For the analysis of settlements, the lesser details at high altitudes does not matter as much, because the main settlements were located in the valleys.

For current land use (2016) based on LISA (Land Information System Austria) data, several datasets were combined [36]. To add detailed differentiation within agricultural land uses, the IACS (Integrated Administration and Control System) dataset was used [37].

Further, the glaciers are based on Buckel and Otto [38], and a dataset on mid-sized rivers and smaller ponds was gathered from OpenStreetMap [39]. For small rivers and streams, data from a project on "Strategic Planning for Alpine River Ecosystems" were used [40]. The different datasets were combined and prepared by Hohensinner et al. [35]. For the recent agricultural land analysis, the IACS spatial dataset of 2018 was used. For the exposure analysis of particularly valuable agricultural land, the results of the BEAT project were used as a spatial dataset, comprising the areas of valuable agricultural land [41]. For the economic evaluation of agricultural land, regional standard output coefficients were considered [42]. The standard output represents the average monetary value of agricultural output at farm-gate price in euro per hectare or per head of livestock. In order to quantify the decline of agricultural land in recent decades, the data of the agricultural structure survey and IACS data from the Green Report 2021 were used [43].

Flooding area combines information derived from the existing flood hazard maps provided by the Federal Water Engineering Administration (Bundeswasserbauverwaltung) and the Austrian Service for Torrent and Avalanche Control (Wildbach- und Lawinenverbauung), as well as data on 200-year flood events from HORA (Austrian Flood Hazard Map). The flood-prone area used for the exposure analysis encompasses all available flooding data in Austria, showing the area potentially affected by a 300-year flood event, which will be referred to as the flood-prone area in this paper. The focus is put on a 300-year flood, which includes areas with lower probability of flooding, in order to integrate areas that might be affected more by flooding in the future due to climate change impacts. Additionally, by covering an area of a 300-year flood in this research, 'protected' areas behind structural flood protection measures are included in the land use analysis of flood-prone areas. These data were also prepared by Hohensinner et al. [35].

2.3. *Methods*

2.3.1. Human Transformation of Alpine Water Bodies

In order to ascertain human modifications of the Alpine channel network, the historical and current active channels, i.e., water-covered areas and adjacent unvegetated sediment bars, were vectorized using ESRI ArcGIS 10.6. A comparative analysis yielded quantitative losses of running waters as a consequence of river channelization and flood prevention measures. The areal changes in standing water bodies (lakes, ponds, reservoirs) were derived analogously [35].

2.3.2. Settlement Development

The settlement area for this research includes buildings and adjacent streets, squares, gardens, and smaller parks. The method applied to research the development of settlements inside flood-prone areas is described as a *"classical approach for flood exposure assessment"* by Papilloud et al. [44], which has been broadly applied. Using ESRI ArcGIS 10.6., the settlement development was calculated by overlapping the historical with the current data; the difference between the datasets represents the development. To calculate the exposure of settlements to flooding, the results of the first step (settlement development) were further crossed with the flood-prone area. This made it possible to see the difference between the development inside and outside flooding areas. Besides the temporal aspect of the analysis, it was also interesting to look at the regional differences. For this purpose, the data were intersected with the municipality borders, showing settlement development inside and outside flood-prone areas per municipality. The results from the GIS analysis were further processed in Excel to calculate the relative development, which is based on the factor calculated by dividing the current settlement area by the historical area.

The historical data in East Tyrol and the eastern part of the province Carinthia was less detailed than in the other regions, and settlements included grassland to some extent [35]. In order to correct the settlement extents, as an estimate, half of the historical settlement area was excluded from the investigation based on a previous analysis of local sample data. The results of some individual municipalities may therefore be less accurate. Even though the analysis was broken down to a municipality level, the spatial distribution was observed on a regional level. In addition, due to some spatial inaccuracy between the historical and the current data [35], the analysis worked with the assumption that settlement areas have not been decreasing in a significant manner since the mid-19th century. Therefore, settlement areas that appear in the historical context but do not in the current data were added to the current settlement.

The flood-prone area is based on current calculations but was also used for the historical analysis. It is likely that a 300-year flood in the historical context covered more ground. The results therefore only show the minimum of settlements exposed to a 300-year flood.

2.3.3. Agricultural Land

In the case of agricultural land, the analysis is based on the IACS dataset of 2018. It allows a detailed differentiation of agricultural land use on field level (18 land use categories). Spatially overlapping the land use data with the flood-prone areas shows the agricultural areas affected by flooding. To calculate the economic value of crop production in flood-prone areas, the spatial land use areas in hectares were multiplied by the regional standard output coefficients (according to crop type). This reveals the economic value of crop production within flood-prone areas (livestock production was not taken into account). It allows regionally differentiated conclusions on how strong the pressure might be for farm enterprises and where the situation might be more conflict-ridden between farming and flood risk management. Results of the research project BEAT categorized agricultural land within agricultural production zones based on their relevance for regional food supply [41]. Intersecting flood-prone areas with these data shows potential pressure in the view of food security. The analysis was carried out for Austria and specifically for the Federal State of Carinthia as an Alpine region.

3. Results

3.1. Human Transformation of Alpine Water Bodies

Since the first half of the 19th century, almost all larger running waters in the Austrian Alps have been severely modified by humans due to river straightening and channel narrowing. The main objective for the comprehensive hydraulic works was on land reclamation, improvement of navigation, and log driving. As a consequence, since 1826–1859, 40% of the former surface area of the Alpine running waters have vanished (see Figure 2). Expressed in a simplified manner: on average, rivers and streams have lost 40% of their former active channel width (omitting reductions in channel length due to river straightening). The data clearly show truncated flood retention and conveyance capacities due to the infilling of former channel areas. Surprisingly, stagnant water bodies have slightly increased in surface area. The reason for this can be found in new glacier lakes that have emerged after glacier decline and—most of all—in the construction of large reservoirs for hydro-energy purposes. Due to their storage capacity, the latter can partly help mitigate flood risks in Alpine environments [45]. Figure 2 also reveals the comprehensive areal reduction in ice-covered areas as a consequence of glacier retreat. Glaciers formerly covered 2.6% of the study area; today, they merely form 0.7%. This equals to an areal reduction of 73%. Because glaciers annually and inter-annually buffer precipitation in form of ice, they significantly influence the flow and flood regime of Alpine rivers. The severely reduced glaciated areas indicate significant losses in runoff storage capacity, resulting in increasing surface runoff during precipitation events [46]. Thus, it amplifies the flood risk in downstream valley sections.

3.2. Land Use in Flood-Prone Areas

3.2.1. Settlement Development

Although the area for running waters and glaciers has been reduced within the time frame of the analysis, settlement areas have expanded. These analyses were conducted for the catchments of the rivers Rhine, Salzach, and Drava, which include main parts of the Austrian federal states of Vorarlberg, East Tyrol, Salzburg, and Carinthia.

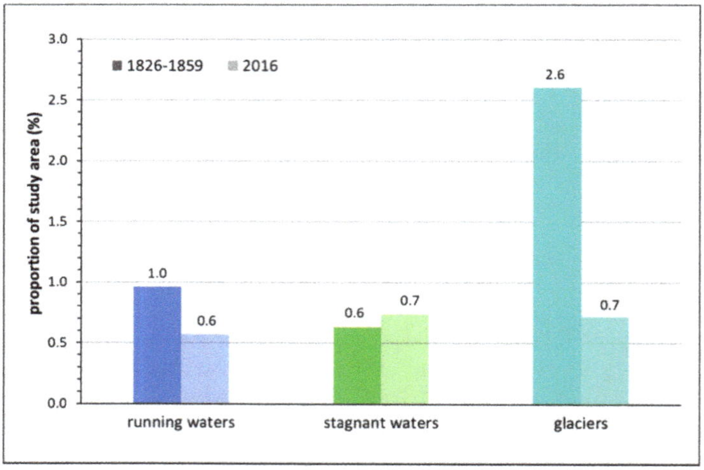

Figure 2. Proportions of the Alpine water bodies and glaciers in the Austrian catchments of the rivers Rhine, Salzach, and Drava (%) (left column: 1826–1859, right: 2016).

First, data on the general settlement development (i.e., inside and outside flooding areas) between 1826–1859 and 2016 were analyzed. The results show that development was primarily taking place in areas around the main cities and in larger valleys, such as the Rhine-Valley in Vorarlberg. Settlement expansion did not occur to the same extent in the mountainous regions. A comparison of the amount of settlement area per region historically and currently was made (see Figure 3). Although in the historical context, the percentage of settlement area per region is below 0.5%, in 2016, it increased to 3.6% in Vorarlberg and to approximately 2% in Carinthia and Salzburg. The settlement development therefore varies between the different Alpine regions but has significantly increased throughout the whole study area.

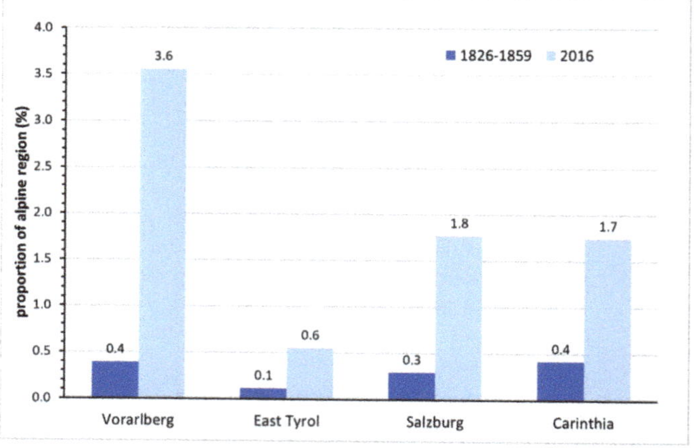

Figure 3. Proportion of settlement area per region.

Besides the general settlement development, the amount of space that is affected by flooding is an important variable. The results show that in Vorarlberg, 10 % of the study area is potentially affected by flooding. In Carinthia, it is 7%, in Salzburg 5%, and in East Tyrol, 3 %. At local level, considerable differences within each federal state can be found, from flood zones covering 97.6% of the surface area of the municipality of Altach to 0.89% in Dünserberg (both located in Vorarlberg). These differences derive from the municipality's location being either in a broad valley or in a very mountainous area. Thus, municipalities were and are affected by flooding in a very different way.

To show the settlement areas exposed to flooding, development inside and outside flood-prone areas was analyzed, as well as the absolute and relative values, which are based on the factor calculated by dividing the current settlement area by the historical area. The results from the historical analysis show that, for the whole study area, 21% of settlement areas were located within flood-prone areas. In 2016, this amounted to 29%; therefore, an increase in development in flood-prone areas can be detected. Differences can be observed when comparing absolute and relative development. The absolute value indicates a higher increase in settlements outside of flooding areas, whereas the relative value is higher for development inside flood-prone areas. On the one hand, this is caused by the limited use of floodplains for development in the 19th century, when adding just a few buildings could result in a high relative value. On the other hand, a shift towards using (former) floodplains for development can be observed.

Besides the temporal aspect of the analysis, the regional differences—based on the municipalities—was explored. According to the absolute numbers, settlement development in flood-prone areas shows a similar dissemination as the general settlement development. Municipalities with a high settlement expansion in flood zones are located in urban areas and the main valleys. Peripheral regions show less development in flood-exposed areas. This demonstrates the link between settlement development in flood-prone areas and the general settlement dynamic within a municipality. Furthermore, a correlation can be found between the number of buildings and infrastructure located in flooding areas and the amount of land potentially affected by flooding within a municipality.

To gain further insight into the spatial distribution of flood damage potential, the amount of settlement area situated in flood-prone areas per municipality was calculated.

Especially in Salzburg and Carinthia, it can be observed that, in more mountainous and peripheral regions, a higher share of settlements is located in flood-prone areas (both in the historical and the current context; see Figure 4).

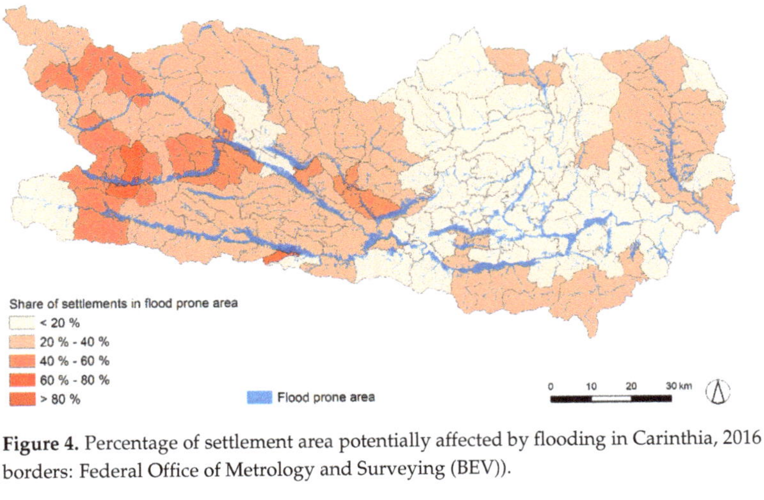

Figure 4. Percentage of settlement area potentially affected by flooding in Carinthia, 2016 (source of borders: Federal Office of Metrology and Surveying (BEV)).

By the results of this exposure analysis, the growing pressure on flood-prone areas and the increase in flood damage potential due to settlement development becomes evident.

3.2.2. Agricultural Land

As settlements are expanding, agricultural land is declining. In the period from 1951 to 2020, the agricultural area in Austria has decreased by 35% (excl. alpine pastures) [43]. Within the last 20 years, agricultural areas have decreased by 14%, from about 3 million ha (year 2000) to 2.6 million ha in 2020 [43]. Figure 5 shows the decline in agricultural land area over the last 20 years, differentiated by Austrian federal states. The significantly higher decline in the Alpine parts of the country is clearly visible. In Carinthia, the southernmost province, the decline in agricultural area was 21%.

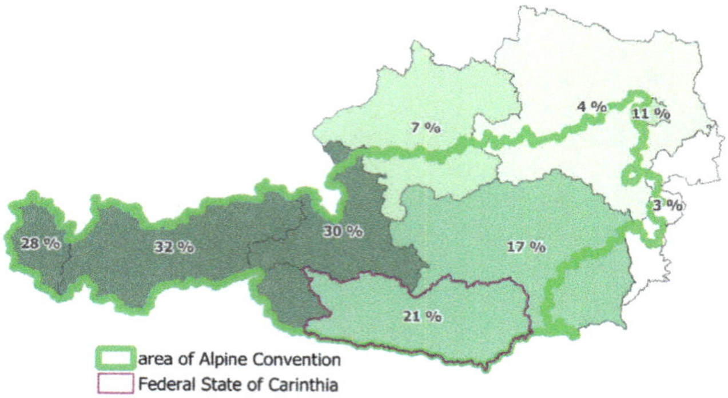

Figure 5. Decline in agricultural land in Austrian federal states between 2000 and 2020 in % (source of borders: Federal Office of Metrology and Surveying (BEV), Alpine Convention).

In total, 38% of Austria was used for agriculture in 2018, which amounts to 3.2 million ha (including 0.6 million ha of alpine pastures). At municipality level, this share of agricultural area varies greatly, for example, in municipalities in Carinthia, it differs between 1% and 56%. According to the spatial intersection of agriculture with flood-prone areas, 7.7% (246,300 ha) of Austrian agricultural areas are situated within flooding areas. In Carinthia, this share is 8.2%. The spatial distribution of these areas is, of course, not uniform, and the share of agricultural land within flood-prone areas varies greatly by municipality. In Carinthia, for example, in the municipality Villach, it is 62%, in four others, it is approximately 50%, and in 12 municipalities, it is still more than 20% (see Figure A1 in Appendix A).

Looking at agricultural land use in flood-prone areas of Carinthia in detail fodder growing (27.5%), extensive grassland (24.2%), feed grain (20.9%), and intensive grassland (17.4%) are the most predominant of 18 land use categories. Within the entire region of Austria, intensive grassland (23.2%), feed grain (21.5 %), extensive grassland (15.2%), fodder growing (10.4%), and oleaginous fruit (8.3%) are the predominant land use categories. In the Alpine region, there is proportionally more field fodder and extensive grassland, whereas bread crop, oleaginous fruit, and intensive grassland are less present (see Figure A2 in Appendix A).

While the first part was looking at agricultural land in general, in the next step a focus is put on land valuable for food production and the economic value of agriculture. About 42% of Austrian agricultural areas are considered particularly valuable for agricultural use—due to their relevance for regional food supply—by the research project BEAT [41]. In order to define valuable agricultural land, the natural characteristics for agricultural production, the resulting production potential—also under climate change conditions—and the regional distribution at the level of small-scale agricultural production areas were considered.

Intersecting these valuable agricultural areas with flood-prone areas reveals that in Austria, 12% (157,200 ha in total) of the total agricultural production areas designated as valuable are located in flood risk areas, whereby their spatial distribution varies greatly

Figure 6 shows the small-scale agricultural production areas (areas of similar agricultural production conditions) and the associated share of the total valuable area in the respective production area that lies within flood-prone areas.

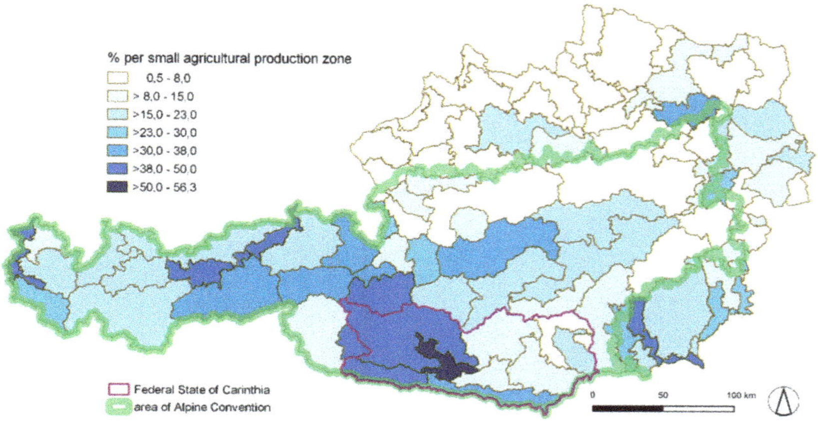

Figure 6. Share of high value agricultural area in Austria situated in flood risk zones per small agricultural production zone (=grey polygons), 2018 (source of borders: Federal Office of Metrology and Surveying (BEV), Alpine Convention).

Particularly in Alpine regions, a greater proportion of soils identified as important for food security tend to be located in flood-prone areas. Whereas the Austrian average of high value agricultural land situated within flood-prone areas is 12%, Carinthia has a share of 21%. Figure 7 shows the distribution of this share for Carinthian municipalities.

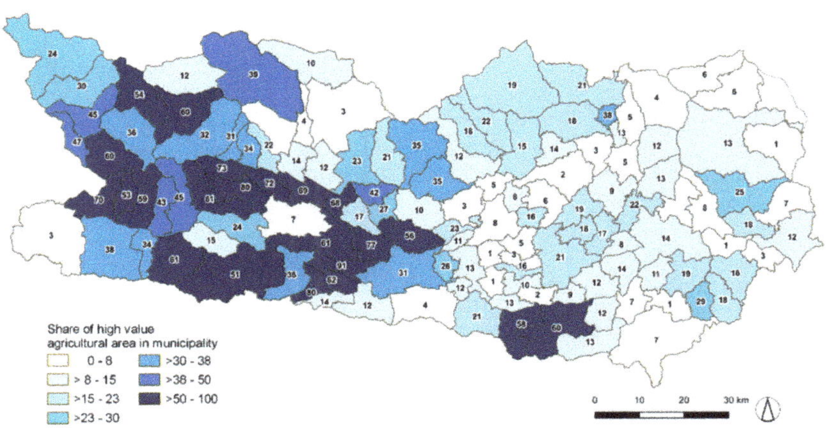

Figure 7. High value agricultural area in Carinthia situated in flood risk zones, share of high value area per municipality, 2018 (source of borders: Federal Office of Metrology and Surveying (BEV)).

The economic value of agricultural areas is calculated by multiplying the hectare of agricultural land in flood-prone areas with an average regional Standard Output (SO). Figure 8 shows the production value of agricultural land in flood-prone areas per municipality. Very high in absolute figures is the agricultural production in flood-prone zones, for example, in the municipalities in St. Andrä, Klagenfurt, Wolfsberg, Hermagor-Pressegger See, and Spittal/Drau. In addition to the absolute numbers, the relative share of agricultural production value at risk is equality important for farm enterprises and food production.

For example, in the communities Freistritz an der Gail, Weißenstein, Baldramsdorf, Nötsch im Gailtal, and Spittal/Drau, the relative share is above 60%. The relative value takes the different municipality sizes into account and shows especially high numbers in the south west of Carinthia.

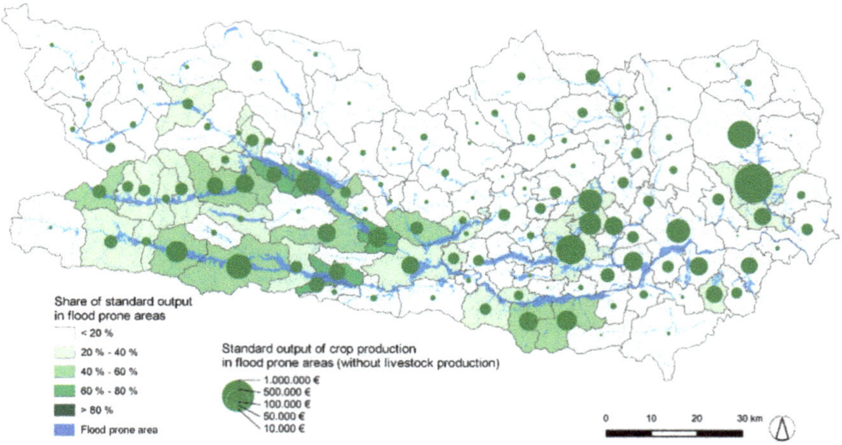

Figure 8. Standard output in EUR of agricultural areas (only value of crop production in intersecting areas—without livestock value) situated in flood-prone areas—sum per municipality and share within flood-prone areas, 2020 (source of borders: Federal Office of Metrology and Surveying (BEV)).

In this calculation, only the standard output values of crop production are considered. In order to be able to assess the full economic exposure, it would also be necessary to include animal production, which is, in part, closely intertwined with crop production.

4. Discussion

This research investigated land use in flood-prone areas to assess the pressure these areas are undergoing and the resulting consequences.

In the case study area, the space for river channels has been reduced by 40% since the middle of the 19th century. In addition, research shows a strong reduction in wetlands there [35]. This significant reduction was caused by river straightening and drainage works in order to decrease the impacts of frequent flood events and to reclaim wetlands for development and intensified agriculture. In the long run, however, these measures led to an increase in flood hazard due to the decreasing space for runoff in the case of floods with low and medium probability [23]. Furthermore, the reduction in glaciers impacts flood runoff. These past developments underline the importance of policies for making space for rivers, not only to reduce flood risks but also to improve the ecological situation of riverine landscapes.

In the case study area, settlement development between the middle of the 19th century and 2016 took place inside and outside flood-prone areas; however, the relative growth is significantly higher inside flooding zones. This result shows a shift towards the use of flood-prone areas for settlement development, which depicts an increase in flood damage potential. This rise in damage potential can also be observed in other exposure-related studies [5,27–29,47]. A multi-hazard exposure analysis for Austria shows a general increase in buildings by 643% between 1919 and 2012, properties exposed to river flooding have grown by 650%, and buildings exposed to torrential flooding have risen by 594% [29]. A general increase in flood damage potential due to settlement development can therefore be observed throughout the scientific literature.

Besides settlement development, our research results also point out that the share of settlements affected by flooding is higher in mountainous regions. In Alpine regions, the space for permanent settlement is limited and, due to the topography, is mainly located in the valleys. At the same time, flooding zones cover big parts of the valley floor. Therefore, there is an overlapping of flood-prone areas and the area for permanent settlement. Land use pressure on flood-prone areas is not limited to Alpine regions in Austria; other mountain areas face similar challenges of settlement expansion and decreasing space for flood runoff [48].

When it comes to the demand of making space for rivers, agricultural areas are the first choice. They provide areas for infiltration and runoff, with potential flood damages being on a lower level compared to settlement areas. The results show that agricultural land is decreasing, with Alpine regions showing a greater decline. In addition, the percentage of agricultural land within flood-prone areas is higher in mountain areas than outside. To gain a better understanding of the impact of flooding on agricultural land, a differentiation between agricultural areas with different characteristics is necessary. According to Grüneis et al. [30], the flooding of agricultural land results in quantifiable economic losses for farmers such as crop damages, erosion, landfall silting, and contamination, as well as hardly quantifiable environmental damages. These economic losses depend on crop species, growth stage, and soil properties and affect not only single-farm enterprises but the whole agricultural sector. This leads to macroeconomic questions of saving regional food supply, which has recently gained more importance, especially due to the increasing land-take by settlements and thus decreasing agricultural land. The results show that, at a national scale, the percentage of highly valuable agricultural land in flood-prone areas is higher in Alpine regions. This is also underlined by regional results. The western part of Carinthia is more mountainous (which is reflected in the limited space for permanent settlement—see Figure A3 in Appendix A) and also displays a higher amount of valuable farmland in flood-prone areas than the eastern part.

In the example of Carinthia, our results show that the share of settlements and valuable agricultural land in flood-prone areas is higher in pronounced mountain regions. The overlapping of flood-prone areas, valuable agricultural land, and settlements is likely to reduce the options of providing land for flood risk management and river restoration.

Research on the current use of former floodplains in Austria in 2003 concludes that over half of them are nowadays used for field crops and grassland [49]. The use of former floodplains for agriculture contributes to the fact that valuable agricultural land is nowadays located in flood-prone areas. The higher amount of valuable agricultural land in flood-prone areas in Alpine regions is further explained by a higher soil quality in the valley areas and the topography, which limits the use of space for agricultural production.

Research by Haidvogl et al. [50] in the area of St. Pölten in Lower Austria discusses the interdependencies of flood protection measures and settlement development. In combination with population growth, the demand for building land, and the construction of flood protection measures, settlement development along the river Traisen in 'flood-protected' areas was encouraged [50]. However, there remains a residual flood risk after the construction of flood protection measures [51]. The so-called 'levee-effect' or 'safe-development paradox' describes an increasing flood damage potential in areas with residual risk [52,53]. This research did not specifically focus on areas with residual risk behind levees due to the scale of the analysis. However, these areas are included because of the extent of the flooding areas applied. The increasing settlement development in flood-prone areas can therefore be connected to the construction of flood protection measures and the creation of 'flood-protected' areas. Besides the 'levee-effect', the limited space for permanent settlement in Alpine areas and the general increase in settlements led to an increase in flood damage potential.

Based on the results, consequences for flood risk management can be derived. Making space for rivers to reduce future flood damage is an important approach that is confronted with limited space and land use pressure in Alpine regions. Many former floodplains, which nowadays could be used for river restoration, e.g., local widening of watercourses or reconnecting rivers with previous river branches, were often claimed by settlement and agriculture. In other countries, the approach of making space for rivers also faces difficulties [19,20,24]. In general, research suggests that a broad inclusion of relevant stakeholders at different governmental levels is necessary to implement a flood risk management that aims to make more room for rivers [16,19]. One of these disciplines is spatial planning, which takes on an important role when it comes to keeping flood runoff and retention areas undeveloped. Another important stakeholder is agriculture. As a major land use agent, it influences the emergence of floods. Agricultural cultivation substantially contributes to flood risk reduction. Reducing surface runoff and maintaining the infiltration capacity of the soil are particularly effective, as developing land for the purpose of retention can additionally reduce floods [43,54]. In general, integrated flood risk management is an interdisciplinary approach including a broad range of technical and non-technical measures to reduce flood risks, in addition to giving space to rivers. This includes emergency measures, flood-adapted building, and hazard zone mapping [12].

The limitations of making room for river are not only connected to the lack of space; the implementation also needs a general shift towards an integrated flood risk management, including a wider range of stakeholders.

5. Conclusions

The aim of this research was to examine the development of land use in flood-prone areas in an Austrian Alpine region and the resulting potential for land use conflicts. Research was guided by two research questions:

(1) How did water bodies and settlements in flood-prone areas change between 1826–1857 and 2016 in Austrian Alpine regions? In the past, one important goal of flood prevention was making space for new settlements and intensifying agricultural production, which led to a decrease in rivers and streams by 40%. Flood-prone areas (potentially affected by a 300-year flood) are nowadays more intensely used by settlements than in the mid-19th century. The amount of settlement areas located in flood-prone areas rose from 21% to 29%; flood damage potential is therefore increasing.

(2) How are flood-prone areas in Alpine regions currently used (focus on settlements and agriculture)? The regional analysis of settlements located in flood-prone areas shows that in pronounced mountain regions, a higher percentage of settlements is affected by flooding. Flood-prone areas are also needed for agricultural production; 7.7% of agricultural areas in Austria are located in flood-prone areas. Looking at valuable agricultural land, 12% is affected by flooding; however, the percentage is higher in Alpine regions. In Carinthia, for example, 21% of valuable agricultural land is located in flood-prone areas.

A lack of space because of the topography, pressure on land due to settlement dynamics, and the importance of land for regional food production are key factors when it comes to the land use pressure in flood-prone areas and the limitations of a making space for rivers policy.

The increasing settlement development in flood-prone areas highlights the importance of spatial planning to preserve areas for flood runoff and retention and thus to reduce flood risk. Further research is needed on spatial planning approaches, particularly in areas with residual risk to address the 'levee-effect' and the increasing flood damage potential. Additionally, a focus in future flood risk management should be put on a more detailed differentiation of agricultural land based on economic and ecologic functions as well as on the flood-adapted cultivation of farmland.

Author Contributions: Conceptualization, L.J.; formal analysis, L.J., S.H., K.S., and K.W.; writing—original draft preparation, L.J., S.H., K.S., and K.W.; writing—review and editing, L.J., S.H., K.S., W.S., and K.W.; visualization, L.J. and K.S.; project administration, W.S. All authors have read and agreed to the published version of the manuscript.

Funding: This research was conducted within the research project "Integrated Flood Risk Management in Mountain Areas: Assessing Sectoral Interdependencies, Conflicts and Options for Policy Coordination" (PoCo-FLOOD) which was funded by the Austrian Academy of Sciences (ÖAW) within the Earth System Sciences (ESS) research program. Lena Junger was supported by the Doctoral

School "Human River Systems in the 21st Century (HR21)" of the University of Natural Resources and Life Sciences, Vienna.

Institutional Review Board Statement: Not applicable.

Informed Consent Statement: Not applicable.

Data Availability Statement: The datasets used are not readily available because the data are partly subjected to third party restriction. Selected data are available on request.

Conflicts of Interest: The authors declare no conflict of interest.

Appendix A

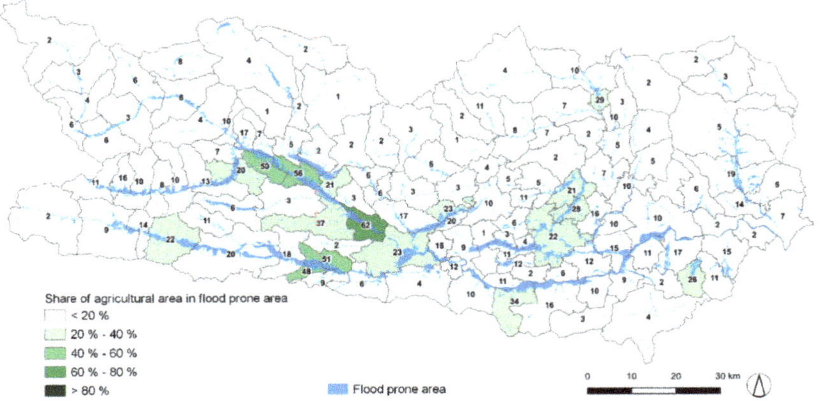

Figure A1. Share of agricultural area in Carinthia situated in flood risk zones per municipality, 2018 (source of borders: Federal Office of Metrology and Surveying (BEV)).

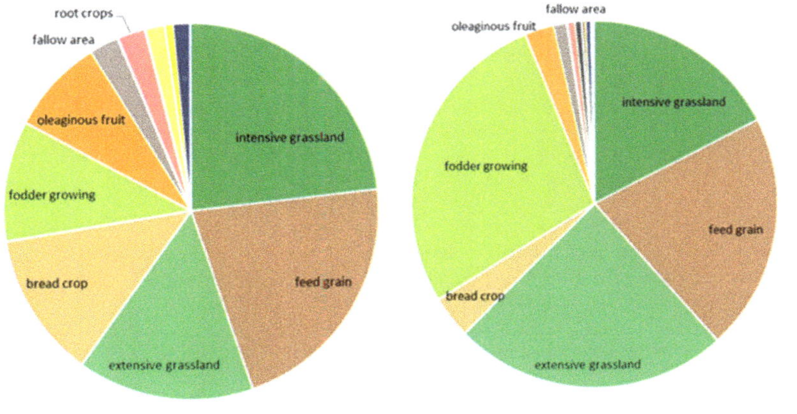

Figure A2. Agricultural land use in flood-prone areas 2018, left side: Austria, right side: Carinthia.

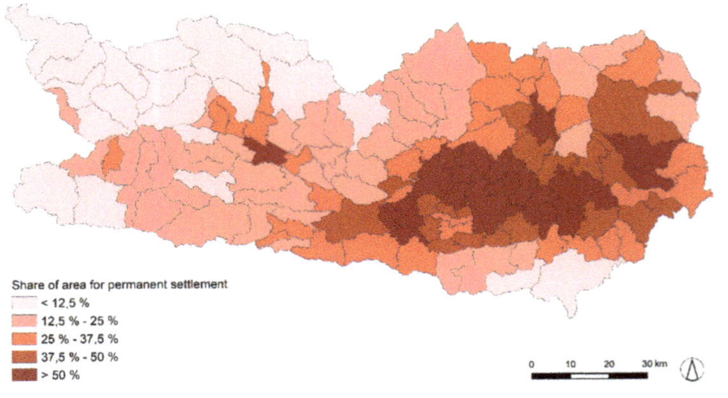

Figure A3. Share of area for permanent settlement per municipality in Carinthia (own presentation; source of borders: Federal Office of Metrology and Surveying (BEV); source of area for permanent settlement: Statistics Austria).

References

1. World Bank. *Cities and Flooding*; The World Bank: Washington, DC, USA, 2012; ISBN 978-0-8213-8866-2.
2. Blöschl, G.; Viglione, A.; Merz, R.; Parajka, J.; Salinas, J.L.; Schöner, W. Auswirkungen des Klimawandels auf Hochwasser und Niederwasser. *Österreichische Wasser und Abfallwirtschaft* **2011**, *63*, 21–30. [CrossRef]
3. Blöschl, G.; Blaschke, A.P.; Haslinger, K.; Hofstätter, M.; Parajka, J.; Salinas, J.; Schöner, W. Auswirkungen der Klimaänderung auf Österreichs Wasserwirtschaft—ein aktualisierter Statusbericht. *Österreichische Wasser und Abfallwirtschaft* **2018**, *70*, 462–473. [CrossRef]
4. Barredo, J.I. Normalised flood losses in Europe: 1970–2006. *Nat. Hazards Earth Syst. Sci.* **2009**, *9*, 97–104. [CrossRef]
5. Löschner, L.; Herrnegger, M.; Apperl, B.; Senoner, T.; Seher, W.; Nachtnebel, H.P. Flood risk, climate change and settlements development: A micro-scale assessment of Austria municipalities. *Reg. Environ. Change* **2017**, *17*, 311–322. [CrossRef]
6. Nachtnebel, H.P.; Apperl, B. Beurteilung des Hochwasser-Schadenspotenzials unter dynamischen Bedingungen. *Österr Wasser- und Abfallw* **2015**, *67*, 120–130. [CrossRef]
7. Schanze, J. The long way towards tolerable flood risks—Initial lessons learnt from the recent Central European floods. *J. Flood Risk Manag.* **2013**, *6*, 167–168. [CrossRef]
8. Klijn, F.; Samuels, P.; van Os, A. Towards flood risk management in the EU: State of affairs with examples from various European countries. *Int. J. River Basin Manag.* **2008**, *6*, 307–321. [CrossRef]
9. Samuels, P.; Klijn, F.; Dijkman, J. An analysis of the current practice of policies on river flood risk management in different countries. *Irrig. Drain.* **2006**, *55*, S141–S150. [CrossRef]
10. Greiving, S. *Räumliche Planung und Risiko*; Gerling Akad. Verl.: Munich, Germany, 2002; ISBN 3-932425-50-2.
11. Sayers, P.; Yuanyuan, L.; Galloway, G.; Penning-Rowsell, E.C.; Fuxin, S.; Kang, W.; Yiwei, C.; LeQuesne, T. *Flood Risk Management A Strategic Approach*; Asian Development Bank: Mandaluyong City, Philippines, 2013.
12. Müller, U. *Hochwasserrisikomanagement: Theorie und Praxis*; Vieweg+Teubner Verlag/GWV Fachverlage GmbH Wiesbaden: Wiesbaden, Germany, 2010; ISBN 978-3-8348-1247-6.
13. Löschner, L. The Spatial Turn in Flood Risk Management: A Case Study of Austria's Changing Flood Policies. Ph.D. Thesis, University of Natural Resources and Life Sciences, Vienna, Austria, 2018.
14. Van Ruiten, L.J.; Hartmann, T. The spatial turn and the scenario approach in flood risk management—Implementing the European Floods Directive in the Netherlands. *AIMS Environ. Sci.* **2016**, *3*, 697–713. [CrossRef]
15. Warner, J.F.; van Buuren, A.; Edelenbos, J. *Making Space for the River: Governance Experiences with Multifunctional River Flood Management in the US and Europe*; IWA Publishing: London, UK, 2012; ISBN 9781780401133.
16. Rijke, J.; van Herk, S.; Zevenbergen, C.; Ashley, R. Room for the River: Delivering integrated river basin management in the Netherlands. *Int. J. River Basin Manag.* **2012**, *10*, 369–382. [CrossRef]
17. Busscher, T.; van den Brink, M.; Verweij, S. Strategies for integrating water management and spatial planning: Organising for spatial quality in the Dutch "Room for the River" program. *J. Flood Risk Manag.* **2019**, *12*, e12448. [CrossRef]
18. Verweij, S.; Busscher, T.; van den Brink, M. Effective policy instrument mixes for implementing integrated flood risk management: An analysis of the 'Room for the River' program. *Environ. Sci. Policy* **2021**, *116*, 204–212. [CrossRef]
19. Hartmann, T. Den Flüssen mehr Raum geben—Umsetzungsrestriktionen in Recht und Praxis. *Raumforschung und Raumordnung* **2011**, *69*, 257–268. [CrossRef]
20. Bogdan, E.A.; Beckie, M.A.; Caine, K.J. Making room for nature? Applying the Dutch Room for the River approach to flood risk management in Alberta, Canada. *Int. J. River Basin Manag.* **2020**, 1–13. [CrossRef]

21. Pardoe, J.; Penning-Rowsell, E.; Tunstall, S. Floodplain conflicts: Regulation and negotiation. *Nat. Hazards Earth Syst. Sci.* **2011**, *11*, 2889–2902. [CrossRef]
22. Hohensinner, S.; Egger, G.; Muhar, S.; Vaudor, L.; Piégay, H. What remains today of pre-industrial Alpine rivers? Census of historical and current channel patterns in the Alps. *River Res. Applic.* **2021**, *37*, 128–149. [CrossRef]
23. Habersack, H.; Bürgel, J.; Kanonier, A. *FloodRisk II: Vertiefung und Vernetzung Zukunftsweisender Umsetzungsstrategien Zum Integrierten Hochwassermanagement*; Synthesebericht: Vienna, Austria, 2009.
24. Rohde, S.; Hostmann, M.; Peter, A.; Ewald, K.C. Room for rivers: An integrative search strategy for floodplain restoration. *Landsc. Urban Plan.* **2006**, *78*, 50–70. [CrossRef]
25. Trautwein, C.; Schinegger, R.; Schmutz, S. Cumulative effects of land use on fish metrics in different types of running waters in Austria. *Aquat. Sci.* **2012**, *74*, 329–341. [CrossRef]
26. Mantyka-Pringle, C.S.; Martin, T.G.; Moffatt, D.B.; Linke, S.; Rhodes, J.R. Understanding and predicting the combined effects of climate change and land-use change on freshwater macroinvertebrates and fish. *J. Appl. Ecol.* **2014**, *51*, 572–581. [CrossRef]
27. Ceola, S.; Laio, F.; Montanari, A. Satellite nighttime lights reveal increasing human exposure to floods worldwide. *Geophys. Res. Lett.* **2014**, *41*, 7184–7190. [CrossRef]
28. Cammerer, H.; Thieken, A.H.; Verburg, P.H. Spatio-temporal dynamics in the flood exposure due to land use changes in the Alpine Lech Valley in Tyrol (Austria). *Nat. Hazards* **2013**, *68*, 1243–1270. [CrossRef]
29. Fuchs, S.; Margreth, K.; Zischg, A. A spatiotemporal multi-hazard exposure assessment based on property data. *Nat. Hazards Earth Syst. Sci.* **2015**, *15*, 2127–2142. [CrossRef]
30. Grüneis, H.; Schroll, K.; Wagner, K. The Role of Agriculture in Flood Risk Management in Austria—Conflicts and Challenges. *JESE-B* **2021**, *10*, 112–127. [CrossRef]
31. Agrawala, S. *Climate Change in the European Alps: Adapting Winter Tourism and Natural Hazards Management*; OECD Organisation for Economic Co-Operation and Development: Paris, France, 2007; ISBN 92-64-03168-5.
32. Alpine Convention. Demographic Changes in the Alps: Report on the State of the Alps. *Alpine Convention Alpine Signals—Special Edition 5, Bozen.* 2015. Available online: https://www.alpconv.org/fileadmin/user_upload/Publications/RSA/RSA5_EN.pdf (accessed on 7 February 2022).
33. Umweltbundesamt. Flächeninanspruchnahme. Available online: https://www.umweltbundesamt.at/umweltthemen/boden/flaecheninanspruchnahme (accessed on 7 February 2022).
34. Löschner, L.; Eder, M.; Herrnegger, M.; Hogl, K.; Nordbeck, R.; Scherhaufer, P.; Schober, B.; Seher, W.; Wesemann, J.; Zahnt, N.; et al. RegioFEM—Informing future-oriented flood risk management at the regional scale (Part I). *J. Flood Risk Manage.* **2021**, *15*, e12754. [CrossRef]
35. Hohensinner, S.; Atzler, U.; Berger, M.; Bozzetta, T.; Höberth, C.; Kofler, M.; Rapottnig, L.; Sterle, Y.; Haidvogl, G. Land Use and Cover Change in the Industrial Era: A Spatial Analysis of Alpine River Catchments and Fluvial Corridors. *Front. Environ. Sci.* **2021**, *9*, 647247. [CrossRef]
36. GeoVille Information Systems. Cadasterenv—Final Report (TD-9): Public Version I 1.0. Available online: https://www.landinformationsystem.at/#/lisa/overview (accessed on 23 December 2021).
37. BMLFUW; BAB. INVEKOS-Datenpool 2017 des BMLFUW. Available online: https://info.bmlrt.gv.at/themen/landwirtschaft/eu-agrarpolitik-foerderungen/direktzahlungen/Invekos.html (accessed on 23 January 2022).
38. Buckel, J.; Otto, J.-C. *The Austrian Glacier Inventory GI 4 (2015) in ArcGis (shapefile) Format*; PANGAEA: Bremen, Germany, 2018.
39. Ramm, F.; OpenStreetMap Data in Layered GIS Format. Technical Report. Available online: https://download.geofabrik.de/europe/austria.html (accessed on 23 December 2021).
40. Muhar, S.; Grüner, B.; Böck, K.; Scheikl, S.; Becsi, R. *SPARE Synthesis Report: "Balancing River Protection Needs, Ecosystem Functions and Human Uses"*; University of Natural Resources and Life ServicesL: Vienna, Austria, 2018.
41. Haslmayr, H.-P.; Baumgarten, A.; Schwarz, M.; Huber, S.; Prokop, G.; Sedy, K.; Krammer, C.; Murer, E.; Pock, H.; Rodlauer, C.; et al. BEAT—Bodenbedarf für die Ernährungssicherung in Österreich: Endbericht zum Forschungsprojekt Nr. 100975; Vienna, Austria. 2018. Available online: https://dafne.at/content/report_release/aa85879d-af0f-4273-a1e2-b7f1d7178d41_0.pdf (accessed on 23 December 2021).
42. BAB. *Standard-Output-Koeffizienten der Pflanzlichen Erzeugung nach Schlagnutzungsarten*; BAB: Vienna, Austria, 2021.
43. BMLRT. *Grüner Bericht 2021 Die Situation der österreichischen Land- und Forstwirtschaft: Gemäß §9 des Landwirtschaftgesetzes*; BMLRT: Vienna, Austria, 2021. Available online: https://gruenerbericht.at/cm4/jdownload/send/2-gr-bericht-terreich/2393-gb2021 (accessed on 23 December 2021).
44. Papilloud, T.; Röthlisberger, V.; Loreti, S.; Keiler, M. Flood exposure analysis of road infrastructure—Comparison of different methods at national level. *Int. J. Disaster Risk Reduct.* **2020**, *47*, 101548. [CrossRef]
45. Lebiedzinski, K.; Fürst, J.; Schulz, K.; Herrnegger, M. Possible impacts of a hydropower reservoir on the flood hazard of an Alpine valley. In Proceedings of the EGU General Assembly 2020, Vienna, Austria, 3–8 May 2020.
46. Jansson, P.; Hock, R.; Schneider, T. The concept of glacier storage: A review. *J. Hydrol.* **2003**, *282*, 116–129. [CrossRef]
47. Früh-Müller, A.; Wegmann, M.; Koellner, T. Flood exposure and settlement expansion since pre-industrial times in 1850 until 2011 in north Bavaria, Germany. *Reg. Environ. Change* **2015**, *15*, 183–193. [CrossRef]
48. Serrano-Notivoli, R.; Mora, D.; Ollero, A.; Sánchez-Fabre, M.; Sanz, P.; Saz, M.Á. Ocupación de la llanura aluvial e inundaciones en el Pirineo Central. *CIG* **2017**, *43*, 309. [CrossRef]

49. Poppe, M.; Muhar, S.; Gregory, E.; Stefan, S. Status quo der österreichischen Flusslandschaften: Erfassung und Bilanzierung der Eingriffe und Nutzungen. *Österreichische Wasser- und Abfallwirtschaft* **2003**, *7–8*, 122–128.
50. Haidvogl, G.; Eberstaller, J.; Eberstaller-Fleischanderl, D.; Fraiß, B.; Gabriel, H.; Hohensinner, S. Historische Landnutzung und Siedlungsentwicklung in Flussauen und Hochwasserschutz: Das Beispiel der Traisen und St. Pöltens 1870–2000. *Österr Wasser- und Abfallw* **2018**, *70*, 305–315. [CrossRef]
51. Merz, B. *Hochwasserrisiken: Grenzen und Möglichkeiten der Risikoabschätzung Mit 33 Tabellen*; Schweizerbart: Stuttgart, Germany, 2006; ISBN 3-510-65220-7.
52. Burby, R.J. Hurricane Katrina and the Paradoxes of Government Disaster Policy: Bringing About Wise Governmental Decisions for Hazardous Areas. *ANNALS Am. Acad. Political Soc. Sci.* **2006**, *604*, 171–191. [CrossRef]
53. Di Baldassarre, G.; Kreibich, H.; Vorogushyn, S.; Aerts, J.; Arnbjerg-Nielsen, K.; Barendrecht, M.; Bates, P.; Borga, M.; Botzen, W.; Bubeck, P.; et al. Hess Opinions: An interdisciplinary research agenda to explore the unintended consequences of structural flood protection. *Hydrol. Earth Syst. Sci.* **2018**, *22*, 5629–5637. [CrossRef]
54. Zischg, A.; Costa, R.; Flury, C.; Schild, A. Einfluß der landwirtschaftlichen Bewirtschaftung auf alpine Naturgefahren eine zusammenfassende Betrachtung. In Proceedings of the 12th Congress INTERPRAEVENT, Grenoble, France, 23–26 April 2012.

land

MDPI

Article

Monitoring of Land Use–Land Cover Change and Potential Causal Factors of Climate Change in Jhelum District, Punjab, Pakistan, through GIS and Multi-Temporal Satellite Data

Muhammad Majeed [1,†], Aqil Tariq [2,†], Muhammad Mushahid Anwar [3], Arshad Mahmood Khan [4], Fahim Arshad [5], Faisal Mumtaz [6,7], Muhammad Farhan [8], Lili Zhang [6,9,*], Aroosa Zafar [10], Marjan Aziz [11], Sanaullah Abbasi [12], Ghani Rahman [3], Sajjad Hussain [13], Muhammad Waheed [5], Kaniz Fatima [5] and Shadab Shaukat [14]

1 Department of Botany, University of Gujrat, Hafiz Hayat Campus, Gujrat 50700, Punjab, Pakistan; m.majeed@uog.edu.pk
2 State Key Laboratory of Information Engineering in Surveying, Mapping and Remote Sensing, Wuhan University, Wuhan 430079, China; aqiltariq@whu.edu.cn
3 Department of Geography, University of Gujrat, Hafiz Hayat Campus, Gujrat 50700, Punjab, Pakistan; mushahid.anwar@uog.edu.pk (M.M.A.); ghani.rahman@uog.edu.pk (G.R.)
4 Department of Botany, Govt. Hashmat Ali Islamia Associate College Rawalpindi, Rawalpindi 46300, Punjab, Pakistan; arshad.mahmood.vt@iiu.edu.pk
5 Department of Botany, University of Okara, Okara 56300, Punjab, Pakistan; fahim.arshad@uo.edu.pk (F.A.); F19-PhD-Bot-5013@uo.edu.pk (M.W.); f19-phd-bot-5005@uo.edu.pk (K.F.)
6 Aerospace Information Research Institute, Chinese Academy of Sciences, Beijing 100101, China; faisal@aircas.ac.cn
7 University of Chinese Academy of Sciences (UCAS), Beijing 101408, China
8 School of Earth Sciences and Engineering, Hohai University, Nanjing 210098, China; m.farhan@hhu.edu.cn
9 Zhongke Langfang Institute of Spatial Information Applications, Langfang 065001, China
10 Department of Pharmacy, Quaid-i-Azam University, Islamabad 45320, Pakistan; aroosa_zafar@cadson.edu.pk
11 Barani Agricultural Research Institute, Chakwal 48650, Punjab, Pakistan; marjan_aziz19@hotmail.com
12 Department of Biochemistry, Shah Abdul Latif University, Khairpur 66020, Sindh, Pakistan; abbasi.sanaullah@salu.edu.pk
13 Department of Environmental Science, Comsats University Islamabad, Vehari Campus, Vehari 61100, Punjab, Pakistan; mc190401094@vu.edu.pk
14 Department of Plant Breeding and Genetics, University of Agriculture, Faisalabad 38000, Punjab, Pakistan; shadab.shaukat@uos.edu.pk
* Correspondence: Zhangll@lreis.ac.cn
† These authors contributed equally to this work.

Citation: Majeed, M.; Tariq, A.; Anwar, M.M.; Khan, A.M.; Arshad, F.; Mumtaz, F.; Farhan, M.; Zhang, L.; Zafar, A.; Aziz, M.; et al. Monitoring of Land Use–Land Cover Change and Potential Causal Factors of Climate Change in Jhelum District, Punjab, Pakistan, through GIS and Multi-Temporal Satellite Data. *Land* **2021**, *10*, 1026. https://doi.org/10.3390/land10101026

Academic Editors: Matej Vojtek, Andrea Petroselli, Raffaele Pelorosso and Le Yu

Received: 12 August 2021
Accepted: 26 September 2021
Published: 30 September 2021

Abstract: Land use–land cover (LULC) alteration is primarily associated with land degradation, especially in recent decades, and has resulted in various harmful changes in the landscape. The normalized difference vegetation index (NDVI) has the prospective capacity to classify the vegetative characteristics of many ecological areas and has proven itself useful as a remote sensing (RS) tool in recording vegetative phenological aspects. Likewise, the normalized difference built-up index (NDBI) is used for quoting built-up areas. The current research objectives include identification of LULC, NDVI, and NDBI changes in Jhelum District, Punjab, Pakistan, during the last 30 years (1990–2020). This study targeted five major LULC classes: water channels, built-up area, barren land, forest, and cultivated land. Satellite imagery classification tools were used to identify LULC changes in Jhelum District, northern Punjab, Pakistan. The perception data about the environmental variations as conveyed by the 500 participants (mainly farmers) were also recorded and analyzed. The results depict that the majority of farmers (54%) believe in the appearance of more drastic changes such as less rainfall, drought, and decreased water availability for irrigation during 2020 compared to 30 years prior. Overall accuracy assessment of imagery classification was 83.2% and 88.8% for 1990, 88.1% and 85.7% for 2000, 86.5% and 86.7% for 2010, and 85.6% and 87.3% for 2020. The NDVI for Jhelum District was the highest in 1990 at +0.86 and the lowest in 2020 at +0.32; similarly, NDBI values were the highest in 2020 at +0.72 and the lowest in 1990 at −0.36. LULC change showed a clear association with temperature, NDBI, and NDVI in the study area. At the same time, variations in the land area of barren soil, vegetation, and built-up from 1990 to 2020 were quite prominent, possibly resulting in temperature increases, reduction in water for irrigation, and changing rainfall patterns. Farmers were found to be quite responsive to such climatic variations, diverting to framing possible mitigation approaches, but they need government assistance. The findings of this study, especially the causes and impacts of rapid LULC variations in the study area, need immediate attention from related government departments and policy makers.

Keywords: farmers' perception; NDVI; climate change; remote sensing; GIS; NDBI

1. Introduction

According to the Fourth Assessment Report of the IPCC, with the end of the current century, the average phase temperature will increase up to 1.1–6.4 °C worldwide [1]. Climate is the main factor for food production, crop growth, and the rural livelihoods of billions of rural communities from different regions of the world [2–5]. In recent decades, global climate change has greatly affected vegetation [6,7]. Pakistan is an agrarian country, and its economy depends on agriculture; it is susceptible to climate fluctuations [8,9]. Land cover degradation is a decrease in the capacity of the land to yield profits in land use (LU) related to quantified control of the land area [10]. Land cover (LC) describes the general characteristics of the land surface, including forest, barren land, water, mountain summits, hill slopes, cropping, and the urban setup [11]. Land cover has the most delicate relation with land use [12,13]. Mapping land use–land cover (LULC) has been completed efficiently with satellite images at many spatial, spectral, and temporal resolutions [1,14,15], while in arid and semi-arid ecology, the application of multi-temporal satellite images has been reduced to develop and assess LULC fluctuations [16,17]. Alterations in LULC could be examined because they affect land ecology permanently, particularly in built-up areas of micro-climate heating [18]. The normalized difference vegetation index (NDVI) estimates the green vegetation density [19] and, in recent decades, has broadly been used for explaining the spatio-temporal features of LULC, with quantitative vegetation cover [20,21].

There are different activities that can alter LULC, which have been extensively studied due to it being the most important part of these types of studies. Various researchers have established an overview of the effects on LULC in different parts of the world, associated with agricultural expansion [22,23], urban expansion [24], and engineering projects such as access and energy [25,26].

The NDVI standards range from "−1.0 to 1.0"; minimum NDVI values are used for whole surface resources, and maximum NDVI standards are used for green flora [27]. Negative NDVI values represent areas with very low or null vegetation cover, such as water bodies and urban areas, whereas positive values concern pixels with vegetation from very low to high cover [28]. When near to "0", NDVI standards are represented by bare soil [29]. The NDVI is extensively applied in remote sensing (RS) investigations because it provides suitable evidence for adding and exploring flora [27,28,30]. The NDVI is applied to determine the combined performance of climatic variation and the vegetation distribution at vast spatial and temporal scales [31] as the biomass of plant diversity is interrelated with precipitation, temperature, and evapotranspiration [32–34]. Geographic information systems (GISs) and remote sensing (RS) are essential tools [35] applied for the investigation of urban dimensions and density with LULC mapping and ecological impacts of urban programming within certain periods [36]. Remote sensing provides on-time availability of LULC and vegetation cover data at specific periods in an economical manner [37–39]. GISs manage and analyze spatial data accurately and are an important and basic need of this area of study [40].

RS data are a helpful tool in the mapping of LULC [41,42]. For LULC mapping, the temporal Landsat sensor data of the Landsat-7 Enhanced Thematic Mapper (ETM), Landsat-5 Thematic Mapper (TM) with ETM+ [43], and Landsat-8 Operational Land Imager (OLI) have been extensively used to discover the variation in the NDVI, NDBI, and

LULC [33]. Assessing variation evaluates the earlier and present situation visually as well as quantitatively and thus supports indenting the fluctuations linked to LULC characteristics depicting different satellite datasets [44]. Proper classifications require previous data from particular regions for collecting recorded data from working areas. Field-recorded data have been used to explain applications for selected classification algorithms [45].

Several such types of attempts to assess and manage LULC changes and degradation using RS data were explored by various researchers from countries around the world [46], such as northwest Ethiopia [4], West Africa [7], Ethiopia [11], Malaysia [12], Zimbabwe [17], Bangladesh [21], Southern Africa [30], Iran [37], Nepal [39], China [40], northern Ethiopia [42], Brazil [47], Iraq [48], and Turkey [49]. Pakistan is regarded as one of the agricultural countries in the world which are directly influenced by climatic fluctuations, which ultimately affect the economy of lowland farmers. From Pakistan, various studies in southern Punjab [1], Islamabad [2], Faisalabad and Multan [27], Vehari [28], Sindh [35], Azad Jammu and Kashmir [50], Multan [51], Lodhran [52], and Khyber Pakhtunkhwa [53] have been conducted for the assessment and management of LULC changes using RS data. In the conducted research, over a longer time period, LULC change assessment was recorded in Jhelum District using RS and GIS tools. Following are the main objectives of the current study:

- To identify temporal LULC changes during the last 30 years and farmers' perception regarding climate change and LULC variations;
- To analyze and map NDVI, NDBI, and LULC changes by using satellite data;
- To compare the various characteristics of LULC, NDBI, and NDVI during the past 30 years.

2. Materials and Methods

2.1. Study Area

Jhelum District is located to the north of the Jhelum River and bounded by Rawalpindi District in the north, Sargodha and Gujrat Districts in the south, Azad Kashmir in the east, and Chakwal District in the west [54] (Figure 1). The total human population of the district is 1.223 million [54]. The climate is a semi-arid, warm subtropical type and recognized as having warm summer and severe winter seasons. Jhelum is a semi-mountainous region; the mean annual rainfall is 880 mm, while the average annual temperature is 23.6 °C [55]. The Jhelum River flows through 247,102 acres of lengthy plain area and 41,207 acres of mountainous zones. The district has the world's second largest salt mine (Khewra salt mine), which covers an area of 2.268 million acres [54]. People of the area have diverse modes of lifestyles, beliefs, and traditions [56]. Some typical landscapes are shown in Figure 2.

2.2. Methods and Materials

2.2.1. Satellite Data

For LULC, NDVI, and NDBI variation assessment in the study area over a temporal gradient of 30 years (1990, 2000, 2010, 2020), Landsat 8 (OLI), Landsat 7 (ETM+), and Landsat 4, 5 (TM) satellite remote sensing imagery data were used and downloaded from the website (http/www.earthexplorer.usgs.gov (accessed on 6 August 2020)) of the United States Geological Survey (USGS). The details of the downloaded satellite images are presented in Table 1.

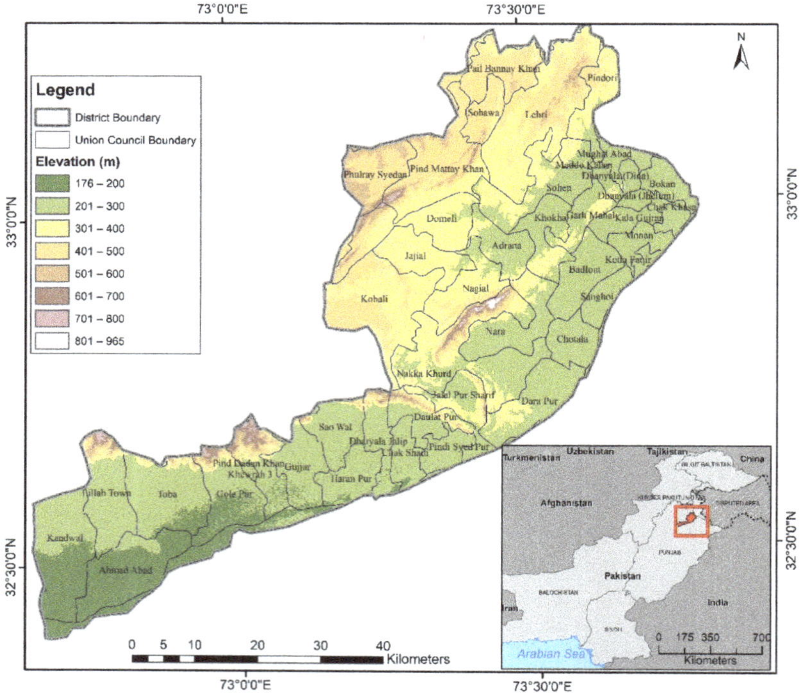

Figure 1. Map of the study area.

Figure 2. View of different land cover types in Jhelum District showing diverse LULC: (**a**) road in the mountains representing a new built-up area in 2020; (**b**) barren land (sandy dune cutting), another important LULC category in the area; (**c**) vegetation cover; (**d**) anthropogenic activities affecting LULC variations; (**e**) deforestation affecting vegetation cover; (**f**) natural disturbance causing LULC variations.

Table 1. Arrangement of Landsat satellite imagery.

Sr. #	Acquirement Date	Data Type	Resolution	Sensors	Path/Rows
1.	17/3/1990	Landsat imagery	30 m	TM	150/037 150/038
2.	25/3/2000	Landsat imagery	30 m	TM	150/037 150/038
3.	12/3/2010	Landsat imagery	30 m	ETM+	150/037 150/038
4.	9/3/2020	Landsat imagery	30 m	OLI	150/037 150/038

2.2.2. Survey Data

In this study, responses of 500 farmers in Jhelum District were recorded by using the snowball sampling method to document perceptions linked to LULC and climate changes. A total of 25 union councils in the study area were targeted, encompassing five villages per union council. The data linked to different climatic variables were recorded during August 2020 to July 2021. Study contributors were selected concentrating on middle-aged and elderly people (range: 30–80 years old), particularly farmers from 75 considered villages of Jhelum District. The investigated variables were composed of climate change records (including rainfall duration, rainfall intensity, and temperature variations) and LULC variations during the past 30 years.

GPS was used to record the sample locations for the considered LULC categories. A mobile-associated tool (Open Data Kit) was used for gathering the digital and geo-referenced field records.

2.2.3. Climatic Data

Climate data (precipitation and temperature) of Jhelum District for the last 30 years (1990 to 2020) were acquired from the Pakistan Metrological Department (PMD). The analysis of variance (ANOVA) test was applied in SPSS version 17 to seek the significant differences among the group means of collected climatic data. Furthermore, the collected climatic data of the study locations were geo-referenced, interpolated, and mapped by using ArcGIS software.

2.3. *Image Classification*

The Landsat images are composed of several bands, where there are 11 bands in the Landsat 8 images. These bands were composited to obtain single-color imagery and to subset the research area, and extraction by the mask tool was conducted in ArcGIS 10.1 software [57]. Digital LULC grouping through the supervised classification technique was used, and field data were employed as ground truth data. LULC maps for the mentioned temporal intervals were developed using supervised classification by centering the research area of focus in the field assessment together with the exercise and authentication portions. Finally, LULC imagery was re-classified in ArcGIS 10.1 to quantify the variations over the indicated study years. ERDAS imagine 15 and ArcGIS 10.1 proved practical tools for assessing the LULC using satellite images. The detailed research framework is presented in Figure 3.

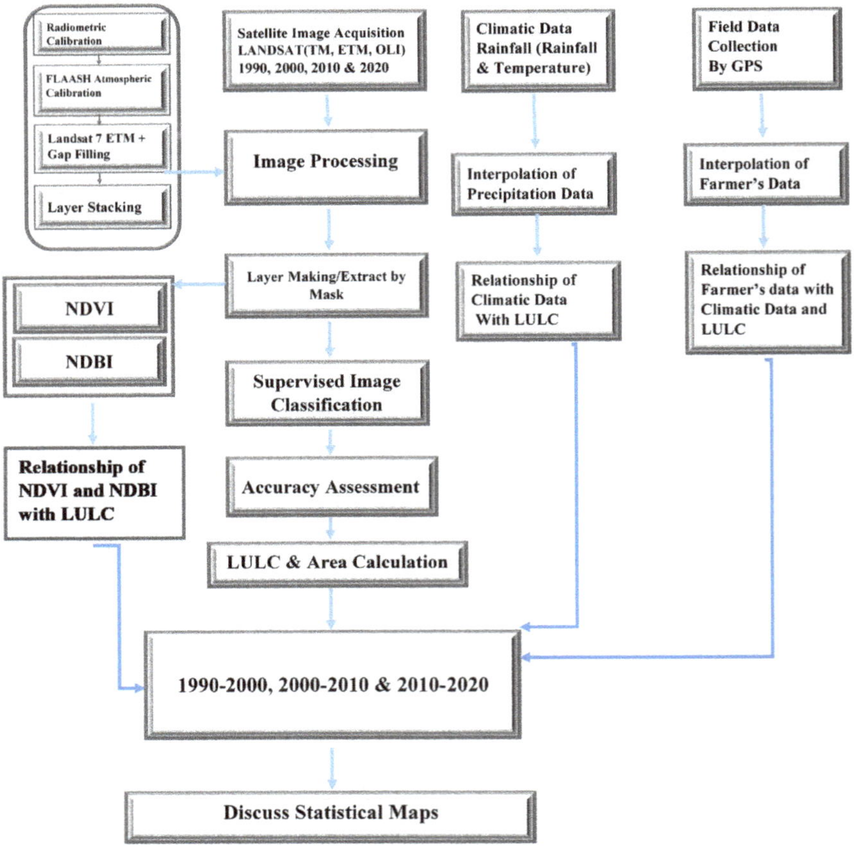

Figure 3. Flow chart for the methodology.

2.4. Assessment of NDVI and NDBI

The variations in the land cover encompassing vegetation for the considered study years (1990, 2000, 2010, 2020) were assessed by calculating and analyzing the NDVI. This remote sensing index to seek vegetation health is calculated as follows [58]:

$$NDVI = \frac{NIR - RED}{NIR + RED} \quad (1)$$

where NIR represents near-infrared radiation band (TM band 4; OLI and ETM band 5), and RED represents the red radiation band (TM band 3; OLI and ETM band 4).

Likewise, the NDBI was used for built-up determination in the study area. The NDBI was obtained by using Arc GIS 10.1 software, and the following formula, as communicated by [59]:

$$NDBI = \frac{MIR - NIR}{MIR + MIR} \quad (2)$$

where MIR represents the central infra-red band (TM band 5; OLI and ETM band 6), and NIR depicts the near-infrared band (ETM and TM band 4; OLI band 5). Therefore, the NDVI was used to assess change in the vegetation cover, and the NDBI was used to find variations in the built-up area by using satellite imagery and show expressive standards between −1 and 1.

2.5. Accuracy Estimation

Accuracy can be considered to determine the effectiveness of several image processing methods in the alignment of imagery [59,60]. The error matrix is the greater grouping and has conjoint meaning to existing accuracy outcomes [61]. Numerous statistical procedures of accuracy assessment can find out of the error matrix including the percentage for producers' accuracy or workers' accuracy as a total accuracy that reveals the error prepared by coincidence [62].

$$\text{Overall accuracy} = \frac{\text{No. of sample classes grouped accurately}}{\text{No. of reference sample classes}} \tag{3}$$

There should be a maximum standard of assurance for any suitable study outcome in the conclusion of the accuracy assessment. The KHAT standards below determine how a good RS group supports, or how far it is accurate in, the reference facts [35]. The mathematical equation of KHAT is

$$K = \frac{\text{observed accuracy } - \text{chances of assessment}}{1 - \text{chances of agreement}} \tag{4}$$

3. Results and Discussion

3.1. Farmers' Perceptions about Temperature and LULC

Farmers were interviewed to obtain their perceptions about climate change, LULC, and their impact on climatic variability, adaptation, and experience stages. Almost all of the farmers (94.5%) stated that climactic change effects were noted and witnessed in Jhelum District (Figure 4).

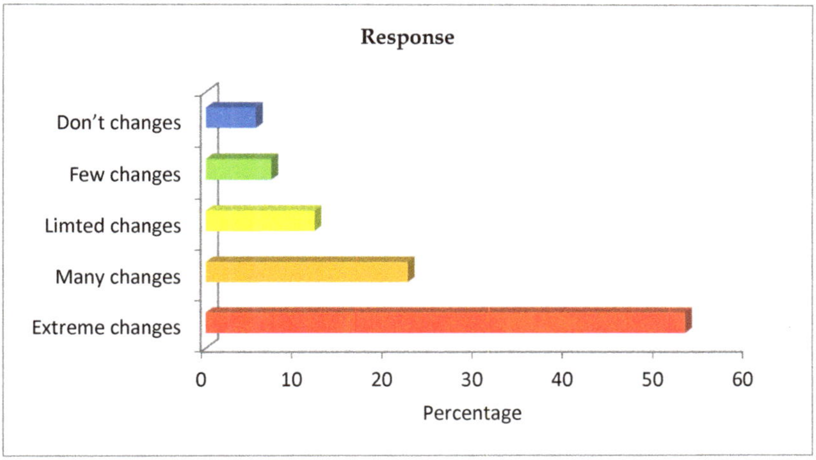

Figure 4. Farmers' opinion of the degree of variations in climate variables in recent years.

A total of 54% of the farmers perceived that significant variations have occurred in the rainfall volume, beginning of the planting period, spreading of the harvesting time, and irregular drought circumstances which occasionally happen in the growth phase. The same decreases in rainfall and increasing drought conditions have been reported in other areas of Pakistan [63,64]. About 25% of the farmers supposed that their livings face many fluctuations due to the rise in the temperature in the study area. Only 5.5% perceived no change in climate in the last 30 years. Concerning the temperature variation, 94.5%

of the farmers stated that they had practiced different crop varieties in recent years, and only 5.5% carried out something different (see Table 2). However, most of the respondents understood that rainfall has been decreasing in terms of volume and the number of rainy days, while the intensity of rainfall has been increasing, which has brought devastation in the form of floods with the passage of time.

Table 2. Farmers' feedback about climatic and LULC variations.

Sr. #	Climatic Variables	Feedback	Percentage
1	LULC variations	Yes	82
		No	18
2	Temperature	Increase	94.5
		Reduction	5.5
		No change	0
3	Irrigation water	Increase	25.3
		Reduction	63.5
		No change	12.2

About 63.5% of the farmers observed that water availability has reduced in recent years, while only 32% of the cropped area is irrigated, and 68% is rain-fed land. However, 72% of the irrigation area would boost fertilizer utilization: 28% of the irrigation types recognized the purpose of changing cropping patterns without using fertilizers. Regarding applications to detect climatic pressures perceived by agriculture (irrigation and rainfall), different investigations led on the agriculture type (rainfall duration, number of occurrences of rainfall and its intensity) for the farmers' actual and scheduled practices found that 36.5% of the informants only observed increases in rainfall, whereas 63.5% observed reductions in rainfall (Table 3). Agricultural practices were documented between two groups of informers, including their association with temperature and rainfall, the ratio of fertilizer usage, and scheduled capitalization on detected occasions.

Table 3. Farmers' observations about rainfall.

Sr. #	Climatic Variables	Feedback	Percentage
1	Rainfall period	High	37.2
		Low	57.6
		No variation	16.2
2	Number of events of rainfall	High	14.3
		Low	80
		No variation	6.7
3	Rainfall density	High	22.5
		Low	72.5
		No variation	5

3.2. Climate Factors of the Research Area

Climate variation showed the most substantial influence on the adaptation of LULC categories in numerous regions of the land [41,65,66]. Similarly, alternation in climatic impacts on the biosphere of the land has a close link with hydrological and energy chains, explaining the effect on the vegetation index (VI) where it increases to its highest quantity [67]. In recent years, universal variation in weather has had many influences on vegetation [7,68].

Climate change has a disproportionate influence on the adaptations of LULC types in different parts of the biosphere. Among different climatic aspects, rainfall and temperature were more associated with LULC. Meanwhile, the recorded data of temperature throughout the field investigations along the coordinates were entered into the ArcGIS 10.1 software and afterward interpreted applying inverse distance weighting (IDW), from which the spatial map of temperature was achieved. These maps indicate the temperature change all over the study area, which indicates the cooler and hotter areas in Jhelum District.

Figure 5 represents the average temperature and rainfall maps for Jhelum District. At the same time, the thematic map (presenting the spatio-temporal variation in temperature) and the central area were acquired. The increase in temperature was recorded up to 28.84 °C, and the decrease in temperature was documented as 27.25 °C. Furthermore, it is assumed that the survey points nearby Jhelum District show maximum temperatures. From these particular points, it is confirmed that the lowest temperature was recorded in the water channels (Jhelum River) and forest parts, and, on the other hand, the average temperature was noted in the barren area and plane part. The highest temperature was noted in the built-up parts. The average rainfall and lowest rainfall in our study area are presented in Figure 5. The highest rainfall increased to 212 mm, while the lowest rainfall value documented was 68.33 mm. The rainfall map remarkably indicates that the maximum rainfall was recorded in Tehsil Pind Dadan Khan. It can be observed that maximum rainfall was recorded in areas such as vegetation area and forest land.

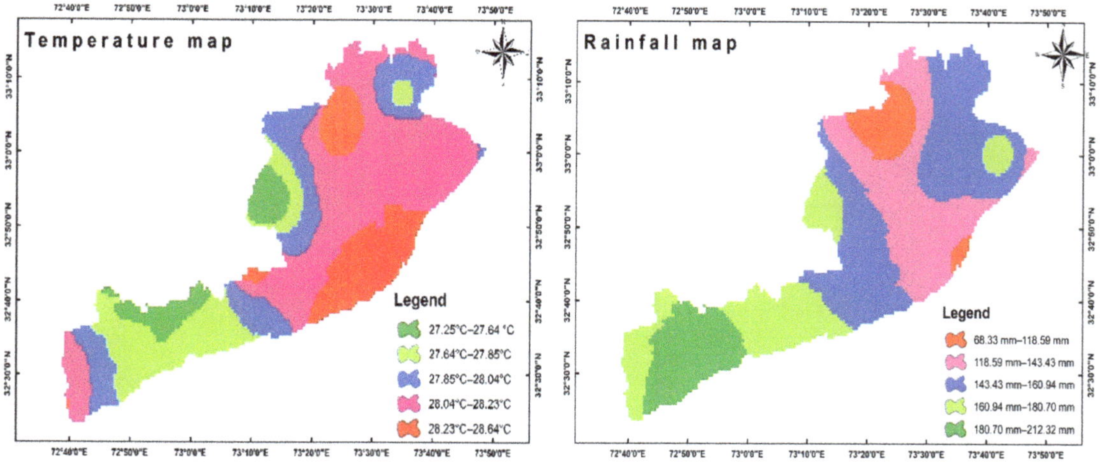

Figure 5. Maps of climate features (rainfall is in millimeters, and temperature is in degrees Celsius) in the research area.

3.3. LULC Change Detection

LULC types with the highest and lowest variations in LULC were assessed at all levels to find out the maximum relative variation over the last 30 years in Jhelum District. The supervised classification analysis indicated that the research district was protected with different land topographies (forest, cultivated area, water, built-up land, and barren land)—the classification arrangement of LULC was carried out by applying surveys with GIS information from Jhelum. Training sites for supervised classification were selected based on different GPS-based samples taken from the field for each land use–land cover class. Then, those GPS locations were plotted on images, and signatures were saved to perform supervised classification using the iso-cluster algorithm. From 1990, cultivated area was 49.54%, followed by water (9.61%); covered area by built-up area was 1.95%; and barren land covered about 31.79%. However, in 2020, forest and cultivated area were 3.36% and 63.39%, followed by water (4.09%); covered area by built-up area was 3.50%; and barren land covered about 25.65%, as presented in Figure 6.

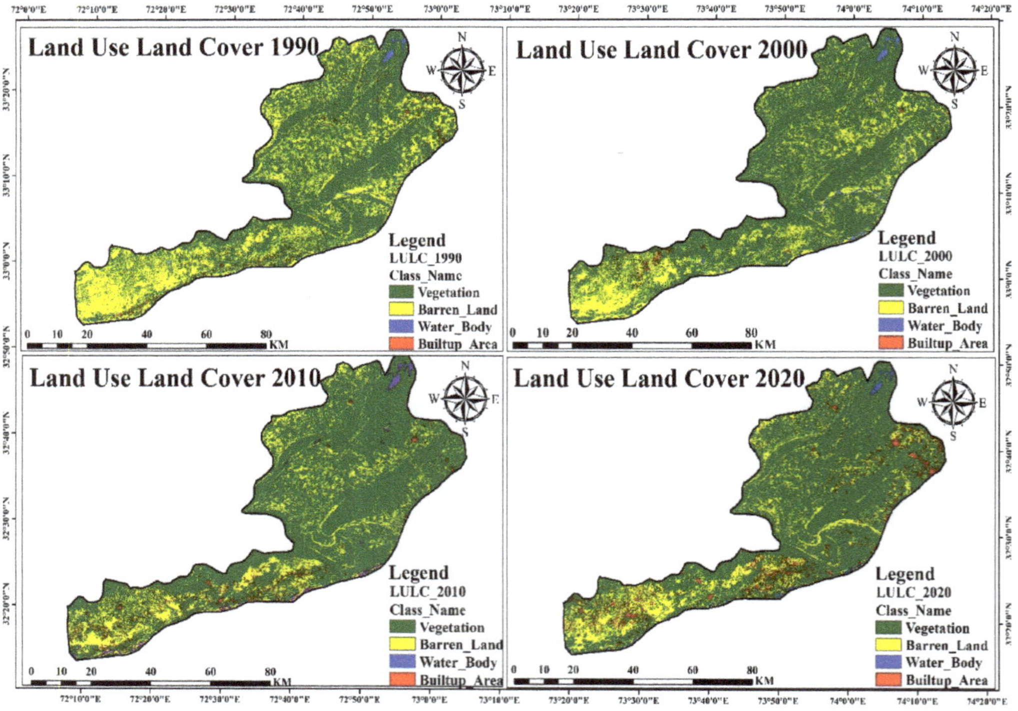

Figure 6. Land use–land cover maps of the district.

Built-up area in the year 1990 covered 1.95% overall, while in the year 2020, built-up area expanded up to 3.5%, compared to the year 1990. However, there was a massive expansion of built-up areas with significant proliferation from 1990 to 2020 (Table 4). Barren area in 1990 covered 31.79%, while in 2020, barren land reduced (25.65%) compared to 1990. It was estimated that bare land changed to housing colonies, commercial parts, and roads. Water channels covered one of the smallest areas among all the categories for Jhelum (9.61%, 8.51%, 6.30%, and 4.09% in 1990, 2000, 2010, and 2020, respectively).

Table 4. Summary of LULC changes from 1990 to 2020.

LULU	1990		2000		2010		2020		1990–2020	
	Ha	%	Ha	%	Ha	%	Ha	%	Ha	%
Forest	25,710.89	7.11	21,313.15	5.90	16,190.64	4.48	12,182.64	3.36	−13,528.25	−3.74
Cultivated area	179,025.57	49.54	192,955.93	53.39	215,529.7	59.64	229,096.76	63.39	50,071.19	13.86
River	34,742.13	9.61	30,742.13	8.51	22,777.12	6.30	14,785.12	4.09	−19,957.01	−5.52
Barren land	114,885.09	31.79	108,358.77	29.98	96,244.73	26.63	92,692.26	25.65	−22,192.83	−6.14
Built-up area	7030.51	1.95	8024.21	2.22	10,652	2.95	12,637.41	3.50	5606.9	1.55
	361,394.19	100	361,394.19	100	361,394.19	100	361,394.19	100		

In the current attempt, LULC categories with the lowest and highest variations in LULC were nominated by minimum and maximum standards to categorize the comprehensive relative change during the past 30 years in Jhelum. There are several smaller colonies that have settled along the central highway and nearby Jhelum District. The total number of these colonies is more than 50 in Jhelum District. The estimated area of the mentioned colonies ranges from 4 to 6 acres. The spread of housing colonies along main roads is an indication of urban expansion in the studied district. Change discovery aims to recognize which LULC expanded or reduced over the past 30 years, and which land uses changed into another LULC category. The results of [69] showed that in recent years, there has been an increase in built-up areas, whereas the ratio of increase for the occupied area was slightly smaller, which is estimated to directly produce a fast rise in the expansion of urban areas in future years, resulting in a reduction in the vegetative area.

3.4. The NDVI and NDBI

The standards of the NDVI are an indication of the volume of chlorophyll content existing in vegetation, where the highest NDVI values depict healthy and thick vegetation, whilst the lowest NDVI values represent thin vegetation. From the study of Jhelum District, NDVI standards in 1990 varied from maximum value of +0.86 to −0.12, whereas during 2000, the same varied from +0.75 to −0.17, and in 2010, the NDVI value showed the minimum value, which was −0.28, while the maximum was +0.62, whilst in 2020, the NDVI displayed the minimum value, of −0.32, and the maximum was +0.56 (Table 4). As averages, NDVI values were detected as 0.37, 0.29, 0.17, and 0.12 for 1990, 2000, 2010, and 2020, respectively. The NDVI of Jhelum District was at its maximum in 1990 at +0.86, and during 2020, it was −0.32, which determines the NDVI classes representing the spatial arrangement of vegetative and green zones found on the map, which displays the creative and best vegetative areas for cultivation as forest and vegetative land (Figure 7).

Figure 8 indicates the extracted NDBI classes demonstrating the spatial arrangement of built-up and water land for 1990, 2000, 2010, and 2020. Average NDBI values were observed to be 0.04, 0.15, 0.19, and 0.27 for 1990, 2000, 2010, and 2020, respectively. Likewise, NDBI standards for Jhelum District were greater in 2020 at +0.72 and lowest in 1990 at −0.36. In Figure 8, the maps show that the red zones were found to be the minimum vegetative land areas, such as water channels, built-up area, and bare land. The NDBI was linked to the temperature, where the values for the NDVI were greater in regions with maximum temperature areas.

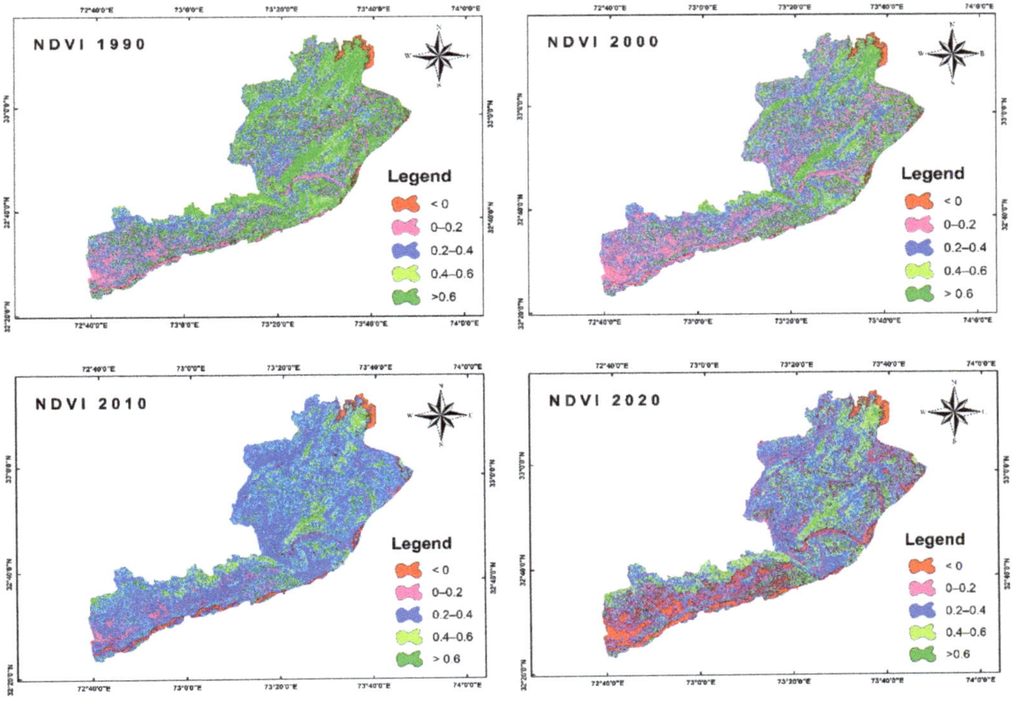

Figure 7. Normalized difference vegetation index maps of Jhelum.

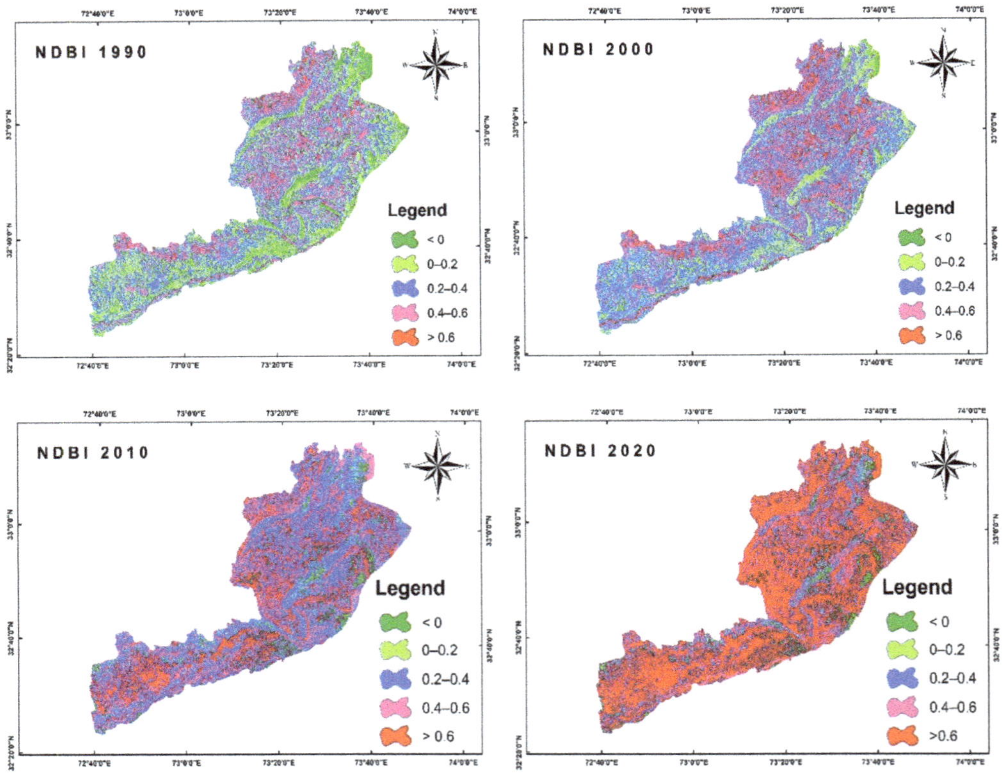

Figure 8. Normalized difference built-up index maps of the research area.

Linear regression analysis was applied to create a link between the NDBI and NDVI. First, regression analysis (R^2) was conducted to determine how variations in the LU intensity within the LULC unit differ over space and pass the intra-LU change of the NDBI. However, a negative association between the NDVI and NDBI can be presented, with a correlation coefficient of $R^2 = 0.82$ for 1990, 0.79 for 2000, 0.76 for 2010, and 0.72 for 2020 shown in all imagery between the vegetation index (NDVI) and NDBI-derived built-up portions, as shown in Figure 9. Furthermore, the regression analysis indicated that in the given areas where the NDBI values were the highest, the NDVI values were the lowest.

The NDVI is generally applied in all vegetation indices established, and its progress is due to random dissimilarity, as stated by [70]. However, due to specific driving factors such as the local temperature, it is recognized that the link to the NDBI powerfully affects the land surface temperature (LST), followed by main roads and LULC [45]. All the calculated NDVI and NDBI values of the considered study area and duration are presented in Table 5.

Table 5. Summary of maximum and minimum values of NDVI and NDBI.

Years	NDVI			NDBI		
	Maximum	Minimum	Average	Maximum	Mininim	Average
1990	0.86	−0.12	0.37	0.45	−0.36	0.045
2000	0.75	−0.17	0.29	0.54	−0.25	0.145
2010	0.62	−0.28	0.17	0.58	−0.2	0.19
2020	0.56	−0.32	0.12	0.72	−0.18	0.27

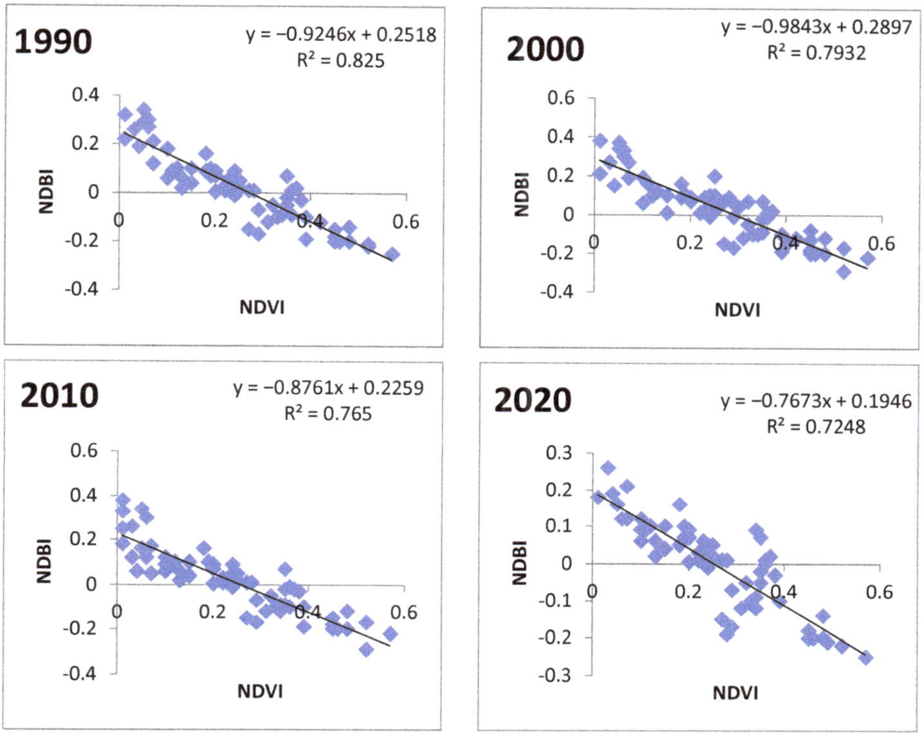

Figure 9. Regression analyses concerning NDBI and NDVI in the study district.

3.5. Accuracy Assessment

Table 6 shows the producer and consumer accuracy outcomes with KHAT (k) values in 1990, 2000, 2010, and 2020 in Jhelum District. The maximum producer and consumer accuracies from the cultivated area remained 88% and 92%, respectively. Additionally, the maximum producer and user accuracies of built-up land were 89.5% and 88.8%, respectively. The average producer and user accuracies were 83.2% and 88.8% for 1990, 88.1% and 85.7% for 2000, 86.5% and 86.7% for 2010, and 85.6% and 87.3% for 2020, respectively, in Jhelum District. The overall accuracy for grouping is 0.93% for 1990, 0.87% for 2000, 0.91% for 2010, and 0.88% for 2020 (Table 6).

Table 6. Kappa (K) and accuracy of producers and consumers.

LULC Classes	Season and Class												Overall Accuracy	K
	Producers' Accuracy (%)						Consumers' Accuracy (%)							
		1	2	3	4	Avg.		1	2	3	4	Avg.		
1990	90.2	85.2	83.7	86.7	81.2	83.2	83.2	88.1	89.7	92.5	90.7	88.8	0.93	0.86
2000	88.1	88	91.3	85.1	88.1	88.1	86.7	86.2	85	88.1	82.5	85.7	0.87	0.82
2010	85.6	84.4	87.4	90.8	84.3	86.5	88.9	86.7	88	82.4	87.5	86.7	0.91	0.89
2020	83.2	80.1	86.5	88.7	89.5	85.6	92	92	85.3	80	89.1	87.3	0.88	0.85

Where: 1 = forest area; 2 = cultivated area; 3 = river; 4 = barren land; 5 = built-up area.

The KHAT (K) coefficients for 1990, 2000, 2010, and 2020 are 0.86, 0.82, 0.89, and 0.85, respectively, in the research area. The accuracy classification was stated as both the consumers' accuracy and producers' accuracy [67,71]. According to [72], the producers' accuracy is described as the quantity of land types properly categorized in the classification of the imagery, whereas the consumers' accuracy is the possibility that a type in the classification of imagery is precise when applied on the land.

4. Conclusions

The current research was conducted in Jhelum District, Punjab (Pakistan), to determine the impact of climatic variations and LULC changes. The livelihood of the farmers in the study area is entirely dependent on agriculture and linked with normal temperatures and rainfall. However, fluctuations in the normal temperature resulted in a shortage of rains, an increase in drought events, and a decrease in water availability for irrigation, hence directly affecting the agrarian community and farming inventions. The increasing temperature and reduced water availability for irrigation due to less rainfall are considered as serious concerns in the study area. Growers are conscious about the climatic fluctuations and familiarize themselves with approaches to manage the impacts but need government support. The outcomes show that the vegetation section contributes an extra grounded constructive link with the NDVI for all the levels, as open area and built-up land negatively associated with LULC and the NDVI during the last 30 years. On average, NDBI and NDVI standards were recorded between 0.37 and 0.12, and 0.04 and 0.27, from 1990 to 2020, respectively, whereas average producer and user accuracies were 83.2% and 88.8% for 1990, 88.1% and 85.7% for 2000, 86.5% and 86.7% for 2010, and 85.6% and 87.3% for 2020. Accordingly, the "Kappa coefficients" for 1990, 2000, 2010, and 2020 were 0.86, 0.82, 0.89, and 0.85, respectively, in the study area. "Barren land" in 1990 occupied the class with 31.79%, but in 2020, it decreased (25.65%) compared to 1990. The outcome indicates that the bare land transformed into housing areas and roads. Water covered 9.61% in 1990, but it remarkably reduced (5.52%) in 2020 compared to 1990 in Jhelum District.

It is concluded that LULC changes are significant for a comprehensive series of uses, comprising temperature, soil destruction, and land planning events. There were main variations in barren land, water channels, and vegetative areas across the studied temporal gradient due to the increasing human influence in acquiring arable lands. The research outcome shows the main observational base for regular inspections of variations in land supervision and will prove helpful for policy makers to improve strategies to manage land capitals efficiently.

Author Contributions: Conceptualization: M.M., A.T., M.M.A. and A.M.K.; methodology: M.M., S.H., F.M., A.T., A.M.K., G.R. and M.M.A.; data curation: M.M., M.F., A.Z., M.A., S.A., A.T., M.W., K.F. and M.F.; validation: A.T., A.M.K. and F.M.; formal analysis: A.T., M.M., S.S. and S.H.; investigation: M.M. and A.M.K.; project administration: A.T., M.M.A., A.M.K. and L.Z.; resources: M.M., A.T., F.A. and L.Z.; software: A.T., M.M., A.Z., M.A., S.A., M.W. and K.F.; visualization: A.T., L.Z., M.F. and G.R.; writing—original draft preparation: A.T., M.M. and A.M.K.; writing—review and editing: A.T., M.M.A. and G.R.; supervision: M.M.A., A.M.K., A.T. and F.A. All authors have read and agreed to the published version of the manuscript.

Funding: This study was supported by the National Natural Science Foundation of China (grant number 41907192) and the Civil aerospace pre-research project (D040102).

Institutional Review Board Statement: Not applicable.

Informed Consent Statement: Not applicable.

Data Availability Statement: The data presented in this study are available on request from the first or corresponding authors.

Acknowledgments: This is part of the PhD research work of the first author. We are thankful to all the study participants who generously shared their perceptions about the climatic variations and their impacts in the study area.

Conflicts of Interest: The authors declare that they have no conflict of interest.

References

1. Hussain, S. Land Use/Land Cover Classification by Using Satellite NDVI Tool for Sustainable Water and Climate Change in Southern Punjab. *COMSATS Univ. Islamabad* **2018**. [CrossRef]
2. Hassan, Z.; Shabbir, R.; Ahmad, S.S.; Malik, A.H.; Aziz, N.; Butt, A.; Erum, S. Dynamics of land use and land cover change (LULCC) using geospatial techniques: A case study of Islamabad Pakistan. *SpringerPlus* **2016**, *5*, 1–11. [CrossRef]
3. Giri, C.; Zhu, Z.; Reed, B. A comparative analysis of the Global Land Cover 2000 and MODIS land cover data sets. *Remote Sens. Environ.* **2005**, *94*, 123–132. [CrossRef]
4. Tewabe, D.; Fentahun, T. Assessing land use and land cover change detection using remote sensing in the Lake Tana Basin, Northwest Ethiopia. *Cogent. Environ. Sci.* **2020**, *6*, 1778998. [CrossRef]
5. Foody, G.M. Status of land cover classification accuracy assessment. *Remote Sens. Environ.* **2002**, *80*, 185–201. [CrossRef]
6. Mannan, A.; Liu, J.; Zhongke, F.; Khan, T.U.; Saeed, S.; Mukete, B.; Shen, C.Y. Application of land-use/land cover changes in monitoring and projecting forest biomass carbon loss in Pakistan. *Glob. Ecol. Conserv.* **2019**, *17*, e00535. [CrossRef]
7. Zoungrana, B.J.; Conrad, C.; Thiel, M.; Amekudzi, L.K.; Da, E.D. MODIS NDVI trends and fractional land cover change for improved assessments of vegetation degradation in Burkina Faso, West Africa. *J. Arid. Environ.* **2018**, *153*, 66–75. [CrossRef]
8. Abid, M.; Schilling, J.; Scheffran, J.; Zulfiqar, F. Climate change vulnerability, adaptation and risk perceptions at farm level in Punjab, Pakistan. *Sci. Total Environ.* **2016**, *547*, 447–460. [CrossRef] [PubMed]
9. Hussain, M.; Liu, G.; Yousaf, B.; Ahmed, R.; Uzma, F.; Ali, M.U.; Ullah, H.; Butt, A.R. Regional and sectorial assessment on climate-change in Pakistan: Social norms and indigenous perceptions on climate-change adaptation and mitigation in relation to global context. *J. Clean. Prod.* **2018**, *200*, 791–808. [CrossRef]
10. Hammad, H.M.; Abbas, F.; Saeed, S.; Fahad, S.; Cerdà, A.; Farhad, W.; Bakhat, H.F. Offsetting land degradation through nitrogen and water management during maize cultivation under arid conditions. *Land Degrad. Dev.* **2018**, *29*, 1366–1375. [CrossRef]
11. Abate, N.; Kibret, K. Detecting and quantifying land use/land cover dynamics in Wadla Delanta Massif, northcentral highlands of Ethiopia. *Ethio. J. Environ. Stud. Manag.* **2016**, *9*, 450–469. [CrossRef]
12. Udin, W.S.; Zahuri, Z.N. Land use and land cover detection by different classification systems using remotely sensed data of Kuala Tiga, Tanah Merah Kelantan, Malaysia. *J. Trop. Resour. Sustain. Sci.* **2017**, *5*, 145–151.
13. Zaidi, S.M.; Akbari, A.; Abu, S.A.; Kong, N.S.; Gisen, A.; Isabella, J. Landsat-5 time series analysis for land use/land cover change detection using NDVI and semi-supervised classification techniques. *Pol. J. Environ. Stud.* **2017**, *26*, 2833–2840. [CrossRef]
14. Solaimani, K.; Arekhi, M.; Tamartash, R.; Miryaghobzadeh, M. Land use/cover change detection based on remote sensing data (a case study; Neka Basin). *Agric. Biol. J. N. Am.* **2010**, *1*, 1148–1157. [CrossRef]
15. Aredehey, G.; Mezgebu, A.; Girma, A. Land-use land-cover classification analysis of Giba catchment using hyper temporal MODIS NDVI satellite images. *Int. J. Remote Sens.* **2018**, *39*, 810–821. [CrossRef]
16. Mohamed, S.A.; El-Raey, M.E. Land cover classification and change detection analysis of Qaroun and Wadi El-Rayyan lakes using multi-temporal remotely sensed imagery. *Environ. Monit. Assess.* **2019**, *191*, 1–9. [CrossRef]
17. Maviza, A.; Ahmed, F. Analysis of past and future multi-temporal land use and land cover changes in the semi-arid Upper-Mzingwane sub-catchment in the Matabeleland south province of Zimbabwe. *Int. J. Remote Sens.* **2020**, *41*, 5206–5227. [CrossRef]
18. Butt, A.; Shabbir, R.; Ahmad, S.S.; Aziz, N. Land use change mapping and analysis using remote sensing and GIS: A case study of Simly watershed, Islamabad, Pakistan. *Egypt J. Remote Sens. Space Sci.* **2015**, *18*, 251–259. [CrossRef]
19. Liang, S.Z.; Ma, W.D.; Sui, X.Y.; Yao, H.M.; Li, H.Z.; Liu, T.; Wang, M. Extracting the spatiotemporal pattern of cropping systems from NDVI time series using a combination of the spline and HANTS Algorithms: A case study for Shandong Province. *Can. J. Remote Sens.* **2017**, *43*, 1–15. [CrossRef]
20. Reddy, A.S.; Reddy, M.J. NDVI based assessment of land use land cover dynamics in a rainfed watershed using remote sensing and GIS. *Int. J. Sci. Eng. Res.* **2013**, *4*, 87–93.
21. Rahman, M.T.U.; Tabassum, F.; Rasheduzzaman, M.; Saba, H.; Sarkar, L.; Ferdous, J.; Uddin, S.Z.; Islam, Z. Temporal dynamics of land use/ land cover change and its prediction using CAANN model for southwestern coastal Bangladesh. *Environ. Monit. Assess.* **2017**, *189*, 565. [CrossRef] [PubMed]
22. Macedo, M.N.; DeFries, R.S.; Morton, D.C.; Stickler, C.M.; Galford, G.L.; Shimabukuro, Y.E. Decoupling of deforestation and soy production in the southern Amazon during the late 2000s. *Proc. Natl. Acad. Sci. USA* **2012**, *109*, 1341–1346. [CrossRef] [PubMed]
23. Hietala-Koivu, R.; Lankoski, J.; Tarmi, S. Loss of biodiversity and its social cost in an agricultural landscape. *Agric. Ecosyst. Environ.* **2004**, *103*, 75–83. [CrossRef]
24. Xu, X.; Xie, Y.; Qi, K.; Luo, Z.; Wang, X. Detecting the response of bird communities and biodiversity to habitat loss and fragmentation due to urbanization. *Sci. Total Environ.* **2018**, *624*, 1561–1576. [CrossRef] [PubMed]
25. Velastegui-Montoya, A.; Lima, A.D.; Adami, M. Multitemporal Analysis of Deforestation in Response to the Construction of the Tucuruí Dam. *ISPRS Int. J. Geo-Inform.* **2020**, *9*, 583. [CrossRef]
26. Llerena-Montoya, S.; Velastegui-Montoya, A.; Zhirzhan-Azanza, B.; Herrera-Matamoros, V.; Adami, M.; de Lima, A.; Moscoso-Silva, F.; Encalada, L. Multitemporal Analysis of Land Use and Land Cover within an Oil Block in the Ecuadorian Amazon. *ISPRS Int. J. Geo-Inform.* **2021**, *10*, 191. [CrossRef]
27. Ahmad, F. A review of remote sensing data change detection: Comparison of Faisalabad and Multan Districts, Punjab Province, Pakistan. *J. Geogr. Reg. Plan.* **2012**, *5*, 236–251. [CrossRef]

28. Ahmad, F. Pixel purity index algorithm and dimensional visualization for ETM+ image analysis: A case of district Vehari. *Glob. J. Hum. Soc. Sci. Arts Hum.* **2012**, *12*, 76–82.
29. Lambin, E.F.; Geist, H.J.; Lepers, E. Dynamics of land-use and landcover change in tropical regions. *Annu. Rev. Environ. Resour.* **2003**, *28*, 205–241. [CrossRef]
30. Harris, A.; Carrb, A.S.; Dashc, J. Remote sensing of vegetation cover dynamics and resilience across Southern Africa. *Int. J. Appl. Earth Obs. Geoinf.* **2014**, *28*, 131–139. [CrossRef]
31. Lunetta, R.S.; Knight, J.F.; Ediriwickrema, J.; Lyon, J.G.; Worthy, L.D. Land-cover change detection using multi-temporal MODIS NDVI data. *Remote Sens. Environ.* **2006**, *105*, 142–154. [CrossRef]
32. Zoran, M.; Anderson, E. The use of multi-temporal and multispectral satellite data for change detection analysis of the Romanian black sea coastal zone. *J. Optoelectron. Adv. Mater.* **2006**, *8*, 252–256.
33. Omran, E.S.E. Detection of land-use and surface temperature change at different resolutions. *J. Geogr. Inf. Syst.* **2012**, *4*, 189–203. [CrossRef]
34. Usman, M.; Liedl, R.; Shahid, M.A.; Abbas, A. Land use/land cover classification and its change detection using multi-temporal MODIS NDVI data. *J. Geogr. Sci.* **2015**, *25*, 1479–1506. [CrossRef]
35. Uddin, K.; Gurung, D.R.; Giriraj, A.; Shrestha, B. Application of remote sensing and GIS for flood hazard management: A case study from Sindh Province, Pakistan. *Am. J. Geog. Inform. Syst.* **2013**, *2*, 1–5.
36. Kumar, K.; Kumar, V.; Kumar, D. Land use and land cover change detection Ingagas river valley watershed using remote sensing and GIS. *Int. J. Res. Eng. Appl. Sci.* **2016**, *6*, 31–37.
37. Tariq, A.; Shu, H. CA-Markov chain analysis of seasonal land surface temperature and land use landcover change using optical multi-temporal satellite data of Faisalabad, Pakistan. *Remote Sens.* **2020**, *12*, 3402. [CrossRef]
38. Tariq, A.; Riaz, I.; Ahmad, Z. Land surface temperature relation with normalized satellite indices for the estimation of spatio-temporal trends in temperature among various land use land cover classes of an arid Potohar region using Landsat data. *Environ. Earth Sci.* **2020**, *79*, 1–15. [CrossRef]
39. Wang, S.W.; Gebru, B.M.; Lamchin, M.; Kayastha, R.B.; Lee, W.K. Land Use and Land Cover Change Detection and Prediction in the Kathmandu District of Nepal Using Remote Sensing and GIS. *Sustainability* **2020**, *12*, 3925. [CrossRef]
40. Fan, F.; Weng, Q.; Wang, Y. Land use and land cover change in Guangzhou, China, from 1998 to 2003, based on Landsat TM/ETM + imagery. *Sensors* **2007**, *7*, 1323–1342. [CrossRef]
41. Li, H.; Zhang, S.; Sun, Y.; Gao, J. Land cover classification with multisource data using evidential reasoning approach. *Chin. Geogr. Sci.* **2011**, *21*, 312–321. [CrossRef]
42. Ayele, G.T.; Tebeje, A.K.; Demissie, S.S.; Belete, M.A.; Jemberrie, M.A.; Teshome, W.M.; Teshale, E.Z. Time series land cover mapping and change detection analysis using geographic information system and remote sensing, Northern Ethiopia. *Air Soil Water Res.* **2018**, *11*. [CrossRef]
43. Amiri, R.; Weng, Q.; Alimohammadi, A.; Alavipanah, S.K. Spatial–temporal dynamics of land surface temperature in relation to fractional vegetation cover and land use/cover in the Tabriz urban area, Iran. *Rem. Sens. Environ.* **2009**, *113*, 2606–2617. [CrossRef]
44. Arora, G.; Wolter, P.T. Tracking land cover change along the western edge of the US Corn Belt from 1984 through 2016 using satellite sensor data: Observed trends and contributing factors. *J. Land Use Sci.* **2018**, *13*, 1–22. [CrossRef]
45. Pal, S.; Ziaul, S. Detection of land use and land cover change and land surface temperature in English Bazar urban centre. *Egypt J. Remote Sens. Space Sci.* **2017**, *20*, 125–145. [CrossRef]
46. Choudhury, D.; Das, K.; Das, A. Assessment of land use land cover changes and its impact on variations of land surface temperature in Asansol-Durgapur Development Region. *Egypt. J. Remote Sens. Space Sci.* **2019**, *22*, 203–218. [CrossRef]
47. Lu, D.; Li, G.; Moran, E. Spatiotemporal analysis of land use and land cover change in the Brazilian Amazon. *Int. J. Remote Sens.* **2013**, *34*, 5953–5978. [CrossRef]
48. Rahman, G.; Rahman, A.-u.; Ullah, S.; Dawood, M.; Moazzam, M.F.U.; Lee, B.G. Spatio-temporal characteristics of meteorological drought in Khyber Pakhtunkhwa, Pakistan. *PLoS ONE* **2021**, *16*, 1–16. [CrossRef]
49. Nowacki, G.J.; Abrams, M.D. Is climate an important driver of post-European vegetation change in the Eastern United States? *Glob. Chang. Biol.* **2015**, *21*, 314–334. [CrossRef]
50. Majeed, M.; Bhatti, K.H.; Pieroni, A.; Sõukand, R.; Bussmann, R.W.; Khan, A.M.; Amjad, M.S. Gathered Wild Food Plants among Diverse Religious Groups in Jhelum District, Punjab, Pakistan. *Foods* **2021**, *10*, 594. [CrossRef]
51. Ibrahim, F.; Gaylan, R. Urban land use land cover changes and their effect on land surface temperature: Case study using Dohuk City in the Kurdistan Region of Iraq. *Climate* **2017**, *5*, 13. [CrossRef]
52. Akar, Ö.Z.; Güngör, O.Ğ. Integrating multiple texture methods and NDVI to the Random Forest classification algorithm to detect tea and hazelnut plantation areas in northeast Turkey. *Int. J. Remote Sens.* **2015**, *36*, 442–464. [CrossRef]
53. Hussain, S.; Mubeen, M.; Akram, W.; Ahmad, A.; Habib-ur-Rahman, M.; Ghaffar, A.; Amin, A.; Awais, M.; Farid, H.U.; Farooq, A.; et al. Study of land cover/land use changes using RS and GIS: A case study of Multan district, Pakistan. *Environ. Monit. Assess.* **2020**, *192*, 2. [CrossRef] [PubMed]
54. Majeed, M.; Bhatti, K.; Amjad, M. Impact of climatic variations on the flowering phenlogy of plant species in Jhelum district Punjab, Pakistan. *Appl. Ecol. Environ. Res.* **2021**, *19*, 3343–3376.
55. Majeed, M.; Bhatti, K.H.; Amjad, M.S.; Abbasi, A.M.; Bussmann, R.W.; Nawaz, F.; Rashid, A.; Mehmood, A.; Mahmood, M. Khan, W.M.; et al. Ethno-veterinary uses of Poaceae in Punjab, Pakistan. *PLoS ONE* **2020**, *15*, 0241705. [CrossRef]

56. Tariq, A.; Shu, H.; Li, Q.; Altan, O.; Khan, M.R.; Baqa, M.F.; Lu, L. Quantitative Analysis of Forest Fires in Southeastern Australia Using SAR Data. *Remote Sens.* **2021**, *13*, 2386. [CrossRef]
57. Huyen, N.T.; Tu, L.H.; Liem, N.D.; Tram, V.N.Q.; Minh, D.N.; Loi, N.K. Assessing impacts of land use and climate change on soil and water resources in the Srepok Watershed, Central Highland of Vietnam. *Policy Brief Ser.* **2016**, *2016*, 1–4. [CrossRef]
58. Tariq, A.; Shu, H.; Siddiqui, S.; Munir, I.; Sharifi, A.; Li, Q.; Lu, L. Spatio-temporal analysis of forest fire events in the Margalla Hills, Islamabad, Pakistan using socio-economic and environmental variable data with machine learning methods. *J. For. Res.* **2021**, *13*, 12. [CrossRef]
59. Lin, C.; Wu, C.C.; Tsogt, K.; Ouyang, Y.C.; Chang, C.I. Effects of atmospheric correction and pansharpening on LULC classification accuracy using WorldView-2 imagery. *Inf. Process Agric.* **2015**, *2*, 25–36. [CrossRef]
60. Zhang, Z.; Liu, S.; Wei, J.; Xu, J.; Guo, W.; Bao, W.; Jiang, Z. Mass change of glaciers in Muztag Ata–Kongur Tagh, Eastern Pamir, China from 1971/76 to 2013/14 as derived from remote sensing data. *PLoS ONE* **2016**, *11*, e0147327. [CrossRef]
61. Rahman, G.; Dawood, M. Spatial and temporal variation of rainfall and drought in Khyber Pakhtunkhwa Province of Pakistan during 1971–2015. *Arab. J. Geosci.* **2018**, *11*, 1–13. [CrossRef]
62. Prestele, R.; Alexander, P.; Rounsevell, M.D.; Arneth, A.; Calvin, K.; Doelman, J.; Verburg, P.H. Hotspots of uncertainty in land-use and land-cover change projections: A global-scale model comparison. *Glob. Chang. Biol.* **2016**, *22*, 3967–3983. [CrossRef]
63. Gao, L.; Xiaofei, W.; Brian, A.J.; Qingjiu, T.; Yu, W.; Jochem, V.; Xihan, M.; Xingfa, G. Remote sensing algorithms for estimation of fractional vegetation cover using pure vegetation index values: A review. *ISPRS J. Photogram Remote Sens.* **2020**, *159*, 364–377. [CrossRef]
64. Shah, S.H.I.A.; Yan, J.; Ullah, I.; Aslam, B.; Tariq, A.; Zhang, L.; Mumtaz, F. Classification of Aquifer Vulnerability by Using the DRASTIC Index and Geo-Electrical Techniques. *Water* **2021**, *13*, 2144. [CrossRef]
65. Rodriguez-Galiano, V.F.; Ghimire, B.; Rogan, J.; Chica-Olmo, M.; Rigol-Sanchez, J.P. An assessment of the effectiveness of a random forest classifier for land-cover classification. *ISPRS J. Photogram. Remote Sens.* **2012**, *67*, 93–104. [CrossRef]
66. Hussain, S.; Mubeen, M.; Ahmad, A.; Akram, W.; Hammad, H.M.; Ali, M.; Masood, N.; Amin, A.; Farid, H.U.; Sultana, S.R.; et al. Using GIS tools to detect the land use/land cover changes during forty years in Lodhran District of Pakistan. *Environ. Sci. Pollut. Res.* **2020**, *27*, 39676–39692. [CrossRef]
67. Mumtaz, F.; Arshad, A.; Mirchi, A.; Tariq, A.; Dilawar, A.; Hussain, S.; Shi, S.; Noor, R.; Noor, R.; Daccache, A.; et al. Impacts of reduced deposition of atmospheric nitrogen on coastal marine eco-system during substantial shift in human activities in the twenty-first century. *Geomat. Nat. Hazards Risk* **2021**, *12*, 2023–2047. [CrossRef]
68. Hu, P.; Sharifi, A.; Tahir, M.N.; Tariq, A.; Zhang, L.; Mumtaz, F.; Shah, S.H.I.A. Evaluation of Vegetation Indices and Phenological Metrics Using Time-Series MODIS Data for Monitoring Vegetation Change in Punjab, Pakistan. *Water* **2021**, *13*, 2550. [CrossRef]
69. Ahmad, A.; Ahmad, S.R.; Gilani, H.; Tariq, A.; Zhao, N.; Aslam, R.W.; Mumtaz, F. A Synthesis of Spatial Forest Assessment Studies Using Remote Sensing Data and Techniques in Pakistan. *Forests* **2021**, *12*, 1211. [CrossRef]
70. Ghaderizadeh, S.; Abbasi-Moghadam, D.; Sharifi, A.; Zhao, N.; Tariq, A. Hyperspectral image classification using a hybrid 3D-2D convolutional neural networks. *IEEE J. Sel. Top. Appl. Earth Obs. Remote Sens.* **2021**, *14*, 7570–7588. [CrossRef]
71. Hamza, S.; Khan, I.; Lu, L.; Liu, H.; Burke, F.; Nawaz-ul-Huda, S.; Baqa, M.F.; Tariq, A. The Relationship between Neighborhood Characteristics and Homicide in Karachi, Pakistan. *Sustainability* **2021**, *13*, 5520. [CrossRef]
72. Tariq, A.; Shu, H.; Siddiqui, S.; Imran, M.; Farhan, M. Monitoring Land Use And Land Cover Changes Using Geospatial Techniques, A Case Study Of Fateh Jang, Attock, Pakistan. *Geogr. Environ. Sustain.* **2021**, *14*, 41–52. [CrossRef]

land

MDPI

Article

Towards the Use of Land Use Legacies in Landslide Modeling: Current Challenges and Future Perspectives in an Austrian Case Study

Raphael Knevels [1,*], Alexander Brenning [1], Simone Gingrich [2], Gerhard Heiss [3], Theresia Lechner [2], Philip Leopold [3], Christoph Plutzar [2,4], Herwig Proske [5] and Helene Petschko [1]

1 Department of Geography, Friedrich Schiller University Jena, 07743 Jena, Germany; alexander.brenning@uni-jena.de (A.B.); helene.petschko@uni-jena.de (H.P.)
2 Department of Economics and Social Sciences, University of Natural Resources and Life Sciences, 1070 Vienna, Austria; simone.gingrich@boku.ac.at (S.G.); t.lechner@mysynergis.com (T.L.); christoph.plutzar@umweltbundesamt.at (C.P.)
3 Center for Low-Emission Transport, AIT Austrian Institute of Technology GmbH, 1210 Vienna, Austria; Gerhard.Heiss@ait.ac.at (G.H.); Philip.Leopold@ait.ac.at (P.L.)
4 Division of Conservation Biology, Vegetation Ecology and Landscape Ecology, University of Vienna, Rennweg 14, 1030 Vienna, Austria
5 Remote Sensing and Geoinformation Department, JOANNEUM RESEARCH Forschungsgesellschaft mbH, 8010 Graz, Austria; herwig.proske@joanneum.at
* Correspondence: raphael.knevels@uni-jena.de

Citation: Knevels, R.; Brenning, A.; Gingrich, S.; Heiss, G.; Lechner, T.; Leopold, P.; Plutzar, C.; Proske, H.; Petschko, H. Towards the Use of Land Use Legacies in Landslide Modeling: Current Challenges and Future Perspectives in an Austrian Case Study. *Land* **2021**, *10*, 954. https://doi.org/10.3390/land10090954

Academic Editors: Matej Vojtek, Andrea Petroselli and Raffaele Pelorosso

Received: 4 August 2021
Accepted: 4 September 2021
Published: 8 September 2021

Publisher's Note: MDPI stays neutral with regard to jurisdictional claims in published maps and institutional affiliations.

Abstract: Land use/land cover (LULC) changes may alter the risk of landslide occurrence. While LULC has often been considered as a static factor representing present-day LULC, historical LULC dynamics have recently begun to attract more attention. The study objective was to assess the effect of LULC legacies of nearly 200 years on landslide susceptibility models in two Austrian municipalities (Waidhofen an der Ybbs and Paldau). We mapped three cuts of LULC patterns from historical cartographic documents in addition to remote-sensing products. Agricultural archival sources were explored to provide also a predictor on cumulative biomass extraction as an indicator of historical land use intensity. We use historical landslide inventories derived from high-resolution digital terrain models (HRDTM) generated using airborne light detection and ranging (LiDAR), which are reported to have a biased landslide distribution on present-day forested areas and agricultural land. We asked (i) if long-term LULC legacies are important and reliable predictors and (ii) if possible inventory biases may be mitigated by LULC legacies. For the assessment of the LULC legacy effect on landslide occurrences, we used generalized additive models (GAM) within a suitable modeling framework considering various settings of LULC as predictor, and evaluated the effect with well-established diagnostic tools. For both municipalities, we identified a high density of landslides on present-day forested areas, confirming the reported drawbacks. With the use of LULC legacy as an additional predictor, it was not only possible to account for this bias, but also to improve model performances.

Keywords: land use/land cover legacy; airborne LiDAR-based HRDTM; generalized additive model; landslide susceptibility modeling; historical landslide inventory bias; biomass extraction

1. Introduction

Landslides pose a threat to human lives and infrastructure. A changing climate and land use/land cover (LULC) alter the landslide risk and thus have societal consequences [1,2]. In Austria, landslides are relevant natural hazards preconditioned by factors such as lithology, geomorphology, tectonic structures and LULC, and are mainly triggered by long-lasting heavy rainfall and rapid snowmelt [3]. Therefore, understanding the factors that increase the chance of landslide occurrence is crucial for spatial planning in the face of ongoing and expected future climate and LULC changes.

LULC types, and their changes, are reported to have different hydrological and geomechanical effects controlling slope stability [4]. While forest is often considered to stabilize slopes [5], forest harvesting or road construction undercutting slopes may reduce slope stability [6,7]. However, in landslide studies, LULC has often been considered as a static factor representing solely the present-day LULC [8] (i.e., latest available LULC). Recently, more studies have begun to account for historical LULC in landslide analysis, assigning LULC dynamics an important role in explaining landslides [9–12]. Beguería [9] and Persichillo et al. [11] discovered a high landslide susceptibility on abandoned cultivated land, even after revegetation by shrubs or trees in the Spanish Pyrenees and in the Oltrepò Pavese (Italy), respectively. Gariano et al. [10] and Pisano et al. [12] found evidence that land management reduced landslide occurrences in Southern Italy (Calabria and Molise), supporting the importance of LULC changes in spatial planning practice. However, due to the availability of mainly remote-sensing products (aerial or satellite imagery), these landslide analyses were only able to consider historical LULC since the mid-20th century (e.g., since 1954 in [12] or since 1957 in [9]). To the authors' knowledge, only one study used historical cartographic documents as additional sources (e.g., Napoleonic cadastral map) to assess the long-term legacy effects of LULC on mass movements [13].

In general, legacy effects describe the influence of past events or processes on later states, often spanning decades to centuries [14,15]. Long-term legacy effects of past LULC have been shown to exist in the context of socio-ecological dimensions such as contemporary forest structure, management and disturbance risk [16] or biodiversity [17,18]. For mass movements, Lopez-Saez et al. [13] revealed the potential of historical LULC changes in explaining the paradoxical observation of reduced rockfall hazards despite an increased urban exposure in the Grenoble conurbation since 1850. Especially the forest densification at the upper part of the slope was considered to contribute to the identified decrease in rockfall frequency and energy for volumes up to 5 m^3 [13].

For landslide analysis and modeling, landslide inventories are a fundamental source to improve the understanding of the factors that precondition and trigger landslides. In the last decade, low-cost, airborne LiDAR-based high-resolution digital terrain models (HRDTM) became available area-wide for all federal provinces of Austria ($\leq$10 m $\times$ 10 m) [19]. Many studies demonstrated the potential of LiDAR HRDTM and its derivatives to identify landslides, and thus to substantially improve conventionally created landslide inventories [20–22], especially underneath the forest cover, where passive remote-sensing sensors are of limited utility [23].

Generally, landslide inventories have an unknown level of incompleteness [24] and thus may be biased. Inventory biases have previously been studied for remotely-sensed, event-based or archival inventories, and can be caused by focusing solely on administrative boundaries, damage reports or on a single triggering event [25,26]. While most authors reported the usefulness of LiDAR-derived historical landslide inventories, only few analyzed drawbacks of the data source [26,27]. For example, it is very challenging or even impossible to determine a landslide's exact extent, absolute age, trigger, and potential for reactivation when using only HRDTM derivatives [28,29]. Additionally, according to Petschko et al. [27], forest cover may have a conservation effect on the landslide morphology (i.e., "young" morphology of a "very old" landslide), while landslides may be underrepresented on agricultural land and near infrastructure due to land rehabilitation (i.e., "very young" morphology but no visibility on orthophoto) resulting in a landslide inventory that is substantially biased towards a high landslide density in recently forested areas. Analyzing the effect of systematically incomplete landslide inventories, Steger et al. [26] discovered that landslide susceptibility models emphasized bias-describing predictors (e.g., larger regression coefficients), and as a consequence the bias was directly propagated into the landslide predictions. While Petschko et al. [27] recommended to exclude inventory-biasing observations (i.e., "old" and "very old" landslides) or to drop the bias-describing predictor (i.e., present-day land cover) from modeling, Steger et al. [26] included the bias-describing predictor (e.g., forested area) as a random effect and used only the fixed effects to make model predictions.

In this study, we investigated the association between LULC legacies and landslide susceptibility using an airborne LiDAR-derived historical landslide inventory in two municipalities in Austria (Waidhofen an der Ybbs and Paldau). We addressed the following main questions: (i) are long-term LULC legacies important and reliable predictors of landslide susceptibility? And (ii) can LULC legacies help to understand and account for possible inventory biases in modeling present-day landslide susceptibility? Additionally, we analyzed the transferability of landslide models between study areas and the effect of dropping inventory-biasing observations.

For the analysis, we digitized and classified LULC patterns for three-time cuts comprising nearly 200 years using various spatial data sources. In addition, yields and livestock statistics were compiled from archival sources and statistical publications, and summarized as socio-ecological variables reflecting plot-level LULC legacies. For the assessment, we used generalized additive models (GAM) within a modeling framework considering different combinations of LULC legacy implementations while also accounting for land surface variables (e.g., slope angle, etc.) and lithological conditions as possible confounders. We evaluated the effect of LULC legacies using well-established diagnostic tools for model assessment and interpretation.

2. Materials and Methods

2.1. Study Area

The study was conducted in two municipalities in Austria: Waidhofen an der Ybbs (referred to as Waidhofen) in Lower Austria, and Paldau in Styria. The two municipalities represent different landscapes (Figure 1A,C).

Waidhofen is located in the Ybbstaler Alps from 14°39′ E 47°52′ N to 14°56′ E 48°01′ N, covers an area of 131 km^2 and has a population of about 13,000 inhabitants [30]. Its elevation rises towards the south with a relative relief of 54–651 m/km^2. In the limestone-dominated south, up to 1205 m above the Adriatic (m AA) are reached; towards the flysch zone in the north the mountains transition into gentle hills (302 m AA) [31]. In contrast, Paldau lies in the East Styrian Basin with a relative relief of 6–138 m/km^2 (282 to 465 m AA), characterized by unconsolidated sediments of the Neogene to Quaternary period [32]. The municipality is an agriculturally favorable region mainly with corn and pig farming. Paldau has a population of 3000 inhabitants [33], it extends from 15°43′ E 46°54′ N to 15°51′ E 46°59′ N and covers 39 km^2.

The geological setting coupled with the characteristic very local, frequent and intense rain events in summertime create conditions that make both study areas particularly prone to landslides of different types and magnitudes [34–36]. In the last decades, landslide occurrences have caused substantial damage to settlements and infrastructure in both study areas [37,38].

2.2. Data

This study is based on land surface and landslide data from various sources and at different spatial resolutions. As our target resolution, we used 10 m × 10 m to account for the dependence of landslide susceptibility on local-scale topography.

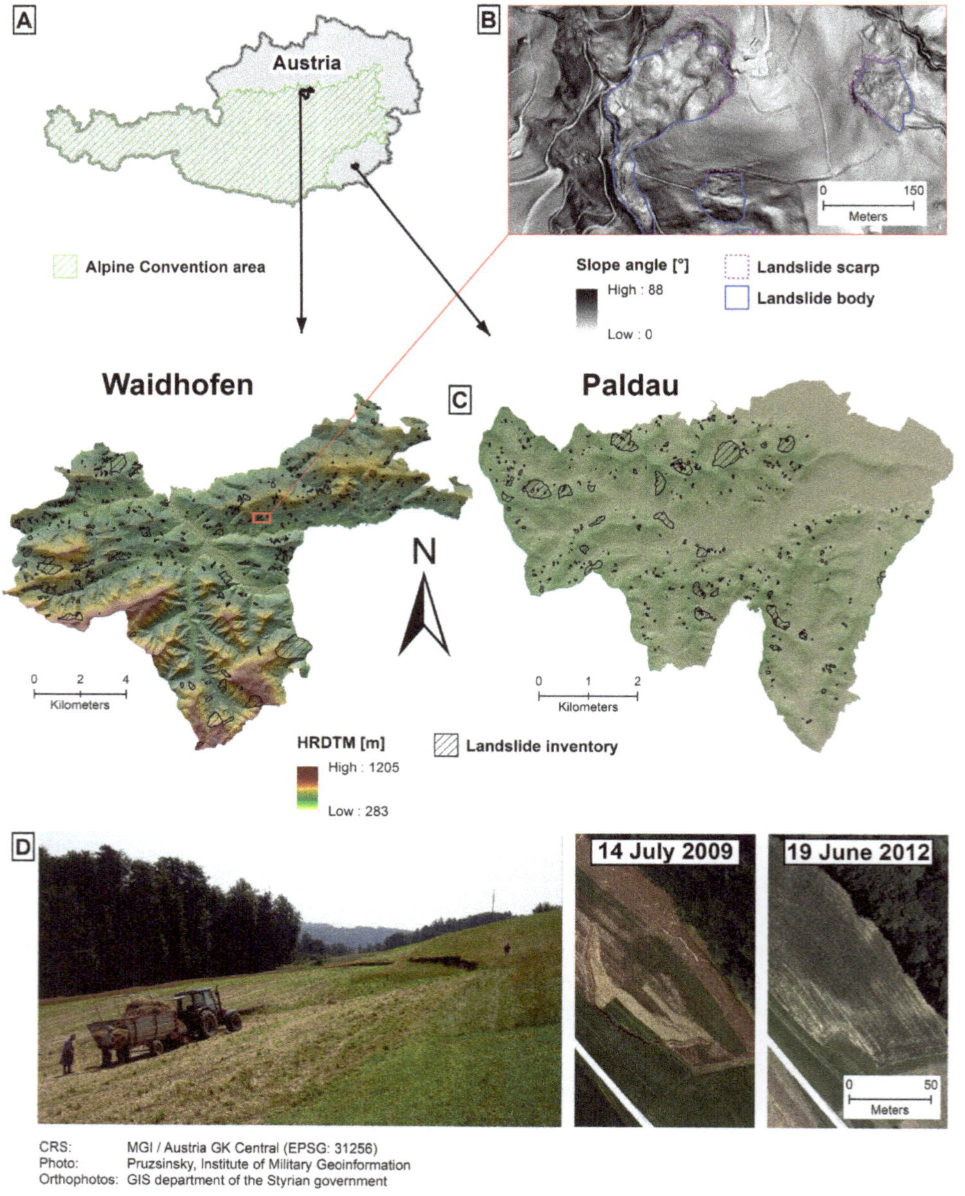

Figure 1. Overview of study area. (**A**): Location of study areas in Austria. (**B**): Landslides visible in Waidhofen's slope map (red rectangle in (**C**)). (**C**): HRDTM and derived landslide inventory. (**D**): A landslide on cropland in the district South East Styria, which occurred after an extreme rainfall event in June 2009, and "disappeared" in the following years. Photo taken on 23 July 2009.

2.2.1. Land Surface Data

For both study areas, airborne LiDAR-based HRDTMs of 1 m × 1 m resolution were available, which were provided by the GIS department of the Styrian government for

Paldau (acquisition year: 2009), and by the provincial government of Lower Austria for Waidhofen (acquisition year: 2017), respectively (Table A1).

Datasets of hydrologic and hydropedologic characteristics were compiled by the Austrian Research Centre for Forests in 2014 for Waidhofen [39] (50 m × 50 m resolution) and in 2017 for Paldau [40] (100 m × 100 m resolution), respectively, and were made available by the respective federal states. The datasets include information on soil parameters such as soil type, total pore volume, and hydraulic conductivity of the topsoil (0–20 cm).

Geological basemaps at a 1:50,000 scale were provided by the Geological Survey of Austria for Waidhofen and by the Styrian GIS department for Paldau, respectively. In both geological basemaps, alluvial deposits were corrected in order to match valley floors visible in the HRDTM. Furthermore, we reclassified the geological units into a smaller number of relevant classes. In Waidhofen the reclassification was based on lithology and geomechanical properties which resulted in seven lithological units of (i) alluvial deposits, (ii) talus and glacial deposits, (iii) Inneralpine Neogene, (iv) Klippen zone, (v) flysch zone, (vi) Upper Austroalpine marls, and (vii, reference level) Upper Austroalpine limestone. In Paldau, we specified five geological units based on grain size distribution and age of origin: (i, reference level) Neogene formations with coarse-grained layers, (ii) Neogene formations dominated by fine-grained sediments, (iii) pre-Würmian Pleistocene formations, (iv) Würmian and Holocene sediments, and (v) other units.

2.2.2. LULC Legacy

By combining unique historical spatially explicit information on LULC in the two study areas and numerical information on land use intensity, we were able to create information on LULC legacies that could be used as an input in landslide susceptibility modeling. This is, to our knowledge, the first study that uses such a long, multi-temporal and spatially explicit LULC record to better understand regional-scale landslide susceptibility.

To generate this dataset, we collected, digitized and harmonized data depicting different spatial patterns for the years 1820, 1960 and 2015 (Table 1). Our data sources included the Franciscan Cadastre of 1820, aerial photographs of 1960, and aerial orthophotos combined with Integrated Administration and Control System (IACS) data of 2015 ("present-day"). Numerical information on LULC intensity (e.g., agricultural yields and machinery numbers) was collected at communal, district or provincial levels from archival sources and both historical and recent statistics, and was up- or downscaled to match the municipal level. The data sources assigned to a time cut slightly differed in their year of origin between the municipalities (Table 1). The selection of the time cuts was based on characteristic socio-economic framework conditions of the Central European cultural landscape following Bender [41] (i.e., time cuts around 1850, 1914, 1960 and present age) as well as on the availability and temporal proximity of data.

We created a vector GIS database with LULC polygons from these sources using ESRI ArcGIS. We distinguished between forest area, cropland and grassland, and classified the remaining LULC as "settlement and other" (for details refer to Table A2 in the Appendix A). Due to varying data quality (Q, Table 1), in the first step, the cadastral maps (1820) and orthophotos (2015) were georeferenced and digitized (EPSG: 32633) since these allowed us to map boundaries between distinct land uses. In a second step, we used these layers as base maps for digitizing the less distinct areas from greyscale aerial photographs (1960). Spatially persistent natural and built structures were kept unchanged for all time cuts. Furthermore, we consistently used a display scale of 1:1000 to keep the digitization error constant. The estimated overall positional error is up to 3–5 m for 1820, 5–10 m for 1960, and less than 3 m for 2015.

Table 1. Sources of land use legacy information of at least acceptable spatial and thematic quality identified for the case study areas.

Year	Study Area	Source	Source Holder	Q	Q-Explanation
			Land use and land cover		
1820	Wh	Maps of Franciscan Cadastre	Provincial Archive of Lower Austria	++	Sharp delimitation of utilization unit
	P		Federal Office for Calibration and Measurement	++	
1962	Wh	Aerial images	Federal Office for Calibration and Measurement	~	Differentiation based on greyscale aerial photography
1953	P			~	
2015	Wh, P	Orthophotos & IACS *	Open Data Austria	++	Parcel-sharp delimitation of arable land and grassland
			Agricultural yields (cereal and grassland)		
1820	Wh	Text records of Franciscan Cadastre	Provincial Archive of Lower Austria	+	based on two cadastral municipalities of Waidhofen
1820	P	Sandgruber [42]		~	average of Styria
1960	Wh, P	Agricultural statistics	Statistics Austria Library	++	data on municipality level
2015	Wh, P	IACS *	Open Data Austria	+	data of farms in municipality
			Wood yields		
1820	Wh	Text records of Franciscan Cadastre	Provincial Archive of Lower Austria	+	based on two cadastral municipalities of Waidhofen
1820	P	Gingrich et al. [43]		~	average of Styria
1965	Wh, P	Weiss et al. [44]		~	Austrian average
2015	Wh, P	Forest inventory	Federal Forest Office	~	state averages

Wh: Waidhofen, P: Paldau; * Data status of temporal extent of 2016; Quality (Q): ++: high spatial and thematic quality, +: high spatial or thematic quality, ~: acceptable spatial and thematic quality. Adapted from Knevels et al. [45].

Based on the numerical information, we quantified biomass extraction as yields (kg fresh weight per ha and year, kg FW/ha/a) at the lowest possible administrative level and allocated it to cropland, grassland and forests. Biomass extraction represents the output from the agricultural production system and is thus an indicator of the LULC intensity during historical time periods [46]. For the biomass extraction from cropland, we employed the cereal yields as proxy; data on harvest from meadows was collected for grassland yields, and wood extraction was derived from forest yields.

We finally intersected the three vector datasets for 1820, 1960 and 2015 successively to one final file, keeping all available attributes, and converted the result to raster format at the uniform target resolution. Biomass extraction was summed over the time cuts to obtain a cumulative land use intensity (Figure A1 in Appendix A). This is an innovative approach for this socio-ecological indicator in the context of landslide science as other authors calculated cumulative materials flows [47] or greenhouse gas emissions [48] in other application contexts. To avoid artifacts due to geometric inaccuracies inherent in the vector files and from sliver polygons, isolated pixels were identified, and the affected grid cells were excluded from the landslide sampling design described below (see Section 2.2.3). The estimated digitization errors (3–10 m) might partly be counteracted by using a target resolution of 10 m × 10 m (i.e., resolution corresponds to largest estimated error).

The created historical LULC legacy dataset was made publicly available [49].

2.2.3. Airborne LiDAR-Derived Landslide Inventories

For both study areas, we derived historical landslide inventories by mapping landslides visible in the HRDTM following the approach proposed by Schulz [22]. We included landslides in earth and debris materials, focusing on the slide type with possible transitions to complex slide flows according to the classification scheme by Cruden and Varnes [50]. Landslides were digitized as polygon features also distinguishing between landslide body and scarp (Figure 1B). Selected mapped landslides were inspected in the field for validation, and we corrected the inventory where necessary. In total, in Waidhofen, 621 landslides were mapped, covering 5.31% of the municipal area, and in Paldau, 418 landslides (4.14% of the area; Figure 1C; Table A3 in the Appendix A).

Following Hussin et al. [51], we randomly sampled landslide presence points in the landslide scarp area using the recommended 50-m distance constraint, and attributed the point to the corresponding grid cell in our target grid (in total 974 and 559 landslide points in Waidhofen and Paldau, respectively). For sampling non-landslide points, we defined the landslide-free area by excluding the mapped landslides and a surrounding 50-m buffer to account for digitization errors. We furthermore excluded so-called trivial areas—areas considered as not susceptible to landsliding (e.g., floodplains, flat areas) [52]. Isolated grid cells (see Section 2.2.2.) and anthropogenic structures with similar geomorphometric characteristics as landslides (e.g., quarries) were also masked. For the landslide absence locations, we distributed random points in a 1:1 sampling ratio using a minimum nearest-neighbor distance of 50 m to reduce spatial autocorrelation.

2.3. Methods

The relationships between LULC legacies and landslide distribution were analyzed using GAMs [53,54] while also accounting for the local topography as an important preparatory factor for landslides [8]. GAMs have become popular in landslide susceptibility studies due to their ability to model nonlinear relationships while allowing for a separate interpretation of additive effects in terms of odds ratios and variable importance [55].

Our analysis was conducted in the free and open source computing environment R (R version 3.5.3) [56]. We used the GAM implementation of the *mgcv* package [54] and the *mlr* package [57] as the modeling framework. Furthermore, for terrain analysis we used System for Automated Geoscientific Analysis (SAGA) GIS 6.3.0 [58] through *RSAGA* [59] and Terrain Analysis Using Digital Elevation Models (TauDEM) 5.3 [60] via R system calls.

For downscaling the input data to the target resolution, we applied bilinear interpolation for resampling using SAGA GIS. However, we acknowledge that we are unable to capture local-scale patterns of geology or soil parameters.

2.3.1. Landslide Susceptibility Modeling Design

For landslide susceptibility modeling, we related land surface variables, soil parameters, lithological units and LULC legacies as predictors to landslide occurrences (Table 2 for overview).

Our model design enabled us to explore relationships between LULC legacy and landslide distribution, and to improve the understanding of the potential biases in airborne LiDAR-derived landslide inventories. We created landslide susceptibility models with different sets of input variables: (i) The baseline model 'GAM-Base' excludes LULC legacy variables; for the assessment of the LULC legacy effect, we built (ii) a GAM using the present-day LULC as an additional predictor (GAM-2015), (iii) a GAM based on the LULC legacy information from 1960 to 2015 (GAM-1960), and (iv) a GAM based on the LULC legacy information since 1820 (GAM-1820). Moreover, we tested (v) a GAM using the setting of GAM-2015, but excluding potentially inventory-biasing observations following the recommendations of Petschko et al. [27] (GAM-2015-Masked; i.e., all observations located in continuously forest-covered areas). Furthermore, we allowed modeled landslide occurrences to be dependent on the combined effect of the historical biomass extraction and present-day LULC class rather than modeling LULC legacy variables as additive terms. Thus, the predictor of LULC legacy information in GAM-1820 and GAM-1960 was implemented as a parametric, linear interaction term between the LULC of 2015 and the historical biomass extraction (sum since 1820 or 1960). Moreover, we tested the model's transferability between study areas, but excluded predictors that are specific to each area (i.e., lithology).

For both study areas, the airborne LiDAR-derived historical landslide inventory and the created landslide susceptibility models are available as Supplementary Materials.

Table 2. Predictor variables for landslide susceptibility modeling.

Variable(s)	Software	Setting	Method
land surface variable			
convergence index (100 m, 500 m)	SAGA GIS	r = 100 m, 500 m	[61]
curvature (plan, profile)	SAGA GIS		[62]
flow accumulation, D-Infinity	TauDEM	log-transformed	[63]
normalized height	SAGA GIS	w = 5; t = 2; e = 2	[64]
slope angle	SAGA GIS		[62]
slope angle, catchment area	SAGA GIS		[65]
slope aspect (S-N, W-E)	SAGA GIS	cosine, sine transformed	[62,66]
topographic position index (TPI)	SAGA GIS	r = 500 m	[67]
topographic wetness index (SWI)	SAGA GIS		[65]
soil			
total pore volume		up to 20 cm depth, median	
hydraulic conductivity		up to 20 cm depth, median	
lithology			
geology		* ref: Waidhofen (vii), Paldau (i)	
land use/land cover legacy			
LULC 2015		ref: 'Forest'	
biomass extraction (1820, 1960)		sum since 1820 and 1960	

Setting, scale-dependent parameters: r: radius; w,t,e: Parameters in SAGA GIS module Relative Heights and Slope Positions; * ref = reference level (see Section 2.2.1).

2.3.2. Assessment of the Effect of Land Use Legacy

The empirical effect of LULC legacy on landslide occurrence was assessed in terms of model performance, variable importance, odds ratios (OR) of the LULC classes, and transformation function plots of the three most important predictor-response relationships.

For the model assessment, we applied a *k*-fold spatial cross-validation (SpCV) to achieve independent test areas and thus a bias-reduced predictive performance as a measure of model generalization [55]. For SpCV, we partitioned the data into five disjoint folds using *k*-means clustering of the coordinates ($k = 5$), and repeated this procedure 100 times. In each repetition, subsequently, four folds were used as training data while the remaining fold was used for validation until each fold was used once for validation (i.e., *k* models per repetition, 500 models in total). Furthermore, we ensured comparability between the models' performance estimates using identical training and validation data for each study area and repetition.

The area under the receiver operating characteristic curve (AUROC) was computed as the performance measure. The AUROC is a quality measure suitable for binary response data and is a common evaluation tool for landslide susceptibility models [55]. AUROC values lie between 0.5 (no discrimination) and 1.0 (perfect discrimination), and were interpreted following the recommendations of Hosmer et al. [68]. Additionally, differences in model performances were tested using Wilcoxon signed-rank tests ($\alpha = 0.05$; R *coin* package) [69,70], and *p*-values were adjusted for multiple comparisons according to Benjamini and Hochberg [71].

As a measure of variable importance, we extracted for each variable the mean decrease in deviance explained (mDD, %) under the consideration of all SpCV models. The mDD indicates the explanatory contribution of a variable to the overall explained deviance of the corresponding model [72,73]. The higher the mDD value, the greater is the contribution of a variable, and thus its importance. To compute the mDD, we left the variable of interest out while fixing the remaining smoothing parameters (*mgcv::gam sp* argument) during model (re-)training, and subsequently measured the percentage differences of the deviance explained [73].

Transformation function and OR plots were used to explore the relationships between landslide distribution and the three most important predictors as well as LULC legacy of each model setting. Additionally, we extracted comparative predictor-response relationships reported in studies of the same area [73,74]. A transformation function plot shows the predictor-response relationship as a parametric (linear) or non-parametric smoothing function by using the additive structure of a GAM. We visualized predictor-response relationships on the logit scale (i.e., linear predictor scale). An OR indicates the chance that an outcome occurs given a specific exposure, relative to a reference exposure [75]. An OR less than one means an exposure with lower odds of the outcome while an OR higher than one shows an exposure with higher odds of the outcome, while accounting for the other variables in the model; an OR of one is associated with no influence of the exposure [75]. We calculated ORs for the LULC classes with 'forest' as reference level. Additionally, we derived ratios of ORs (rOR) by dividing a model's predictor-response relationships by the corresponding relationship in GAM-Base. rOR enables a more sensitive comparison between models.

3. Results

3.1. LULC Change

The analysis of almost 200 years of changes in LULC displayed distinct trends in the two municipalities (Figure 2).

Even though forest covered more than one third of the total land area in both case studies throughout the investigation period (Figure 2A), in Paldau, its area decreased by 113.2 ha (8%) from 1820 to 2015, while in Waidhofen, the forest area increased by more than 1000 ha (18%) from 1820 to 2015. In both areas, cropland extent declined while grassland and "settlement and other" expanded. In Waidhofen, in 2015 only 1% of the area was used as cropland (1820: 26%), while in Paldau, the share diminished from 43% to 37%. Areas classified as "settlement and other" increased in Waidhofen more than fivefold from 1820 to 2015 (246 ha to 1431 ha) and in Paldau more than sevenfold (104 ha to 750 ha).

Biomass extraction nearly doubled in both municipalities from one time cut to the next (Figure 2A; Paldau: 4969 to 10,630 to 18,674 kg/ha/a, Waidhofen: 4661 to 8410 to 16,773 kg/ha/a). Cereal yields increased most strongly in both municipalities (factors 9.8 and 7.9 in Waidhofen and Paldau, respectively). Grassland yields increased less strongly and reached their maximum in 1960. In both study areas, wood extraction declined slightly from 1820 to 1960 (Waidhofen: −6%, Paldau: −2%), but it more than doubled from 1960 to 2015 (Waidhofen: 5067 kg/ha/a, Paldau: 5667 kg/ha/a). In 1820, wood extraction had the largest share in biomass extraction in both municipalities.

The landslide distribution for the different LULC classes and time cuts showed contrasting patterns in Waidhofen and Paldau (Figure 2C). While in Waidhofen, the change pattern in landslide distribution and LULC distribution by LULC category was similar (Figure 2A,C), in Paldau the change pattern showed clear differences. There, 82.4% of the landslides were located in present-day (2015) forest areas, which also showed the highest landslide frequency ratio (FR) of 2.5 (Figure 2C); in Waidhofen, 42.2% of the landslides were found in present-day forested areas with a FR of nearly 1. In Waidhofen in 2015, grassland showed the highest FR of 1.14 (Paldau: FR of 0.78). Furthermore, in Paldau, 69% of the landslides were located in continuously forest-covered areas (28.7% for LULC change pattern) and in Waidhofen 20.7% (27.5% for LULC change pattern), respectively. In both study areas, landslides were least frequent on present-day cropland (Paldau: FR 0.05, Waidhofen: 0.31).

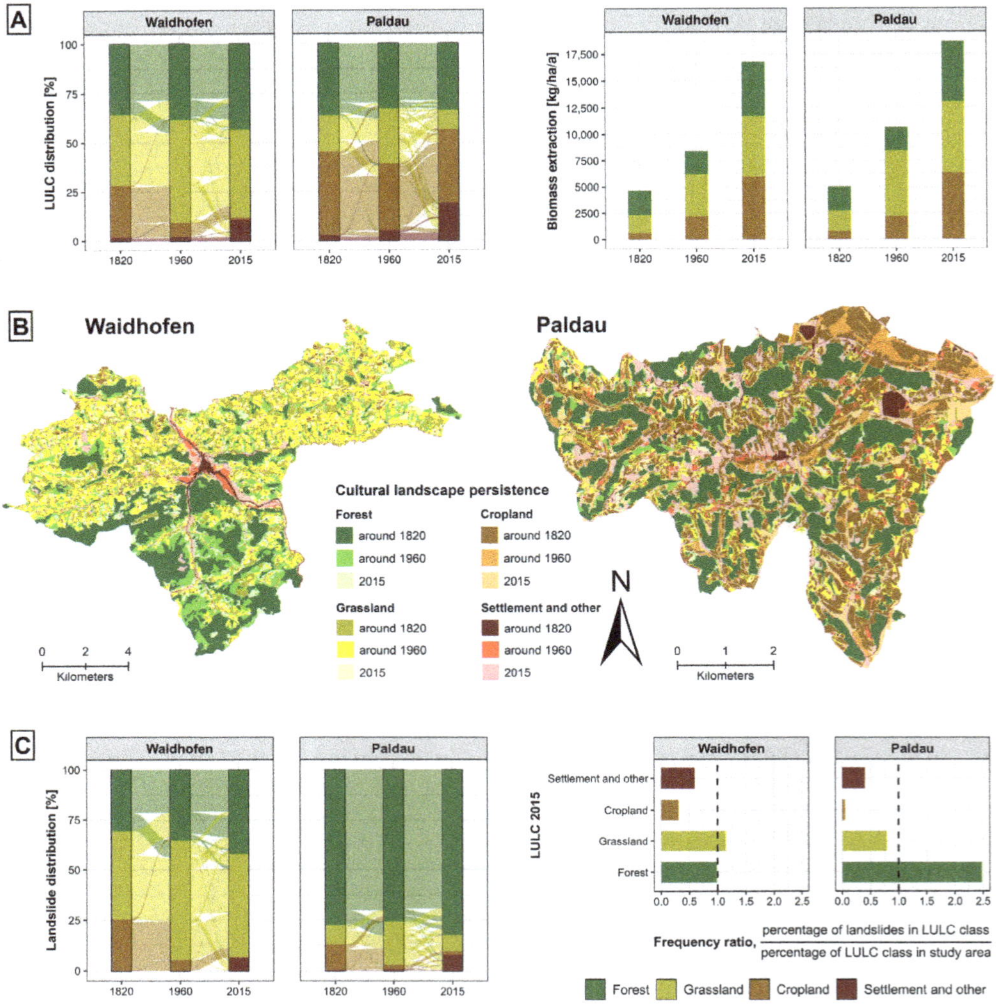

Figure 2. LULC change in the study areas for different time cuts. (**A**): Changes of LULC patterns (in %, left) and changes of biomass extraction (in kg/ha/a, right). (**B**): Cultural landscape persistence maps showing persistent historical structures in present-day LULC. (**C**): Landslide distribution for different LULC patterns (in %) and their changes (left), and landslide frequency ratios for the LULC pattern in 2015 (right).

3.2. *LULC Legacy Effects on Landslide Occurrence*

3.2.1. Model Performance and Transferability

We applied SpCV with the AUROC as performance measurement to gain information on the model's capabilities to discriminate landslides from non-landslides observations. The performance assessment showed distinct differences between the study areas (Figure 3, Table A4 in the Appendix B).

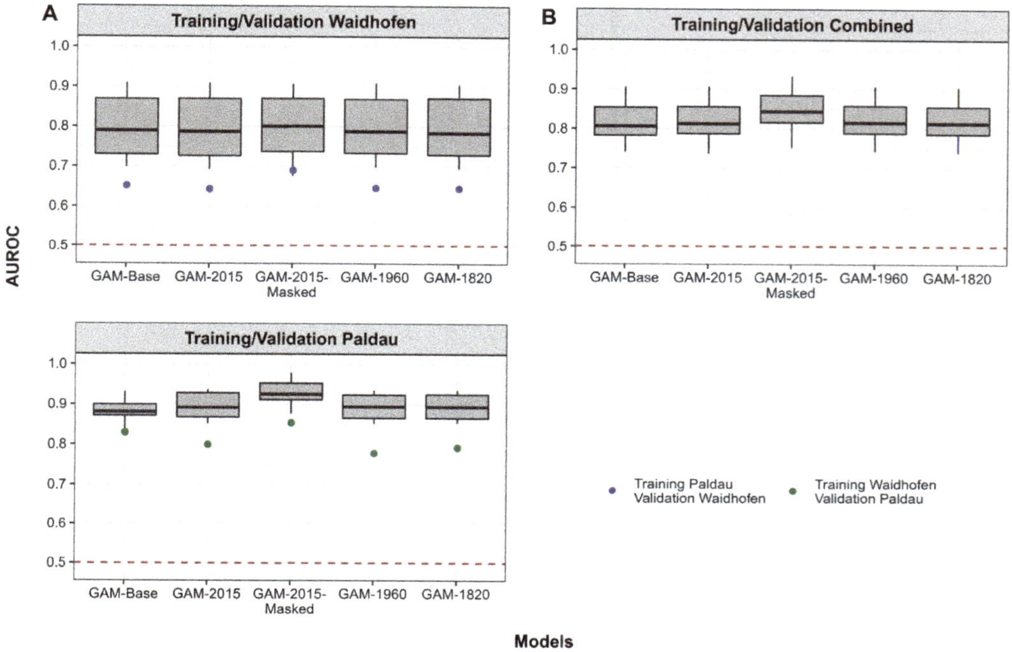

Figure 3. Model performance in SpCV and transferability assessment. (**A**): Study-area specific SpCV and transferability estimates. (**B**): SpCV estimates using the combined data from both municipalities.

For Waidhofen the median AUROC (mAUROC) in SpCV ranged between 0.78 for GAM-1820 and 0.80 for GAM-2015-Masked, i.e., acceptable to excellent discrimination capabilities. In contrast, in Paldau, GAM-Base had the lowest mAUROC of 0.88 and GAM-2015-Masked the highest mAUROC of 0.93, i.e., excellent to outstanding discrimination. AUROC estimates for Waidhofen were substantially more variable than for Paldau (interquartile ranges, IQR, 0.13 to 0.14 versus 0.03 to 0.06). Model transfer to the other study area resulted in a strong drop in model performance, with a decrease of 0.11–0.14 for model transfer to Waidhofen and 0.05–0.12 in Paldau. As in SpCV, GAM-2015-Masked performed best (Waidhofen: 0.69, Paldau: 0.85), followed by GAM-Base (Waidhofen: 0.65, Paldau: 0.83) in both study areas.

In the performance assessments for the combined data, the mAUROC and IQR values of the combined validation data fall in between the estimates of Paldau and Waidhofen (Figure 3B, Table A4 in the Appendix B). GAM-Base had the lowest mAUROC of 0.81 and GAM-2015-Masked the highest mAUROC of 0.84.

Regarding the effect of LULC legacy predictors on the performance, for Paldau the performance estimates were marginally, but yet significantly higher with the inclusion of legacy information (order of significant "<" and non-significant "=" differences in AUROCs: GAM-Base < GAM-1960 < GAM-1820 = GAM-2015 < GAM-2015-Masked, Table A5 in the Appendix B). However, for Waidhofen, such a tendency was not identifiable (GAM-2015 < GAM-1820 = GAM-Base < GAM-1960 < GAM-2015-Masked, Table A5 in the Appendix B). Furthermore, for both study areas the landslide model excluding observations located in continuously forest-covered areas (i.e., GAM-2015-Masked) showed the highest estimates in SpCV (combined and non-combined data) and transferability assessment.

3.2.2. Variable Importance

The assessment of variable importance showed differences in the variable ranking of the study areas, although slope angle was shared as a top-tier predictor (Figure 4, Table A6 in the Appendix B). In Waidhofen, the top three ranks were identical across all settings: 1. slope angle (mDD 7.45–8.63%), 2. lithology (mDD 3.72–6.38%), and 3. plan curvature (mDD 2.09–2.83%). In Paldau and across all settings, four different variables occurred in the top three ranks: slope angle (mDD 4.38–7.68%), profile curvature (mDD 3.05–6.37%), slope aspect (S-N, mDD 3.94%), and a LULC legacy variable (LULC 2015 or LULC legacy 1820/1960, mDD 3.64–3.92%). In all models, slope angle was the most important variable, except for Paldau's GAM-2015-Masked, where profile curvature was more important (mDD 6.37%).

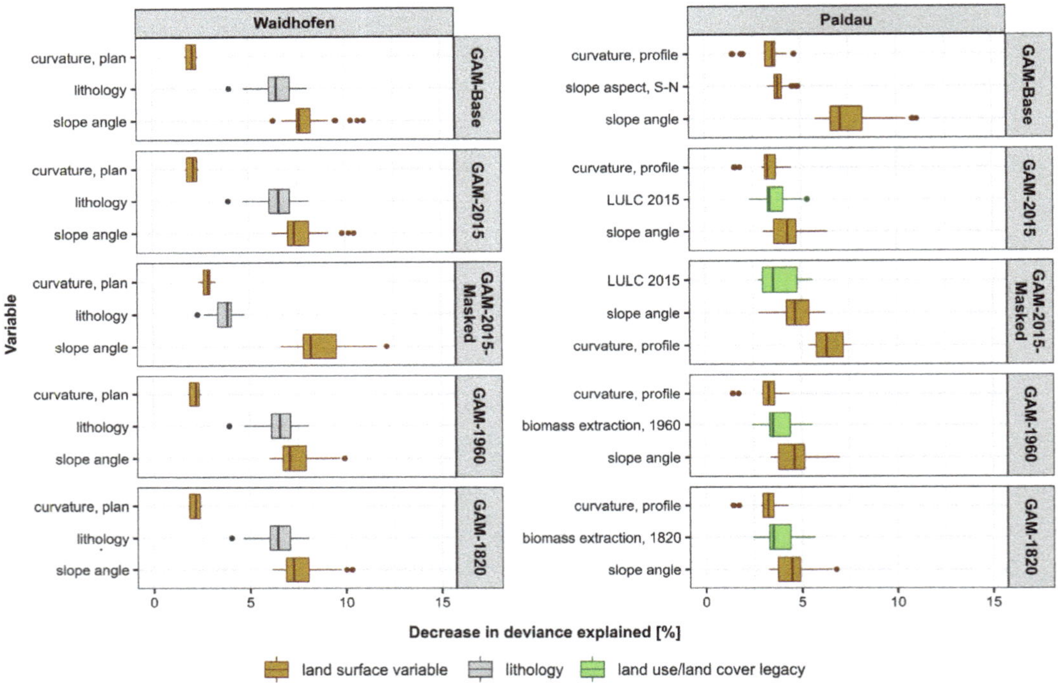

Figure 4. The three most important variables in terms of decrease in deviance explained [%]. Refer to Table A6 in the Appendix B for an overview of variable importance for all predictor variables.

The top-ranked variable was always a land surface variable, specifically slope angle and in one setting profile curvature (Table A6 in the Appendix B). Soil variables were less important with a highest rank of 11 with 0.62% mDD in Waidhofen (hydraulic conductivity in GAM-2015-Masked), and of 7 with 1.3% mDD in Paldau (hydraulic conductivity in GAM-2015-Masked), respectively. LULC legacy variables were less important in Waidhofen (mDD 0.26–1.07%), while for Paldau, they showed a high mDD of 3.64–3.92%. Lithology, in contrast, was only important in Waidhofen (mDD 6.34–6.38%; Paldau: 0.25–0.39%).

3.2.3. Predictor-Response Relationships

In both study areas, based on the transformation function plots in logit scale, we found a general agreement on the predictor-response relationships among all model settings, except for the GAM-2015-Masked (Waidhofen: Figure A2, Paldau: Figure A3). However, some differences in the predictor-response relationships for the three most important variables could be identified (Figure 5).

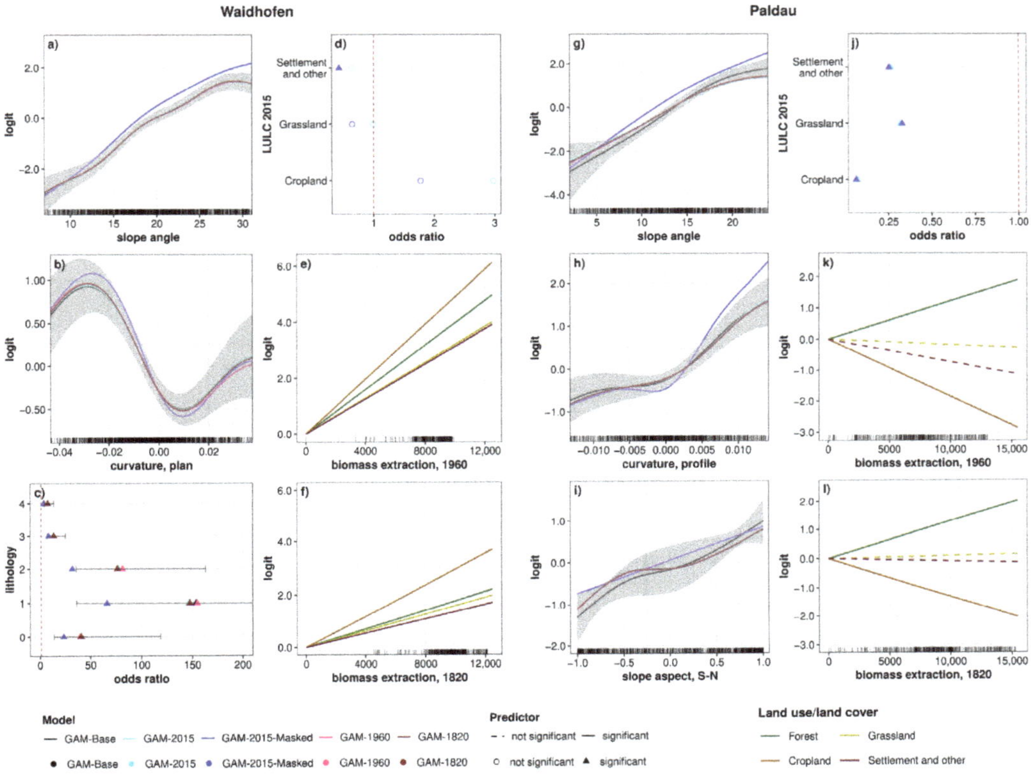

Figure 5. Comparison of predictor-response relationships of the three most important predictors (left column, (**a**–**c**) and (**g**–**i**)) and the LULC legacy variables (right column, (**d**–**f**) and (**j**–**l**)) for each study area. Grey: 95% Bayesian credible interval of GAM-Base. Lithological units in Waidhofen: Reference unit: 'Upper Austroalpine limestone', 0: talus and glacial deposits, 1: Inneralpine Neogene, 2: Klippen zone, 3: flysch zone, 4: Upper Austroalpine marls. Note: the y axes are plot-dependent, and the x axes of non-parametric transformation functions are limited to the 5th to 95th percentile range.

In both study areas, the chance of landslide occurrence was higher on steeper slopes. Among the model settings, the predictor-response relationship of GAM-Masked-2015 indicated the comparatively highest rOR compared to GAM-Base (e.g., for 15°, Paldau: 1.4, Waidhofen: 1.6). Minor differences of the other models in reference to GAM-Base were mostly at lower and upper quartiles of the value ranges (Figure 5a,g), but fell within its 95% Bayesian credible interval.

In Waidhofen, for the variable plan curvature, the chance of landslide occurrence was higher on concave than on convex surfaces, but without substantial differences between the models (Figure 5b). Regarding the variable lithology (Figure 5c), each lithological unit showed a higher chance of landslide occurrences relative to Upper Austroalpine limestone (e.g., GAM-Base: OR of 13 for flysch zone). However, among the model settings, the rORs of GAM-Masked-2015 were generally lower by a factor of 0.6, and fell partially outside the Bayesian credible interval of GAM-Base.

In Paldau, for the variable profile curvature, the chance of landslide occurrence was lower on concave than on convex surfaces (Figure 5h). Substantial differences between the models were identifiable only for GAM-2015-Masked on convex surfaces with a curvature greater than 0.005 m^{-1} (e.g., landslides were 1.3 times more likely relative to GAM-Base). Among the model settings, south-exposed slopes were less susceptible to landslides than north-exposed slopes, with only marginal differences between models (Figure 5i).

Regarding the LULC variables, in Waidhofen LULC 2015 was significant only in the GAM-2015-Masked setting for "settlement and other" with an OR of 0.43 relative to forest areas (Figure 5d). Using additionally the biomass extraction as LULC legacy interaction term (i.e., GAM-1960 and GAM-1820), the predictor was significant and showed a higher chance of landslide occurrence with higher biomass extraction for each LULC class (Figure 5e,f). Compared to forest areas, landslides were more likely on cropland and less on grassland and "settlement and other". However, using the long-term legacy in GAM-1820, the contrast relative to forest areas showed higher ORs for cropland (e.g., for 10,000 kg FW/ha/a: OR of 2.95 for GAM-1820 and of 2.53 for GAM-1960), but lower OR for grassland (OR of 0.83 for GAM-1820 and of 0.46 for GAM-1960) and "settlement and other" (OR of 0.67 for GAM-1820 and of 0.43 for GAM-1960) compared to GAM-1960.

In Paldau, for the variable LULC 2015 in GAM-2015 and GAM-2015-Masked, all LULC classes were significant terms and also showed a similar pattern with minor differences with respect to forest area (reference level; Figure 5j): For GAM-2015, the chance of landslide occurrence on cropland was only 0.07 times as high as in forest areas (OR of 0.06 for GAM-2015-Masked), 0.32 times as high on grassland (OR of 0.33 for GAM-2015-Masked), and 0.26 times as high on "settlement and other" (OR of 0.25 for GAM-2015-Masked). Including the biomass extraction as LULC legacy interaction term (i.e., GAM-1960 and GAM-1820; Figure 5k,l), for both models, the coefficients of grassland and "settlement and other" were not significant anymore, while with a higher biomass extraction the chance of landslide occurrences on cropland was lower (e.g., for 10,000 kg FW/ha/a: OR of 0.16 for GAM-1960 and of 0.27 for GAM-1820) and higher on forest areas (OR of 3.4 for GAM-1960 and of 3.7 for GAM-1820), respectively. Using the long-term legacy in GAM-1820, the ORs of the LULC classes relative to forest areas were generally lower compared to GAM-1960 (e.g., for 10,000 kg FW/ha/a on cropland: OR of 0.05 for GAM-1960 and of 0.07 for GAM-1820).

We extracted reported relationships of comparative studies, in which the landslide dates were known (for Paldau: event-based landslide inventory of southeast Styria in [73]) or the relative landslide age could be approximated (Waidhofen: remotely-sensed landslide inventory based on [21] used e.g., by [74]). For the land surface variables, we generally found a good match in the shape of the predictor-response relationships in both study areas (Figure 6), with the exception of slope aspect (S-N) in Paldau. Contrary to our results, in Knevels et al. [73] the predictor slope aspect was less influential with south-exposed slopes being slightly more susceptible to landsliding than north-exposed slopes.

Regarding the LULC variables, for Waidhofen, Steger et al. [74] reported that landslides were 1.7 to 2.0 times more likely on arable land or pastures compared to forests, while in our study, LULC was only significant when biomass extraction was included (i.e., GAM-1960 and GAM-1820). For southeast Styria, Knevels et al. [73] found non-forest areas more than three times as susceptible to landsliding as forest areas (OR > 3.6). Yet, in Paldau's GAM-2015 and GAM-2015-Masked, we found contrasting results (OR < 0.33 outside forest).

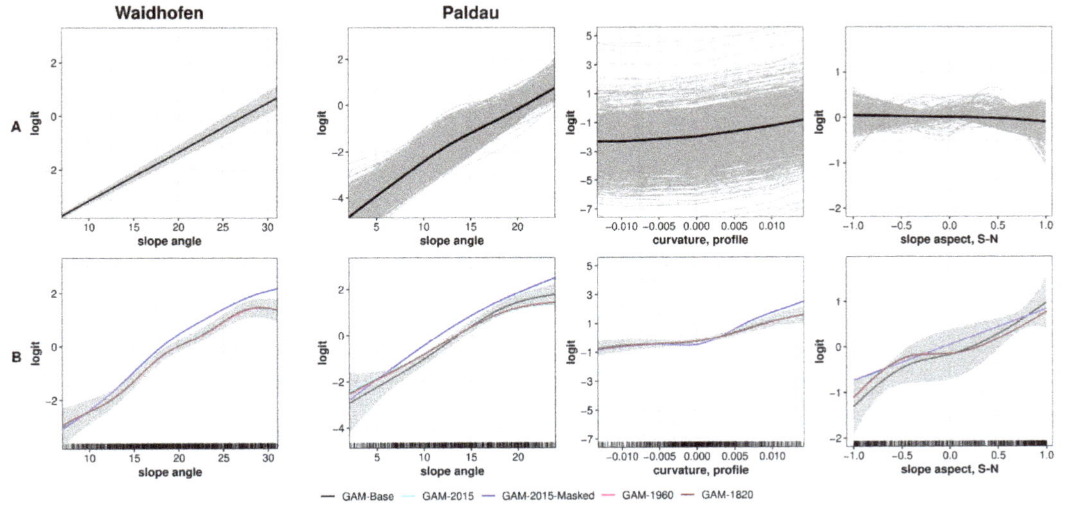

Figure 6. Comparison of predictor-response relationships between event-based landslide and historical LiDAR-derived landslide inventories. (**A**): Event-based inventory of Waidhofen (left, Table 1 in [74]) and southeast Styria (right, [73]). (**B**): Historical LiDAR-derived inventory of Waidhofen (left) and Paldau (right). Note: the logit scale is variable-dependent, the intercept for slope angle in Waidhofen was artificially set, and the x axes are limited to the 5th to 95th percentile range.

4. Discussion

4.1. Initial Objective: Effect of LULC Legacy on Modeling and Biases

We were able to successfully link the historical LULC characteristics covering a period of almost 200 years in two Austrian municipalities to landslide occurrences identified in an airborne LiDAR-derived HRDTM. The established landslide susceptibility models showed performances with acceptable to outstanding discrimination capabilities, confirming the proposed modeling approach. Landslide models including LULC legacy predictors had significantly higher (for Paldau) or at least equal (for Waidhofen) performances compared to a reference model without LULC (GAM-Base, Table A5 in the Appendix B), demonstrating the potential of LULC legacies for explaining landslide susceptibility today.

The use of LULC legacies may improve the understanding of LULC dynamics and landslide occurrences. The LULC legacy categories were generally more susceptible to landsliding with a higher biomass extraction (excluding cropland in Paldau; Figure 5e,f,k,l). In particular, present-day forested areas previously used for agricultural activities were more prone to landsliding than continuously forested areas, confirming findings for less extensive historical time periods by Beguería [9] and Persichillo et al. [11]. Thus, the effect of LULC legacies on landslide occurrences might be a good explanation of contradictory empirical observations such as landslide events in forests [23]. Additionally, for Waidhofen, we conclude that forest areas are more susceptible to landsliding than grassland, after accounting for biomass extraction. While Goetz et al. [7] reported that forest harvesting may temporally lead to open or semi-open forest types and thus reduced slope stability due to different soil hydrological and mechanical conditions (e.g., reduced rainfall interception and root cohesion), Tasser et al. [76] found managed grassland to be significantly less erodible than abandoned areas. However, we acknowledge that potentially landslide mitigating land management strategies are not incorporated in the analysis, and that the difference between forest and grassland may also be related to inventory biases.

Inventory biases are a known issue in landslide modeling [25,26]. We suggest that the identified substantial difference in the predictor-response relationship for LULC compared to other studies, is due to the underlying airborne LiDAR-derived historical landslide inventory. In particular, we assume that in Paldau, the high number of landslides in continuously forest-covered areas (69%, FR of 2.5 in 2015) can be explained by an inventory bias, which ultimately produces biased predictor-response relationships (OR of 0.33 for outside forest relative to forest areas vs. OR of >3.6 in Knevels et al. [73]). A bias was also detectable in Waidhofen's inventory, albeit weaker (about 20% of the landslides in continuously forested-covered areas). This is consistent with the reported drawbacks of such landslide inventories [26,27]. Unreflecting model interpretation might thus lead to the conclusion that forest areas are more susceptible for landslides than other LULC categories (i.e., contrasting the findings of [4,5,13,73]). We assume that the inventory bias results from the fact that landslide traces on agricultural land are quickly "tilled away" (Figure 1D), and that landslides affecting the built environment are removed during reconstruction, while in forests, they are preserved for an extended period of time (see Figure 12 in [28]).

For bias mitigation, we explored three approaches. Following the recommendations of Petschko et al. [27], (i) dropping the bias-describing predictor from modeling (i.e., LULC; GAM-Base) had only minor effects on the predictor-response relationships compared to the other models (excluding GAM-2015-Masked). We assume that confounding effects may explain these minor differences (e.g., slope angle: forest area more likely located on steeper slopes [26]). Besides, (ii) excluding inventory-biasing observations (i.e., continuously forest-covered areas; GAM-2015-Masked) led to substantial differences of predictor-response relationships for some variables (e.g., slope angle in both study areas, profile curvature in Paldau, lithology in Waidhofen, Figure 5), yet the OR of the LULC categories in reference to forest were similar to GAM-2015. However, since the relationships are in general agreement with other research results [73,74] (Figure 6), further research might help to evaluate and identify the "true" relationship. Moreover, reducing the models' training data may not always be suitable (e.g., sparse inventories), and might even decrease model performances [77,78]. (iii) Using LULC legacy information (i.e., biomass extraction, GAM-1820 and GAM-1960), the predictor-response relationships matched GAM-Base except for the minor differences mentioned above, and we could reduce the bias of a higher landslide chance on forest area relative to other LULC categories. Additionally, we identified a tendency towards relatively lower OR in forests using long-term legacies. Albeit the landslide inventory bias still had an effect on the landslide models for all approaches, for studies based on airborne LiDAR-derived landslide inventories we recommend to exclude bias-describing predictors (e.g., LULC) in the first place, or to enrich the feature space with a bias-reducing information predictor (e.g., LULC legacies).

Moreover, when considering LULC as a predictor, it should be kept in mind that each study area has its unique legacy. The transferability assessment showed a poorer transferability of landslide models that were fit only in one study area (AUROC loss up to 0.14, Figure 3). While the transformation functions of slope angle, plan curvature and the LULC legacy category forest matched between study areas, for other important variables they were partly (e.g., concave profile curvature) or even completely opposed (e.g., slope aspect, S-N and other LULC legacy categories; Figures A2 and A3). Moreover, landslide models using the combined data did not improve performances compared to both area-specific models. However, for event-based landslide modeling with especially sparse inventories, the opposite may be true (see Figure 4 in [73]).

Regarding the variable importance assessment, the results are in agreement with findings in other studies despite different landslide inventories: For Waidhofen, Steger et al. [74] also identified the slope angle as the most influential predictor, while present-day land cover ranked last. In the landslide model of Knevels et al. [73] fitted in southeast Styria (where Paldau is located), the present-day LULC (categorized into forest classes) and slope angle ranked second and third, respectively. We conjecture that the lower influence and different shape of the slope aspect relationship (Figure 5) could be due to effects such as pre- or post-failure HRDTM or the specific rainfall event in the study of [73].

4.2. Study Data: Challenges and Requirements

We encourage the use of long-term LULC legacy data in landslide studies. The presented LULC legacy predictor is not only a simple and comprehensible way to highlight historical LULC dynamics but may also support spatial planning in preventive disaster reduction regarding future LULC changes. With historical cartographic documents of several countries (e.g., Habsburg, Franciscan or Napoleonic cadastral map) [79] and information on long-term biomass extraction [80,81] becoming increasingly available in a pre-processed form, there is a real opportunity to further explore LULC legacies in future studies.

For the creation of LULC legacy data, we used data sources of different spatial, temporal and thematic quality and resolution (Table 1). For the time cut of 1960, the available greyscale aerial photographs for the municipalities were nine years apart (Paldau 1953 and Waidhofen 1962). Thus, LULC changes in the context of the "economic miracle" (e.g., mechanized agriculture, urban growth and sprawl) beginning in 1955 may not yet be present in Paldau [82]. Also, the interpretation of the greyscale aerial photographs of low image quality was particularly challenging (especially grassland vs. cropland). Additionally, the thematic resolution of the data sources used for the biomass extraction (i.e., wood and agricultural yields) was very heterogeneous in terms of the aggregated spatial statistical reference unit (e.g., the data availability for wood yields was for time cut 1960 on a national scale and for 2015 on for federal states). Nevertheless, we suggest that the data still allows conclusions to be drawn about long-term trends [43,45,81]. We acknowledge that the digitization of LULC legacy data was a time-consuming process and historical imagery at an appropriate scale may not be available everywhere.

The landslide inventories were separately created for each study area by local experts. There is evidence that landslide inventories are not only incomplete to an unknown degree [24], but also that landslide inventories compiled by different experts may have positional mismatches up to 70% [23]. Therefore, in this study, consistent rules were agreed upon to ensure or at least strive for a homogeneous inventory quality (e.g., uniform digitization scale, order of digitization: first landslide scarp, then body, etc.). To overcome the drawbacks of the inventory (e.g., bias in forest areas) and to enhance the explanatory power of the LULC legacy variables in modeling landslides, we encourage to collect and analyze additional historical aerial photographs, orthophotos or satellite images in order to estimate landslide ages [4,11,27].

5. Conclusions and Outlook

In this study, LULC and legacies of its change could successfully improve the explanation of landslide distribution as we identified the LULC variable as a meaningful predictor in landslide susceptibility modeling. Higher biomass extraction resulted in higher landslide susceptibility (excluding cropland in Paldau), explaining different risk levels in areas with the same present-day LULC. Furthermore, we could confirm that airborne LiDAR-derived inventories may be biased towards currently forested areas (e.g., high landslide density in Paldau with a FR of 2.5). Using long-term LULC legacy variables accounting for changes for almost 200 years, we could successfully reduce the effects of LULC-related inventory bias. However, without any information on the failure date or at least an approximated time slice, the additional landslides preserved in forest areas may still lead to an unknown bias in the model, and thus lead to potentially contradictory relationships between landslide occurrences and LULC legacies (e.g., OR < 0.33 outside forest relative to forest areas in Paldau). Other bias-avoiding strategies such as removing inventory-biasing observation led to improved model performances, but also to different predictor-response relationships. Thus, the implementation and interpretation of LULC as a bias-describing predictor must be carefully considered. We highly encourage further research using event-based landslide inventories with known landslide ages or age ranges to avoid such biases.

The construction of historical LULC datasets is relevant for future use both from methodological and empirical perspectives. Methodologically, the approach chosen here can be used in future analyses in cases where both geographic and statistical information on LULC are available and can be combined. Empirically, the dataset established may be used for future analyses of different issues related to long-term LULC change, in particular those related to sustainable LULC intensification (e.g., cultural landscape change [45]). With the publication of the historical LULC legacy dataset [49], we highly encourage future research with this data and a replication of the analysis, especially once a reliable event-based landslide inventory for the study areas is available.

Supplementary Materials: The following are available online at https://www.mdpi.com/2073-445X/10/9/954/s1: Landslide inventories and landslide susceptibility models of both study areas.

Author Contributions: Conceptualization, R.K., A.B. and H.P. (Helene Petschko); data curation, T.L., H.P. (Herwig Proske), G.H., R.K., C.P. and S.G. formal analysis, R.K.; funding acquisition, P.L., S.G. and H.P. (Helene Petschko); methodology, R.K., A.B. and H.P. (Helene Petschko); supervision, H.P. (Helene Petschko) and A.B.; writing—original draft, R.K.; writing—review & editing, R.K., A.B., S.G., G.H., T.L., P.L., C.P., H.P. (Herwig Proske) and H.P. (Helene Petschko). All authors have read and agreed to the published version of the manuscript.

Funding: This research was conducted within the Integrating Land use Legacies in Landslide Risk Assessment to support Spatial Planning (ILLAS) project funded by the Austrian Climate Research Program (ACRP), grant number KR16AC0K13226. The contribution of S. Gingrich was also financially supported by the European Research Council (ERC-2017-StG 757995 HEFT). We acknowledge support by the German Research Foundation and the Open Access Publication Fund of the Thueringer Universitaets- und Landesbibliothek Jena Projekt-Nr. 433052568.

Institutional Review Board Statement: Not applicable.

Informed Consent Statement: Not applicable.

Data Availability Statement: The long-term spatially explicit information on land use 1830–1960–2015 are available at https://doi.org/10.5281/zenodo.4896571, accessed on 30 June 2021. The landslide inventories and susceptibility models of both study areas are accessible through the Supplementary Materials.

Acknowledgments: We are grateful to the GIS department of the Styrian government and the Provincial Government of Lower Austria for providing the high-resolution digital elevation models, orthophotos, and the datasets of hydrologic and hydropedologic characteristics. Furthermore, we thank the Geological Survey of Austria and the Styrian GIS department for providing the geological base maps.

Conflicts of Interest: The authors declare no conflict of interest. The funders had no role in the design of the study; in the collection, analyses, or interpretation of data; in the writing of the manuscript, or in the decision to publish the results.

Appendix A. Descriptive Summary of Input Data

Table A1. Sources of land surface data.

Study Area	Source Holder	Resolution
	airborne LiDAR-based high-resolution digital terrain model	
Waidhofen	provincial government of Lower Austria	1 m × 1 m, acquisition year: 2014
Paldau	GIS department of the Styrian government	1 m × 1 m, acquisition year: 2009
	hydrologic and hydropedologic parameters	
Waidhofen Paldau	Austrian Research Centre for Forests	50 m × 50 m, year: 2014 100 m × 100 m, year 2017
	geological basemaps	
Waidhofen	Geological Survey of Austria	1:50,000
Paldau	GIS department of the Styrian government	1:50,000

For downscaling to the target resolution of 10 m × 10 m, bilinear interpolation was applied.

Table A2. Overview of the classification and recording of land use classes according to time cut.

land Use/Land Cover Category	Time Cut 1820 LULC Types in the Franciscan Cadastre *	Time Cut 2015 IACS, Orthophotos
forest (including forest pasture)	Hardwood forests, Coniferous forests, Mixed forests, Chestnut forests, Meadows with fruit trees	All forest types digitized from orthophotos
grassland	Dry meadows, wet meadows, pastures, community pastures, shrubs	IACS agricultural parcel: Grassland, alpine pastures, pasture
cropland	Orchards, vegetable gardens, vineyards, arable land (with fruit trees, trees and vines)	IACS agricultural parcel: arable land
settlement and other	Marshes, lakes, ponds, rivers and streams, wastelands and bare rocks, buildings (all types), trails (all types)	Remaining area, which includes, e.g., buildings, impervious surfaces, water bodies, excavation pits and quarries, urban green, near-natural areas

* LULC types not listed here are not present in the study area; LULC classes of the 1960 time cut were digitized in aerial photographs, modified from Knevels et al. [45].

Table A3. Summary of the landslide inventory data.

		Waidhofen			Paldau		
Number	landslides		621			418	
	scarps		829			469	
	bodies		663			348	
	samples		974			559	
Total Area [m^2] (%)	landslides		6,976,638 (5.31)			1,621,250 (4.14)	
		min	*mean*	*max*	*min*	*mean*	*max*
Area [m^2]	landslides	113	11,235	1,163,088	30	3879	206,842
	scarps	2	517	79,640	12	140	1518
	bodies	52	9876	1,083,448	24	3949	206,842
Perimeter [m]	landslides	44	451	7250	26	163	1821
	scarps	7	101	3013	18	71	373
	bodies	29	297	4548	28	168	1821

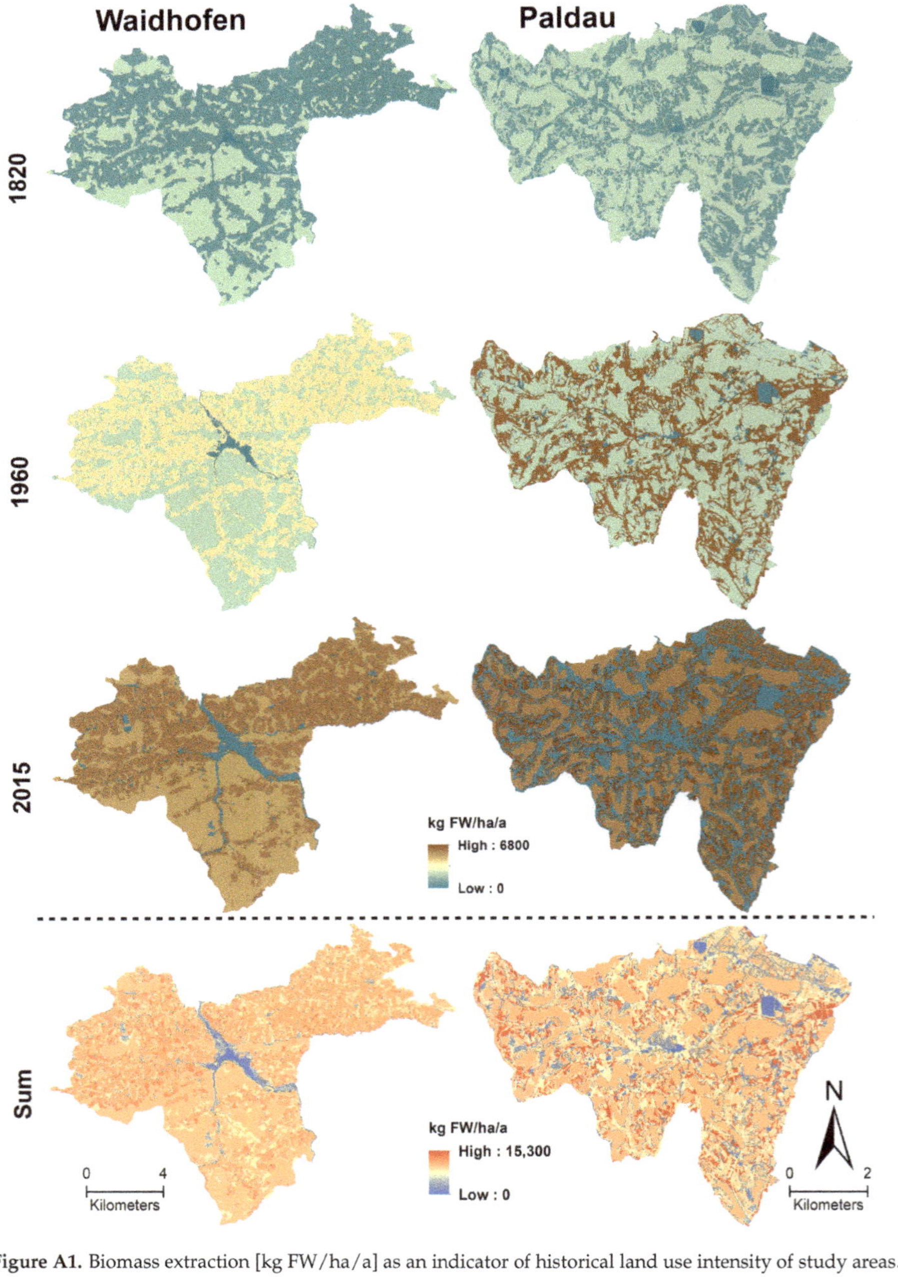

Figure A1. Biomass extraction [kg FW/ha/a] as an indicator of historical land use intensity of study areas.

Appendix B. Summary of Model Assessment Results

Table A4. Results of the model assessments within the study area with SpCV, and assessment of model transferability between both areas.

Model	$\tilde{x}$	Min	Max	IQR	Transfer
		A: Waidhofen			
GAM-Base	0.79	0.7	0.91	0.14	0.65
GAM-2015	0.79	0.69	0.91	0.14	0.64
GAM-2015-Masked	0.8	0.68	0.91	0.13	0.69
GAM-1960	0.79	0.7	0.91	0.14	0.65
GAM-1820	0.78	0.7	0.91	0.14	0.65
		B: Paldau			
GAM-Base	0.88	0.83	0.93	0.03	0.83
GAM-2015	0.89	0.85	0.94	0.06	0.8
GAM-2015-Masked	0.93	0.88	0.98	0.04	0.85
GAM-1960	0.89	0.85	0.94	0.06	0.78
GAM-1820	0.89	0.85	0.94	0.06	0.79
		C: Combined			
GAM-Base	0.81	0.74	0.91	0.07	
GAM-2015	0.81	0.74	0.91	0.07	
GAM-2015-Masked	0.84	0.75	0.93	0.07	
GAM-1960	0.82	0.74	0.91	0.07	
GAM-1820	0.81	0.74	0.91	0.07	

AUROC Statistic: median ($\tilde{x}$), minimum (**Min**), maximum (**Max**), interquartile range (**IQR**), transferability estimate (**Transfer**).

Table A5. Wilcoxon signed-rank tests for AUROC performance differences within each study area estimated with SpCV.

Model	mAUROC	N *	Z	*p*-Values	*r*
		A: Waidhofen			
GAM-2015	0.79				
<GAM-1820	0.78	36	2.7	0.01	0.45
=GAM-Base	0.79	35	0.93	0.18	0.16
<GAM-1960	0.79	36	4.24	<0.001	0.72
<GAM-2015-Masked	0.80	36	2.24	0.02	0.37
		B: Paldau			
GAM-Base	0.88				
<GAM-1960	0.89	66	5.01	<0.001	0.62
<GAM-1820	0.89	66	4.35	<0.001	0.54
=GAM-2015	0.89	66	1.54	0.06	0.19
< GAM-2015-Masked	0.93	66	7.03	<0.001	0.87

mAUROC: median AUROC of SpCV; **N:** Number of observations per group; * tied observations were removed; **Z**: Z score;alternative hypothesis: *greater*, $\alpha = 0.05$; <: Differences significantly greater from previous model, =: No significant difference to the previous model.

Table A6. Variable importance measured as mean decrease in deviance explained (%), rank of variable in parentheses.

Variable	Study Area	GAM-Base	GAM-2015	GAM-2015-Masked	GAM-1960	GAM-1820
land surface variable						
convergence index, 100 m	Wh	1.23 (5)	1.27 (5)	1.3 (6)	1.15 (5)	1.17 (5)
	P	0.91 (7)	0.97 (8)	0.71 (11)	0.96 (8)	0.98 (7)
convergence index, 500 m	Wh	0.87 (8)	0.86 (9)	0.99 (9)	0.83 (10)	0.86 (9)
	P	0.93 (6)	1.16 (6)	3.25 (4)	1.2 (6)	1.18 (6)
curvature, plan	Wh	2.09 (3)	2.1 (3)	2.83 (3)	2.17 (3)	2.17 (3)
	P	2.38 (4)	1.43 (5)	0.84 (10)	1.46 (5)	1.49 (5)
curvature, profile	Wh	1.45 (4)	1.58 (4)	1.78 (4)	1.65 (4)	1.57 (4)
	P	3.32 (3)	3.2 (3)	6.37 (1)	3.08 (3)	3.05 (3)
flow accumulation	Wh	0.42 (13)	0.43 (12)	0.48 (14)	0.5 (13)	0.46 (12)
	P	0.07 (15)	0.2 (16)	0.46 (13)	0.22 (16)	0.21 (16)
normalized height	Wh	0.02 (15)	0.02 (16)	0.05 (16)	0.04 (16)	0.03 (16)
	P	0.95 (5)	0.95 (9)	3.15 (5)	0.93 (9)	0.93 (9)
slope angle	Wh	8.01 (1)	7.7 (1)	8.63 (1)	7.45 (1)	7.61 (1)
	P	7.68 (1)	4.38 (1)	4.62 (2)	4.77 (1)	4.64 (1)
slope angle, catchment area	Wh	1.11 (7)	1.15 (6)	1.1 (8)	1.07 (7)	1.15 (6)
	P	0.55 (11)	0.47 (12)	0.2 (15)	0.44 (12)	0.43 (12)
slope aspect, S-N	Wh	1.12 (6)	1.07 (7)	1.11 (7)	1.06 (8)	1.12 (7)
	P	3.94 (2)	2.3 (4)	1.25 (8)	2.36 (4)	2.33 (4)
slope aspect, W-E	Wh	0.61 (11)	0.58 (11)	0.84 (10)	0.52 (12)	0.56 (11)
	P	0.32 (13)	0.29 (14)	1.68 (6)	0.27 (15)	0.27 (15)
TPI	Wh	0.86 (9)	0.89 (8)	1.64 (5)	0.89 (9)	0.91 (8)
	P	0.7 (9)	0.84 (11)	0.14 (16)	0.84 (11)	0.84 (11)
SWI	Wh	0.66 (10)	0.66 (10)	0.39 (15)	0.68 (11)	0.67 (10)
	P	0.66 (10)	0.39 (13)	0.95 (9)	0.35 (13)	0.35 (13)
soil						
total pore volume	Wh	0.35 (14)	0.31 (14)	0.51 (13)	0.29 (15)	0.31 (15)
	P	0.47 (12)	0.86 (10)	0.68 (12)	0.87 (10)	0.87 (10)
hydraulic conductivity	Wh	0.44 (12)	0.41 (13)	0.62 (11)	0.35 (14)	0.38 (14)
	P	0.9 (8)	0.98 (7)	1.3 (7)	0.98 (7)	0.97 (8)
lithology						
lithology/geology	Wh	6.38 (2)	6.37 (2)	3.72 (2)	6.37 (2)	6.34 (2)
	P	0.25 (14)	0.27 (15)	0.39 (14)	0.27 (14)	0.28 (14)
land use/land cover legacy						
LULC 2015	Wh		0.26 (15)	0.6 (12)		
	P		3.64 (2)	3.92 (3)		
biomass extraction, 1960 *	Wh				1.07 (6)	
	P				3.81 (2)	
biomass extraction, 1820 *	Wh					0.43 (13)
	P					3.85 (2)

* Biomass extraction in interaction with LULC 2015; Wh: Waidhofen, P: Paldau.

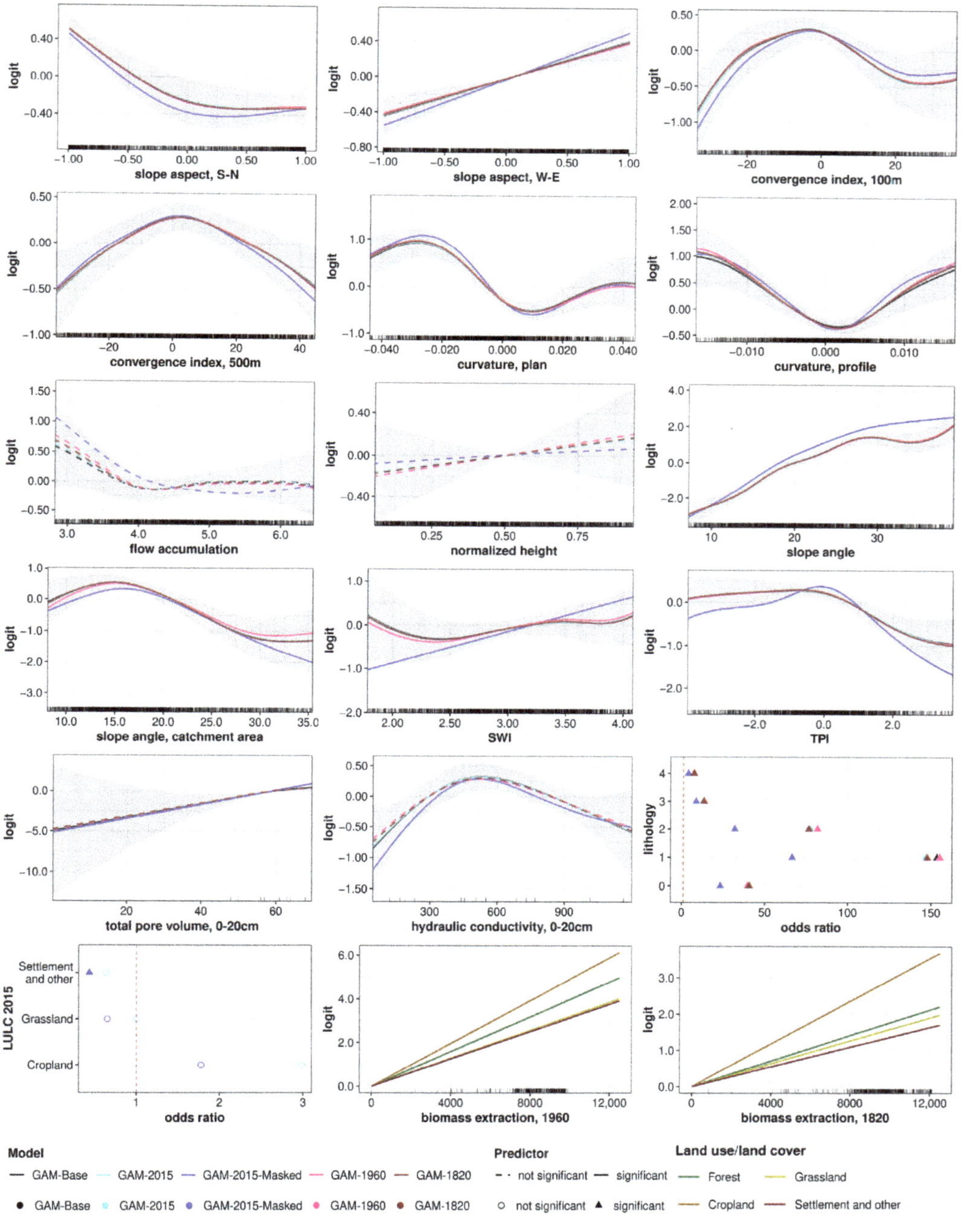

Figure A2. Predictor-response relationships of landslide models in Waidhofen. Grey: 95% Bayesian credible interval of GAM-Base. Reference level of LULC 2015: 'Forest'; Lithological units: Reference: 'Upper Austroalpine limestone', 0: talus and glacial deposits, 1: Inneralpine Neogene, 2: Klippen zone, 3: flysch zone, 4: Upper Austroalpine marls. Note: the y axes are plot-dependent, and the x axes of non-parametric transformation functions are limited to the 5th to 95th percentile range.

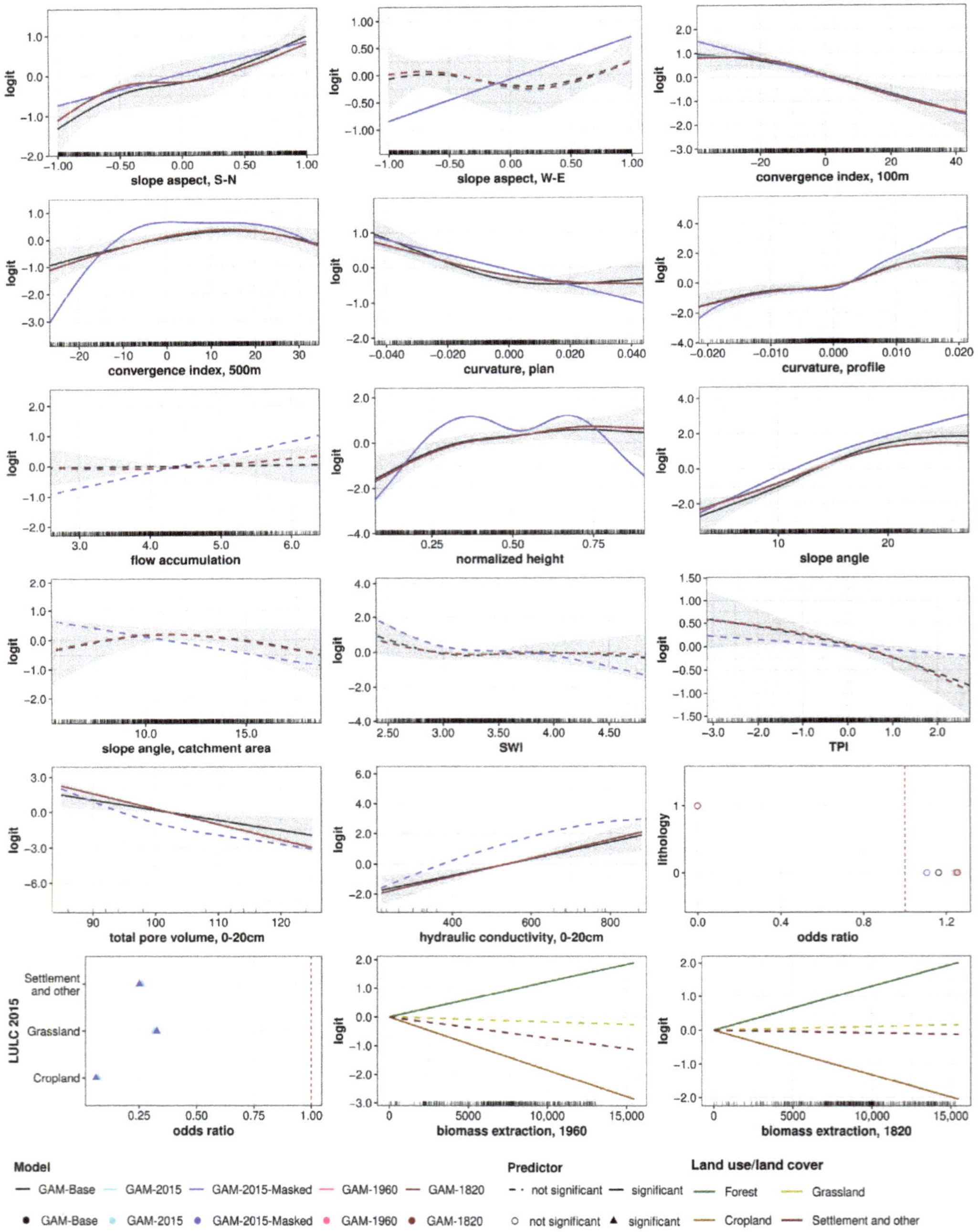

Figure A3. Predictor-response relationships of landslide models in Paldau. Grey: 95% Bayesian credible interval of GAM-Base. Reference level of LULC 2015: 'Forest'; Lithological units: Reference: 'Neogene formations with coarse-grained layers', 0: 'Neogene formations dominated by fine-grained sediments', 1: 'pre-Würmian Pleistocene formations'. Note: the y axes are plot-dependent, and the x axes of non-parametric transformation functions are limited to the 5th to 95th percentile range.

References

1. Crozier, M.J.; Glade, T. Landslide Hazard and Risk: Issues, Concepts and Approach. In *Landslide Hazard and Risk*; Glade, T., Anderson, M., Crozier, M.J., Eds.; John Wiley & Sons: Chichester, UK, 2005; pp. 1–40.
2. Gariano, S.L.; Guzzetti, F. Landslides in a changing climate. *Earth-Sci. Rev.* **2016**, *162*, 227–252. [CrossRef]
3. Schweigl, J.; Hervás, J. *Landslide Mapping in Austria*; European Commission, Joint Research Centre: Luxembourg; Ispra, Italy, 2009. [CrossRef]
4. Reichenbach, P.; Busca, C.; Mondini, A.C.; Rossi, M. The Influence of Land Use Change on Landslide Susceptibility Zonation: The Briga Catchment Test Site (Messina, Italy). *Environ. Manag.* **2014**, *54*, 1372–1384. [CrossRef] [PubMed]
5. Papathoma-Köhle, M.; Glade, T. The role of vegetation cover change for landslide hazard and risk. In *The Role of Ecosystems in Disaster Risk Reduction*; UNU-Press: Tokyo, Japan, 2013; pp. 293–320.
6. Brenning, A.; Schwinn, M.; Ruiz-Páez, A.P.; Muenchow, J. Landslide susceptibility near highways is increased by 1 order of magnitude in the Andes of southern Ecuador, Loja province. *Nat. Hazards Earth Syst. Sci.* **2015**, *15*, 45–57. [CrossRef]
7. Goetz, J.; Guthrie, R.H.; Brenning, A. Forest harvesting is associated with increased landslide activity during an extreme rainstorm on Vancouver Island, Canada. *Nat. Hazards Earth Syst. Sci.* **2015**, *15*, 1311–1330. [CrossRef]
8. van Westen, C.J.; Castellanos, E.; Kuriakose, S.L. Spatial data for landslide susceptibility, hazard, and vulnerability assessment: An overview. *Eng. Geol.* **2008**, *102*, 112–131. [CrossRef]
9. Beguería, S. Changes in land cover and shallow landslide activity: A case study in the Spanish Pyrenees. *Geomorphology* **2006**, *74*, 196–206. [CrossRef]
10. Gariano, S.L.; Petrucci, O.; Rianna, G.; Santini, M.; Guzzetti, F. Impacts of past and future land changes on landslides in southern Italy. *Reg. Environ. Chang.* **2018**, *18*, 437–449. [CrossRef]
11. Persichillo, M.G.; Bordoni, M.; Meisina, C. The role of land use changes in the distribution of shallow landslides. *Sci. Total Environ.* **2017**, *574*, 924–937. [CrossRef] [PubMed]
12. Pisano, L.; Zumpano, V.; Malek, Ž.; Rosskopf, C.M.; Parise, M. Variations in the susceptibility to landslides, as a consequence of land cover changes: A look to the past, and another towards the future. *Sci. Total Environ.* **2017**, *601–602*, 1147–1159. [CrossRef]
13. Lopez-Saez, J.; Corona, C.; Eckert, N.; Stoffel, M.; Bourrier, F.; Berger, F. Impacts of land-use and land-cover changes on rockfall propagation: Insights from the Grenoble conurbation. *Sci. Total Environ.* **2016**, *547*, 345–355. [CrossRef]
14. Cuddington, K. Legacy Effects: The Persistent Impact of Ecological Interactions. *Biol. Theory* **2011**, *6*, 203–210. [CrossRef]
15. Foster, D.; Swanson, F.; Aber, J.; Burke, I.; Brokaw, N.; Tilman, D.; Knapp, A. The Importance of Land-Use Legacies to Ecology and Conservation. *BioScience* **2003**, *53*, 77–88. [CrossRef]
16. Munteanu, C.; Kuemmerle, T.; Keuler, N.S.; Müller, D.; Balázs, P.; Dobosz, M.; Griffiths, P.; Halada, L.; Kaim, D.; Király, G.; et al. Legacies of 19th century land use shape contemporary forest cover. *Glob. Environ. Chang.* **2015**, *34*, 83–94. [CrossRef]
17. Kuussaari, M.; Bommarco, R.; Heikkinen, R.K.; Helm, A.; Krauss, J.; Lindborg, R.; Öckinger, E.; Pärtel, M.; Pino, J.; Rodà, F.; et al. Extinction debt: A challenge for biodiversity conservation. *Trends Ecol. Evol.* **2009**, *24*, 564–571. [CrossRef]
18. Perring, M.P.; Frenne, P.D.; Baeten, L.; Maes, S.L.; Depauw, L.; Blondeel, H.; Carón, M.M.; Verheyen, K. Global environmental change effects on ecosystems: The importance of land-use legacies. *Glob. Chang. Biol.* **2016**, *22*, 1361–1371. [CrossRef]
19. Sonny Digital LiDAR-Terrain Models of Austria (last change 02 March 2021 16:01:16) 2016. Available online: https://data.opendataportal.at/dataset/dtm-austria (accessed on 30 June 2021).
20. Haugerud, R.; Harding, D.; Johnson, S.; Harless, J.; Weaver, C.; Sherrod, B. High-Resolution Lidar Topography of the Puget Lowland, Washington —A Bonanza for Earth Science. *GSA Today* **2003**, *13*. [CrossRef]
21. Petschko, H.; Bell, R.; Glade, T. Effectiveness of visually analyzing LiDAR DTM derivatives for earth and debris slide inventory mapping for statistical susceptibility modeling. *Landslides* **2016**, *13*, 857–872. [CrossRef]
22. Schulz, W.H. Landslides mapped using LIDAR imagery, Seattle, Washington. *US Geol. Surv. Open-File Rep.* **2004**, *1396*. [CrossRef]
23. Eeckhaut, M.V.D.; Poesen, J.; Verstraeten, G.; Vanacker, V.; Nyssen, J.; Moeyersons, J.; Beek, L.P.H.v.; Vandekerckhove, L. Use of LIDAR-derived images for mapping old landslides under forest. *Earth Surf. Process. Landf.* **2007**, *32*, 754–769. [CrossRef]
24. Malamud, B.D.; Turcotte, D.L.; Guzzetti, F.; Reichenbach, P. Landslide inventories and their statistical properties. *Earth Surf. Process. Landf.* **2004**, *29*, 687–711. [CrossRef]
25. Stanley, T.A.; Kirschbaum, D.B. Effects of inventory bias on landslide susceptibility calculations. In Proceedings of the Landslides: Putting Experience, Knowledge and Emerging Technologies into Practice; Association of Environmental & Engineering Geologists (AEG), Roanoke, VA, USA, 4–8 June 2017; pp. 794–806.
26. Steger, S.; Brenning, A.; Bell, R.; Glade, T. The influence of systematically incomplete shallow landslide inventories on statistical susceptibility models and suggestions for improvements. *Landslides* **2017**, *14*, 1767–1781. [CrossRef]
27. Petschko, H.; Bell, R.; Glade, T. Relative Age Estimation at Landslide Mapping on LiDAR Derivatives: Revealing the Applicability of Land Cover Data in Statistical Susceptibility Modelling. In *Landslide Science for a Safer Geoenvironment*; Springer Nature: Cham, Switzerland, 2014; pp. 337–343.
28. Bell, R.; Petschko, H.; Röhrs, M.; Dix, A. Assessment of landslide age, landslide persistence and human impact using airborne laser scanning digital terrain models. *Geogr. Ann. Ser. A Phys. Geogr.* **2012**, *94*, 135–156. [CrossRef]
29. van Westen, C.J.; van Asch, T.W.J.; Soeters, R. Landslide hazard and risk zonation—why is it still so difficult? *Bull. Eng. Geol. Environ.* **2006**, *65*, 167–184. [CrossRef]

30. STATISTIK AUSTRIA Ein Blick auf die Gemeinde Waidhofen an der Ybbs (30301). Österreich Besser Verstehen. Bevölkerungsentwicklung 1869–2019 [A Look at the Municipality of Waidhofen an der Ybbs (30301). Understanding Austria Better. Population Development 1869–2019] 2011. Available online: https://www.statistik.at/blickgem/G0201/g30301.pdf (accessed on 30 June 2021).
31. Wessely, G. *Geologie der Österreichischen Bundesländer-Niederösterreich [Geology of the Austrian Federal Provinces-Lower Austria]*; Verlag der Geologischen Bundesanstalt: Wien, Österreich, 2006.
32. Gross, M. Beitrag zur Lithostratigraphie des Oststeirischen Beckens (Neogen/Pannonium; Österreich). In *Stratigraphia Austriaca*; Piller, W.E., Ed.; Schriftenreihe der Erdwissenschaftlichen Kommissionen/Österreichische Akademie der Wissenschaften; Österreichische Akademie der Wissenschaften: Wien, Austria, 2003; pp. 11–62.
33. STATISTIK AUSTRIA Ein Blick auf die Gemeinde Paldau (62384). Österreich Besser Verstehen. Bevölkerungsentwicklung 1869—2019 [A Look at the Municipality of Paldau (62384). Understanding Austria Better. Population Development 1869—2019] 2011. Available online: https://www.statistik.at/blickgem/G0201/g62384.pdf (accessed on 30 June 2021).
34. Bell, R.; Glade, T.; Granica, K.; Heiss, G.; Leopold, P.; Petschko, H.; Pomaroli, G.; Proske, H.; Schweigl, J. Landslide Susceptibility Maps for Spatial Planning in Lower Austria. In *Landslide Science and Practice. Volume 1: Landslide Inventory and Susceptibility and Hazard Zoning*; Margottini, C., Canuti, P., Sassa, K., Eds.; Springer: Berlin/Heidelberg, Germany, 2013; pp. 467–472.
35. Proske, H.; Bauer, C. Methodik zur Erstellung einer Gefahrenhinweiskarte für Rutschungen in der Steiermark [Methodology of the generation of an indicative hazard map for landslides in Styria]. *Torrent Avalanche Landslide Rock Fall* **2015**, *175*, 84–95.
36. Schwenk, H.; Spendlingwimmer, R.; Salzer, F. Massenbewegungen in Niederösterreich 1953—1990 [Mass Movements in Lower Austria 1953—1990]. *Jahrb. Der Geol. Bundesanst.* **1992**, *132*, 597–660.
37. Hornich, R.; Adelwöhrer, R. Landslides in Styria in 2009. *Geomech. Tunn.* **2010**, *3*, 455–461. [CrossRef]
38. Lemenkova, P.; Glade, T.; Promper, C. Economic Assessment of Landslide Risk for the Waidhofen a.d. Ybbs Region, Alpine Foreland, Lower Austria. Protecting Society through Improved Understanding. In Proceedings of the 11th International Symposium on Landslides and the 2nd NorthAmerican Symposium on Landslides and Engineered Slopes, Banff, AB, Canada, 2–8 June 2012; pp. 279–285. [CrossRef]
39. Eder, A.; Sotier, B.; Klebinder, K.; Sturmlechner, R.; Dorner, J.; Markat, G.; Schmid, G.; Strauss, P. *Hydrologische Bodenkenndaten der Böden Niederösterreichs (HydroBodNÖ) [Data on Hydrological Soil Characteristics of Soils in Lower Austria]*; Bundesamt für Wasserwirtschaft, Institut für Kulturtechnik und Bodenwasserhaushalt; Bundesforschungszentrum für Wald, Institut für Naturgefahren: Innsbruck, Austria, 2011.
40. Klebinder, K.; Sotier, B.; Lechner, V.; Strauss, P. *Hydrologische und Hydropedologische Kenndaten. Raabgebiet und Region. Südoststeiermark [Hydrologic and Hydropedologic Characteristics. Region. of Raab and Southeast. Styria]*; Bundesforschungszentrum für Wald, Bundesamt für Wasserwirtschaft: Innsbruck, Austria, 2017; p. 30. Available online: https://wegenernet.org/downloads/Klebinder-etal_HydroBod-SOStmk-Projbericht_Jul2017.pdf (accessed on 30 June 2021).
41. Bender, O. *Analyse der Kulturlandschaftsentwicklung der Nördlichen Fränkischen Alb Anhand Eines Katasterbasierten Geoinformationssystems [Analysis of the Cultural Landscape Change of the Northern Franconian Alb Using a Cadastre-Based Geoinformation System]*; Forschungen zur Deutschen Landeskunde; Deutsche Akademie für Landeskunde: Leipzig, Germany, 2007.
42. Sandgruber, R. *Österreichische Agrarstatistik 1750—1918*; Hoffmann, A., Matis, H., Eds.; Wirtschafts- und Sozialstatistik Österreich-Ungarns; Verlag für Geschichte und Politik: Wien, Austria, 1978.
43. Gingrich, S.; Erb, K.-H.; Krausmann, F.; Gaube, V.; Haberl, H. Long-term dynamics of terrestrial carbon stocks in Austria: A comprehensive assessment of the time period from 1830 to 2000. *Reg. Environ. Chang.* **2007**, *7*, 37–47. [CrossRef]
44. Weiss, P.; Schieler, K.; Schadauer, K.; Englisch, M. *Die Kohlenstoffbilanz des Österreichischen Waldes und Betrachtungen Zum Kyoto-Protokoll*; Monographien; Umweltbundesamt: Wien, Austria, 2000.
45. Knevels, R.; Brenning, A.; Gingrich, S.; Gruber, E.; Lechner, T.; Leopold, P.; Petschko, H.; Plutzar, C. Kulturlandschaft im Wandel: Ein indikatorenbasierter Rückblick bis in das 19. Jahrhundert. Fallstudie anhand der Gemeinden Waidhofen/Ybbs und Paldau [Cultural Landscape Change: An Indicator-Based Retrospect into the 19th Century. Case Study of the Municipalities Waidhofen/Ybbs and Paldau]. *Mitt. Der Osterreichischen Geogr. Ges.* **2021**, *162*, 255–285. [CrossRef]
46. Erb, K.-H.; Haberl, H.; Jepsen, M.R.; Kuemmerle, T.; Lindner, M.; Müller, D.; Verburg, P.H.; Reenberg, A. A conceptual framework for analysing and measuring land-use intensity. *Curr. Opin. Environ. Sustain.* **2013**, *5*, 464–470. [CrossRef]
47. Mayer, A.; Haas, W.; Wiedenhofer, D. How Countries' Resource Use History Matters for Human Well-being—An Investigation of Global Patterns in Cumulative Material Flows from 1950 to 2010. *Ecol. Econ.* **2017**, *134*, 1–10. [CrossRef]
48. Matthews, H.D.; Zickfeld, K.; Knutti, R.; Allen, M.R. Focus on cumulative emissions, global carbon budgets and the implications for climate mitigation targets. *Environ. Res. Lett.* **2018**, *13*, 010201. [CrossRef]
49. Lechner, T.; Plutzar, C.; Knevels, R.; Gingrich, S. Long-Term Spatially Explicit Information on Land Use 1830–1960–2015: Case Study of the Municipalities Waidhofen/Ybbs and Paldau, Austria [Data set] 2021. 1.0.0. Available online: https://doi.org/10.5281/zenodo.4896571 (accessed on 30 June 2021).
50. Cruden, D.M.; Varnes, D.J. Landslide Types and Processes. In *Landslides Investigation and Mitigation. Transportation Research Board, US National Research Council*; Turner, A.K., Schuster, R.L., Eds.; Special Report 247; U.S. National Academy of Sciences: Washington, DC, USA, 1996; pp. 36–75.
51. Hussin, H.Y.; Zumpano, V.; Reichenbach, P.; Sterlacchini, S.; Micu, M.; van Westen, C.; Bălteanu, D. Different landslide sampling strategies in a grid-based bi-variate statistical susceptibility model. *Geomorphology* **2016**, *253*, 508–523. [CrossRef]

52. Steger, S.; Glade, T. The Challenge of "Trivial Areas" in Statistical Landslide Susceptibility Modelling. In Proceedings of the Advancing Culture of Living with Landslides, Ljubljana, Slovenia, 29 May–2 June 2017; Mikos, M., Tiwari, B., Yin, Y., Sassa, K., Eds.; Springer International Publishing: Cham, Switzerland, 2017; pp. 803–808. [CrossRef]
53. Hastie, T.; Tibshirani, R. Generalized Additive Models. *Statist. Sci.* **1986**, *1*, 297–310. [CrossRef]
54. Wood, S.N. *Generalized Additive Models: An Introduction with R*, 2nd ed.; Chapman and Hall/CRC: Boca Raton, FL, USA, 2017.
55. Brenning, A. Improved spatial analysis and prediction of landslide susceptibility: Practical recommendations. In *Proceedings of the Landslides and Engineered Slopes: Protecting Society through Improved Understanding*; Eberhardt, E., Froese, C., Turner, A.K., Leroueil, S., Eds.; CRC Press/Balkema: Banff, Canada, 2012; Volume 1, pp. 789–794.
56. R Core Team. *R: A Language and Environment for Statistical Computing*, R version 3.5.2.; R Foundation for Statistical Computing: Vienna, Austria, 2018. Available online: https://www.R-project.org/ (accessed on 30 June 2021).
57. Bischl, B.; Lang, M.; Kotthoff, L.; Schiffner, J.; Richter, J.; Studerus, E.; Casalicchio, G.; Jones, Z.M. mlr: Machine Learning in R. *J. Mach. Learn. Res.* **2016**, *17*, 1–5.
58. Conrad, O.; Bechtel, B.; Dietrich, H.; Fischer, E.; Gerlitz, L.; Wehberg, J.; Wichmann, V.; Böhner, J. System for Automated Geoscientific Analyses (SAGA) v. 2.1.4. *Geosci. Model. Dev.* **2015**, *8*, 1991–2007. [CrossRef]
59. Brenning, A.; Bangs, D.; Becker, M. RSAGA: SAGA Geoprocessing and Terrain Analysis; 2018. R Package Version 1.3.0. Available online: https://CRAN.R-project.org/package=RSAGA (accessed on 30 June 2021).
60. Tarboton, D.G.; Dash, P.; Sazib, N. *TauDEM 5.3: Guide to using the TauDEM command line functions 2015*; Utah State University: Logan, UT, USA, 2015. Available online: http://hydrology.usu.edu/taudem/taudem5/TauDEM53CommandLineGuide.pdf (accessed on 30 June 2021).
61. Koethe, R.; Lehmeier, F. *SARA—System Zur Automatischen Relief-Analyse. User Manual*, 2nd ed.; Department of Geography, University of Göttingen: Göttingen, Germany, 1996.
62. Zevenbergen, L.W.; Thorne, C.R. Quantitative Analysis of Land Surface Topography. *Earth Surf. Process. Landf.* **1987**, *12*, 47–56. [CrossRef]
63. Tarboton, D.G. A new method for the determination of flow directions and upslope areas in grid digital elevation models. *Water Resour. Res.* **1997**, *33*, 309–319. [CrossRef]
64. Dietrich, H.; Böhner, J. Cold Air Production and Flow in a Low Mountain Range Landscape in Hessia. In *SAGA–Seconds Out*; Böhner, J., Blaschke, T., Montanarella, L., Eds.; Hamburger Beiträge zur Physischen Geographie und Landschaftsökologie; Universität Hamburg, Institut für Geographie: Hamburg, Germany, 2008; Volume 19, pp. 37–48.
65. Böhner, J.; Selige, T. Spatial Prediction of Soil Attributes Using Terrain Analysis and Climate Regionalisation. *Göttinger Geogr. Abh.* **2006**, *115*, 13–27.
66. Brenning, A.; Trombotto, D. Logistic regression modeling of rock glacier and glacier distribution: Topographic and climatic controls in the semi-arid Andes. *Geomorphology* **2006**, *81*, 141–154. [CrossRef]
67. Guisan, A.; Weiss, S.B.; Weiss, A.D. GLM versus CCA spatial modeling of plant species distribution. *Plant Ecol.* **1999**, *143*, 107–122. [CrossRef]
68. Hosmer, D.W.; Lemeshow, S.; Sturdivant, R.X. Wiley Series in Probability and Statistics. In *Applied Logistic Regression*; John Wiley & Sons: Hoboken, NJ, USA, 2013.
69. Demšar, J. Statistical Comparisons of Classifiers over Multiple Data Sets. *J. Mach. Learn. Res.* **2006**, *7*, 1–30.
70. Hothorn, T.; Hornik, K.; Wiel, M.A.v.d.; Zeileis, A. A Lego System for Conditional Inference. *Am. Stat.* **2006**, *60*, 257–263. [CrossRef]
71. Benjamini, Y.; Hochberg, Y. Controlling the False Discovery Rate: A Practical and Powerful Approach to Multiple Testing. *J. R. Stat. Society. Ser. B (Methodol.)* **1995**, *57*, 289–300. [CrossRef]
72. Goetz, J.; Brenning, A.; Marcer, M.; Bodin, X. Modeling the precision of structure-from-motion multi-view stereo digital elevation models from repeated close-range aerial surveys. *Remote. Sens. Environ.* **2018**, *210*, 208–216. [CrossRef]
73. Knevels, R.; Petschko, H.; Proske, H.; Leopold, P.; Maraun, D.; Brenning, A. Event-Based Landslide Modeling in the Styrian Basin, Austria: Accounting for Time-Varying Rainfall and Land Cover. *Geosciences* **2020**, *10*, 217. [CrossRef]
74. Steger, S.; Brenning, A.; Bell, R.; Glade, T. The propagation of inventory-based positional errors into statistical landslide susceptibility models. *Nat. Hazards Earth Syst. Sci.* **2016**, *16*, 2729–2745. [CrossRef]
75. Szumilas, M. Explaining Odds Ratios. *J. Can. Acad. Child Adolesc. Psychiatry* **2010**, *19*, 227–229.
76. Tasser, E.; Mader, M.; Tappeiner, U. Effects of land use in alpine grasslands on the probability of landslides. *Basic Appl. Ecol.* **2003**, *4*, 271–280. [CrossRef]
77. Petschko, H.; Brenning, A.; Bell, R.; Goetz, J.; Glade, T. Assessing the quality of landslide susceptibility maps—Case study Lower Austria. *Nat. Hazards Earth Syst. Sci.* **2014**, *14*, 95–118. [CrossRef]
78. Shirzadi, A.; Solaimani, K.; Roshan, M.H.; Kavian, A.; Chapi, K.; Shahabi, H.; Keesstra, S.; Ahmad, B.B.; Bui, D.T. Uncertainties of prediction accuracy in shallow landslide modeling: Sample size and raster resolution. *CATENA* **2019**, *178*, 172–188. [CrossRef]
79. Biszak, E.; Biszak, S.; Timár, G.; Nagy, D.; Molnár, G. Historical topographic and cadastral maps of Europe in spotlight—Evolution of the MAPIRE web portal. In Proceedings of the 12th International Workshop on Digital Approaches to Cartographic Heritage, Venice, Italy, 26–28 April 2017; Volume 12, pp. 204–208.

80. Gingrich, S.; Niedertscheider, M.; Kastner, T.; Haberl, H.; Cosor, G.; Krausmann, F.; Kuemmerle, T.; Müller, D.; Reith-Musel, A.; Jepsen, M.R.; et al. Exploring long-term trends in land use change and aboveground human appropriation of net primary production in nine European countries. *Land Use Policy* **2015**, *47*, 426–438. [CrossRef]
81. Krausmann, F.; Gingrich, S.; Haberl, H.; Erb, K.-H.; Musel, A.; Kastner, T.; Kohlheb, N.; Niedertscheider, M.; Schwarzlmüller, E. Long-term trajectories of the human appropriation of net primary production: Lessons from six national case studies. *Ecol. Econ.* **2012**, *77*, 129–138. [CrossRef] [PubMed]
82. Bätzing, W. Die Alpen—Tiefgreifende Nutzungsveränderungen als Herausforderung für den Naturschutz [The Alps—Profound changes in human uses present challenges to nature conservation]. *Nat. Und Landsch.* **2017**, *92*, 398–406. [CrossRef]

land

MDPI

Article

Impact of Land Use Change Due to Urbanisation on Surface Runoff Using GIS-Based SCS–CN Method: A Case Study of Xiamen City, China

Sabita Shrestha [1,2,3], Shenghui Cui [1,3,*], Lilai Xu [4], Lihong Wang [1,3], Bikram Manandhar [1,2,3,5] and Shengping Ding [1,3]

1 Key Laboratory of Urban Environment and Health, Institute of Urban Environment, Chinese Academy of Sciences, Xiamen 361021, China; sabita@iue.ac.cn (S.S.); lhwang@iue.ac.cn (L.W.); bikram@iue.ac.cn (B.M.); spding@iue.ac.cn (S.D.)
2 University of Chinese Academy of Sciences, Beijing 100049, China
3 Xiamen Key Lab of Urban Metabolism, Institute of Urban Environment, Chinese Academy of Sciences, Xiamen 361021, China
4 Research Center for Integrated Disaster Risk Reduction and Emergency Management, Institute for Disaster Management and Reconstruction, Sichuan University, Chengdu 610065, China; lilaixu@scu.edu.cn
5 Institute of Forestry, Tribhuwan University, Hetauda 44107, Nepal
* Correspondence: shcui@iue.ac.cn

Citation: Shrestha, S.; Cui, S.; Xu, L.; Wang, L.; Manandhar, B.; Ding, S. Impact of Land Use Change Due to Urbanisation on Surface Runoff Using GIS-Based SCS–CN Method: A Case Study of Xiamen City, China. *Land* **2021**, *10*, 839. https://doi.org/10.3390/land10080839

Academic Editors: Matej Vojtek, Andrea Petroselli and Raffaele Pelorosso

Received: 14 July 2021
Accepted: 9 August 2021
Published: 11 August 2021

Abstract: Rapid urban development results in visible changes in land use due to increase in impervious surfaces from human construction and decrease in pervious areas. Urbanisation influences the hydrological cycle of an area, resulting in less infiltration, higher flood peak, and surface runoff. This study analysed the impact of land use change due to urbanisation on surface runoff, using the geographic information system (GIS)-based soil conservation service curve number (SCS–CN) method, during the period of rapid urban development from 1980 to 2015 in Xiamen, located in south-eastern China. Land use change was analysed from the data obtained by classifying Landsat images from 1980, 1990, 2005, and 2015. Results indicated that farmland decreased the most by 14.01%, while built-up areas increased the most by 15.7%, from 1980 to 2015. Surface runoff was simulated using the GIS-based SCS–CN method for the rainfall return periods of 5, 10, 20, and 50 years. The spatial and temporal variation of runoff was obtained for each land use period. Results indicate that the increase in surface runoff was highest in the period of 1990–2005, with an increase of 10.63%. The effect of urbanisation can be realised from the amount of runoff, contributed by built-up land use type in the study area, that increased from 14.2% to 27.9% with the rise of urban expansion from 1980 to 2015. The relationship between land use and surface runoff showed that the rapid increase in constructed land has significantly influenced the surface runoff of the area. Therefore, the introduction of nature-based solutions such as green infrastructure could be a potential solution for runoff mitigation and reducing urban flood risks in the context of increasing urbanization.

Keywords: land use change; surface runoff; urban development; green infrastructure; Xiamen

1. Introduction

Urbanisation is a growing concern in the present world. With over 55% of the world's population living in urban areas [1], urban expansion has dramatically influenced the change in urban land use. Land use change due to urbanisation results in more impervious surfaces that have considerable impacts on urban hydrology [2]. Urban expansion leads to large impervious surfaces that reduces rainwater infiltration, generating high surface runoff and peak flow, thus increasing the risks of urban flooding and waterlogging [3]. Urban development is one of the major causes of urban pluvial flooding, aggravated by poor urban drainage systems that become more severe with increasing frequency and magnitude of extreme precipitations as a consequence of climate change, affecting the urban population, infrastructure, and economy [4–6].

The effect of land use change on surface runoff depends on many factors such as spatial characteristics of land use, rainfall, and soil [7]. The effect is most prominent in areas undergoing rapid urbanisation where natural vegetation has been removed and replaced by constructed land [8]. Removal of trees and vegetation and construction of commercial and residential buildings, streets, and parking lots alters the water balance of an area by changing the balance between rainfall, infiltration, evapotranspiration, and runoff [9,10]. Constructed areas such as concrete buildings, roofs, paved roads, and sidewalks lead to an increase in impervious surfaces, which reduces the time of concentration for runoff, leading to higher storm runoff and peak discharges [11,12]. As a result, flooding and waterlogging are common issues in highly urbanised areas [13].

The effect on hydrological processes, particularly surface runoff characteristics, of land use change due to urbanisation has been studied by many researchers [14–16]. Specifically, Hu, Fan, and Zhang [2] investigated the impact of land use change on the distribution of surface runoff in the highly urbanised city of Beijing. Results indicated that the change in surface runoff is strongly associated with change in impervious areas. Chen et al. [17] showed that non-uniform urban expansion and intensification are the major driving forces for changing surface runoff. The studies mentioned above indicate that changes in urban land use with more impervious surfaces reduce the ability to intercept rainfall and exacerbate surface runoff process.

The SCS–CN method, developed by the Natural Resources Conservation Service (NRCS), U.S. Department of Agriculture (USDA), is one of the most widely used methods in many hydrological studies, particularly for estimating surface runoff, accounting several factors such as soil, land use treatment, and topographical features, and incorporating these factors into a single CN parameter [18–21]. However, the conventional method is more time consuming and error prone. Remote sensing (RS) and a geographic information system (GIS) can be used effectively to manage spatial and non-spatial databases that represent the hydrologic characteristics of the watershed [22]. Therefore, the application of GIS and remote sensing with hydrological models yield results with high reliability and accuracy over conventional methods [23]. Many hydrological studies are being conducted with the integration of GIS and RS techniques for simulating surface runoff. For instance, Shadeed and Almasri [24] demonstrated that the integration of GIS with SCS–CN for simulating runoff volume proved to be an effective tool in arid to semi-arid catchments [24]. Similarly, Liu and Li [25] used a GIS-based SCS model for runoff generation in Loess plateau of China and found that the calculated and observed runoff processes were well correlated.

Recently, the GIS-based SCS–CN method has been used widely to study the impacts of land use change and urban growth on surface runoff [2,14,26]. For example, Jahan et al. [27] investigated the impacts of suburban growth on surface runoff using GIS-based SCS–CN, and concluded that the integrated approach is useful for land use change detection and analysis of impact of suburban growth on surface runoff. Furthermore, Vojtek and Vojtekova [14] applied the SCS–CN method in a GIS environment to study the impact of land use change on surface runoff. Results indicated that the method is suitable for spatial modelling of runoff characteristics of a basin where few data are available. Therefore, GIS and remote sensing in the SCS–CN model is a more efficient and cost-effective method that produces adequate results without using complex data in analysing the impact of land use change and urban growth [27–30]. In this study, the GIS and SCS–CN method is integrated to analyse the effect of land use change, and examine the spatial-temporal variation in surface runoff at an urban scale.

Firstly, the study assessed land use change in Xiamen city from 1980 to 2015 by analysing land use images in GIS environment. Then, surface runoff depth was estimated using SCS–CN method. We used GIS-based SCS–CN method to evaluate the effect of land use change on surface runoff. The main objective of this research is to identify the characteristics of land use change from 1980 to 2015 and analyse the impact of land use change due to urbanisation on the temporal and spatial distribution of surface runoff under the rainfall return periods of 5, 10, 20, and 50 years.

2. Materials and Methods

2.1. Study Area

Xiamen City is located on the west coast of the Taiwan Strait, composed of the mainland area along Xiamen Bay, Xiamen Island, Gulangyu, and other islands, including 6 administrative districts: Huli, Siming, Haicang, Jimei, Tong'an, and Xiang'an, as shown in Figure 1. The city has a total land area of 1699 km^2 and a sea area of 324 km^2. The study area is predominantly flat, low relief, with medium-low mountains, plains, and tidal flats. The slope of the terrain descends from northwest to southeast. The northwest part is mountainous, with the highest elevation of 1175 m above sea level, located on Yunding mountain [31]. Xiamen has a subtropical monsoon climate, with humid and mild climates throughout the year. According to the *Xiamen Statistical Yearbook*, annual average temperature and rainfall are approximately 21 °C and 1200 mm, respectively, with most rainfall in June, July, and August, accounting for more than half of the annual rainfall. Xiamen is one of the special economic zones in China and has experienced a rapid economic development and urbanisation [31]. According to Xiamen Municipal Bureau of Statistics [32], urban built-up area of the city has expanded from 38.5 km^2 in 1985 to 348.23 km^2 in 2017. The permanent population of the city was 4.29 million in 2019 [32]. As of 2019, the urbanisation rate of the population in Xiamen is 89.2% [33]. Xiamen has experienced several natural disasters, especially flooding and waterlogging induced by sea-level rise and storm surges, which is further intensified by rapid urban growth in past decades. Hence, there is an urgent need for mitigation of the potential hazards induced by urban development and climate change.

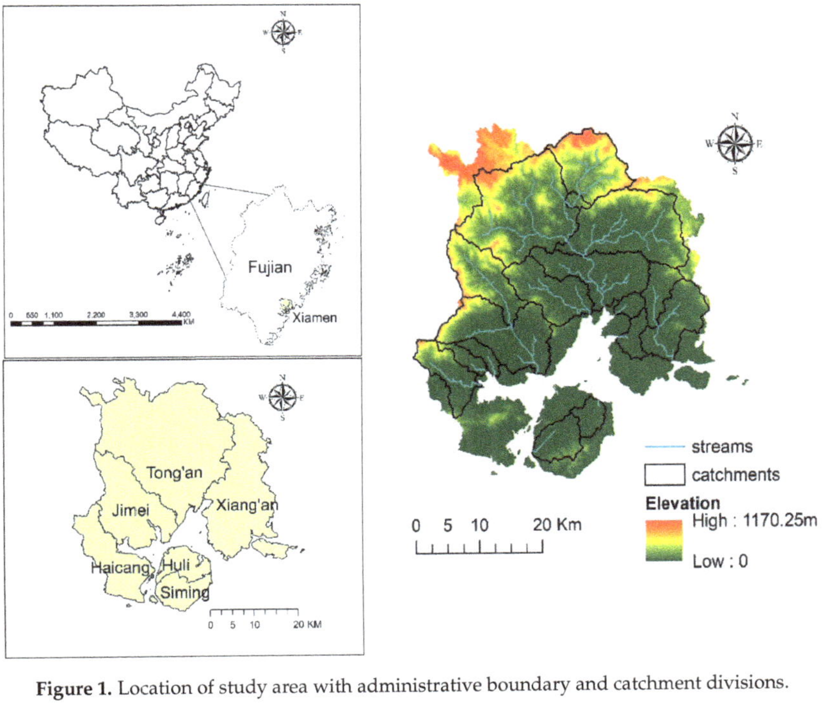

Figure 1. Location of study area with administrative boundary and catchment divisions.

Figure 2 shows the overall methodology of the study conducted for analysing impact of land use change on surface runoff with the application of GIS and SCS–CN method.

The GIS environment was used to build and intersect land use and HGS shapefiles. The land use and soil complex were used to obtain the weighted CN. The CN value of each polygon was estimated using the USDA table. Finally, raster calculator in the GIS was used to calculate runoff depth from CN values for the four rainfall return periods. The methodology was applied to four different years of 1980, 1990, 2005, and 2015, respectively.

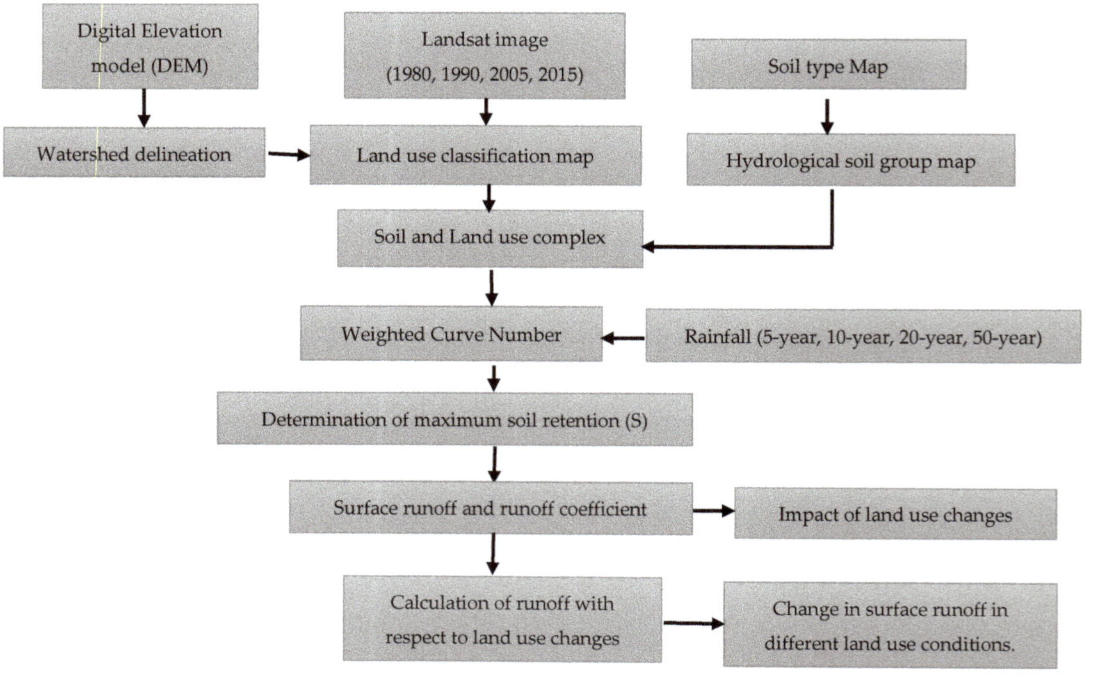

Figure 2. Flowchart showing the methodology of the study.

2.2. Data Source and Methods

(a) Land use data

Multispectral satellite images from Landsat 3 MSS, Landsat 5 TM, and Landsat 7 ETM+ were obtained to create a land use map of the study area for 1980, 1990, 2005, and 2015. The images were downloaded from Centre for Earth Observation and Digital Earth (CEODE), Chinese Academy of Sciences, and the United States Geological Survey (USGS) [34]. Images were georeferenced to the UTM, Zone 50 North, WGS-84 projection, and Beijing 1954 coordinate systems. The spatial resolution is set to 30 m. The land use was classified according to the National Land Cover Data Sets (NLCD) of China generated by Liu et al. [35] in the construction of the *China 20th Century LUCC Spatio-temporal Platform*. The method of unsupervised classification with Iterative Self Organizing Data Analysis Technique Algorithm (ISODATA) was used for image classification. Based on the national land use data product by Geographical Information Monitoring Cloud Platform, the land use was initially classified as six first level categories and twenty-five second level categories, which were reclassified into eight classes for the study. High resolution satellite imagery from Google Earth was used for visual interpretation and accuracy assessment. The overall accuracy and kappa coefficient of classified land use types is more than 85%. The high overall accuracy and kappa coefficient suggests a good relationship between classified image and reference image. The detailed descriptions of land use classes are given in Table 1. ArcGIS 10.5 and ENVI 5.3 software were used to generate various layers and land use maps.

Table 1. Land use classes and their description.

Land Use	Description
Farmland	Areas for growing crops, mainly including paddy fields and arable lands for vegetable farming, with or without regular irrigation facilities. It includes farmland where rice and dry land crops are rotated.
Woodland	Areas referring to forestry land for growing trees, shrubs, bamboos, and coastal mangroves, including trees and shrubs with canopy density more than 30%.
Grassland	Areas of all kinds of grassland, mainly with herbaceous plants covering more than 5%, including shrub grassland with grassland and canopy density.
Water	Areas of natural land waters and water conservancy facilities, including natural and artificial river canals, lakes, and reservoir ponds.
Coastal wetlands	Areas of tidal flats and beach lands, including lands near water level of the rivers and lakes.
Built-up land	Urban land refers to land in large, medium, and small cities, residential areas, and built-up areas above county towns. It also includes construction sites such as factories and mines, large-scale industrial areas, oil fields, salt fields, and quarries, as well as roads, airports, and special sites.
Rural settlements	Refers to rural settlements independent of cities and towns.
Unused land	Areas of bare land, lands covered with gravel, sand, rocks, and saline-alkali and marsh lands. Generally, vegetation coverage is less than 5%.

(b) Soil data

Soil data were obtained from Harmonized World Soil Database (HWSD) and Food and Agriculture Organization (FAO). Soil information for Xiamen city was obtained from 1:1 million soil map of China with a resolution of 1 km (30 × 30 arc seconds) [36]. Hydrological soil groups are classified into A, B, C, and D on the basis of water transmission and infiltration rate of soil when the soil is thoroughly wetted [29]. Soil group A has a high infiltration rate and the lowest runoff potential. These soils consist of sands or gravels that have a high rate of water transmission. Group B and C soils have moderate runoff potential and infiltration rate with a slower rate of water transmission. Soil group D has the lowest infiltration rate and highest runoff potential composed of clay or clayey loam soils with a very slow rate of water transmission [29]. Based on soil texture and soil type, the study area is classified into three HSG types, B, C, and D, as can be seen in Figure 3, which mostly contribute to large surface runoff and less infiltration.

(c) Rainfall data

Long-time series rainfall data from 1985 to 2015, collected at the Xiamen meteorological stations, were acquired from the China Meteorological Data Service Centre of the China Meteorological Administration. The rainfall amount for the return periods of 5, 10, 20, and 50 years were obtained from maximum daily rainfall data for the hydrological analysis, as seen in Table 2. The rainfall for corresponding return periods was determined by using Log Pearson Type III distribution, which is the common probability distribution method used in China [37]. The study assumed that the climatic and soil conditions are constant. To investigate spatial heterogeneity, the study required uniformly distributed station data, which are not available. Therefore, for the temporal analysis of rainfall from the stations, the station with the most complete data was selected as a representative of the study area's rainfall. Furthermore, the study area is small, and changes in rainfall variation is considered insignificant.

(d) Digital Elevation Model (DEM)

A digital elevation model (DEM) with a 30 m resolution was downloaded from the USGS website. Using hydrology tools in ArcGIS 10.8, 19 catchments were obtained, as seen in Figure 1. The sinks in the DEM data were filled, water flow direction was estimated, and flow accumulation was set with a threshold of 10,000 and 60,000 to obtain the catchments. As the study area has short streams and not many large rivers, areas in the southwest do not have adequate streams to obtain a catchment. Hence, runoff obtained from these areas was not significant, and only those areas with enough streams within the threshold are selected as catchments for runoff analysis.

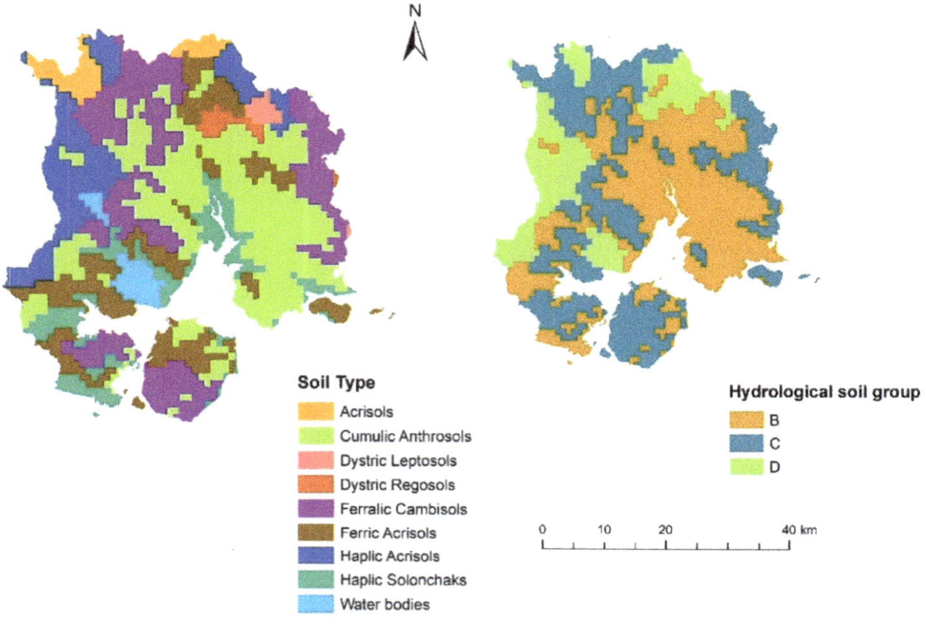

Figure 3. Soil type and hydrological soil group in Xiamen.

Table 2. Rainfall depth for 24 h maximum daily rainfall in different return periods.

Return period (years)	5	10	20	50
Rainfall (mm)	194.3	237.7	280.1	335.3

Source: Xiamen Meteorological Bureau.

2.3. SCS–CN Method

The SCS–CN method is the most commonly used empirical hydrological method developed by the NRCS, USDA, and is widely used to simulate runoff [19,29]. The curve number method has been successfully adopted in many ungauged watersheds and has expanded its scope of application in urbanised catchments and forested watersheds [29]. The surface runoff model uses the curve number approach of the US Soil Conservation Service [29], based on combinations of land use, hydrological soil group, and antecedent moisture condition (AMC) for the estimation of runoff. The amount of runoff was estimated using the SCS–CN method in presence of GIS and RS. The curve number is the most important factor in determining runoff via the SCS based method. The runoff of the soil and land use complex is represented by CN, which is a function of soil type, moisture conditions, and land use type [38].

The SCS–CN model is based on the water balance equation as shown in the Equations (1)–(3).

$$P = Ia + F + Q \quad (1)$$

$$\frac{Q}{P - Ia} = \frac{F}{S} \quad (2)$$

$$Ia = \lambda \times S \quad (3)$$

where P is the rainfall depth (mm), Ia is the initial abstraction of the rainfall (mm), F is infiltration, Q is surface runoff depth (mm), S is the potential maximum soil retention, and λ is abstraction coefficient that ranges between 0.0 to 0.2 or 0.05 for urbanised catchments [39]. The value of 0.2 as mentioned by Natural Resources Conservation Service (NRCS) was used in the study [29]. The runoff depth can be obtained for two conditions from the Equations (1) and (2):

For $P > Ia$ and

$$Q = \frac{(P - Ia)^2}{P - Ia + S} \tag{4}$$

If $P < Ia$, $Q = 0$ and Q from Equation (4) is expressed as follows:

$$Q = \frac{(P - 0.2S)^2}{P + 0.8S} \tag{5}$$

In Equation (5), S was obtained from the dimensionless parameter CN. CN is runoff curve number that ranges from 0 to 100.

$$S = \frac{25,400}{CN} - 254 \tag{6}$$

In the SCS–CN method, curve number plays an important role in determining the surface runoff of an area, and its value depends on the corresponding soil type and AMC. AMC is antecedent moisture condition present in the soil at the beginning of the rainfall. Areas with a higher curve number represent higher runoff generated from the surface. In our study, we chose B, C, and D as three hydrological soil groups (HSG) present in the study area, and the soil moisture condition (AMC II) was set as moderate according to average runoff condition. A combined map of land use and HSG was generated by combining land use and soil maps in ArcGIS using overlay analysis. Then, the CN values were assigned for each polygon based on the information of land use and soil. The CN values for each land use type under AMC II is obtained from the TR-55 lookup Table 3 [29,40].

Table 3. CN numbers for corresponding land use types.

SN	Land Use and Cover Type	Hydrological Soil Group Type		
		B	C	D
1.	Farmland	71	78	81
2.	Forestland	58	72	79
3.	Grassland	56	70	77
4.	Water Body	100	100	100
5.	Coastal wetlands	89	93	95
6.	Built-up land	98	98	98
7.	Rural settlements	71	79	83
8.	Unused land	86	91	94

Area weighted CN was obtained to simulate the runoff of the whole area using the initial curve numbers from the table as in Equation (8). Combining the CN values of different land use and soil complex polygons, weighted CN was calculated for each catchment. The weighted CN is calculated by taking the sum of each CN value multiplied by its fraction of the total area of each land use type [41]. The Equation is given below:

$$CNw = \frac{\sum(CNi \times Ai)}{A} \tag{7}$$

where CNw is the weighted curve number; CNi is the curve number for each land use type; Ai is area of land use with respective curve number; and A is the total area of each land use type. Finally, surface runoff depth was estimated, and runoff coefficient, i.e., the ratio of runoff to rainfall, was calculated.

2.4. Analysing Impact of Land Use Change on Surface Runoff

Impact of land use change on surface runoff was analysed by comparing the difference in runoff variables. Runoff depth and runoff coefficient was used as two variables to assess land use change on surface runoff. Surface runoff from the catchments were obtained for the different land use conditions and runoff was calculated under different rainfall return periods. The difference of runoff and runoff coefficient was obtained for the land use period of 1980–1990, 1990–2005, and 2005–2015 by using Equations (8)–(10):

$$\Delta Q = Qa - Qb \quad (8)$$

$$\Delta \propto = \frac{\Delta Q}{P} \quad (9)$$

$$\Delta \beta = \frac{\Delta Q}{Qb} \times 100\% \quad (10)$$

where Qa and Qb denotes surface runoff depths (mm) of the initial and final land use, respectively, P is the rainfall depth (mm), ΔQ is the change in runoff depth between two periods of land use conditions, $\Delta\alpha$ is the absolute change in the runoff coefficient, and $\Delta\beta$ represent relative change in runoff. Land use change leads to an increase in surface runoff if the values of ΔQ and $\Delta\alpha$ are positive in the above Equation (10).

A relationship between land use and surface runoff was determined by using Pearson's product–moment correlation coefficient. A positive and larger correlation coefficient suggests that the factor is more significant in the change in surface runoff.

2.5. Validation of SCS–CN Model

For analysing the performance of the model, the model was validated using observed flow and simulated runoff between 1981 and 2015. Four statistical indices, as shown in Table 4, were used for testing the goodness of fit.

Table 4. Statistical indices used for model validity.

Coefficient	Description	Optimal Value
Percent bias (PBIAS)	measures the average tendency of the simulated values to be larger or smaller than their observed ones.	0—Optimal, Negative—underestimation, Positive—overestimation
Nash-Sutcliffe efficiency (NSE)	a normalised statistic that calculates the relative magnitude of the simulated flow variance compared to the observed flow variance.	NSE = 1—perfect match, NSE = 0—model predictions accurate as the mean of the observed data, −Inf < NSE < 0—observed mean is a better predictor than the model
Correlation coefficient (r)	statistical measure of the strength of the relationship between the relative movements of simulated and observed flow.	Ranges from −1 to 1 −1—perfect negative correlation 1-perfect positive correlation 0—no correlation
Volumetric efficiency (VE)	represents the fraction of water delivered at the proper time	$-\text{Inf} \leq \text{VE} \leq 1$ close to 1—efficient

3. Results

3.1. Land Use Change in the Study Area

Land use types for four different years can be seen in Figure 4. The spatial distribution of land use types show that the area has experienced significant change, especially due to urban expansion. There was a remarkable expansion of built-up areas from 62.85 to 307.54 km^2 between 1980 and 2015. It was the largest gain, with a net increase of 244.69 km^2. The largest net loss from 1980 to 2015 was observed in farmland and coastal wetlands with 218.2 km^2 and 46.65 km^2, respectively. However, constructed land that includes built-up land and rural settlements underwent the largest net increase. The forest area reduced substantially with a change of 23.93 km^2.

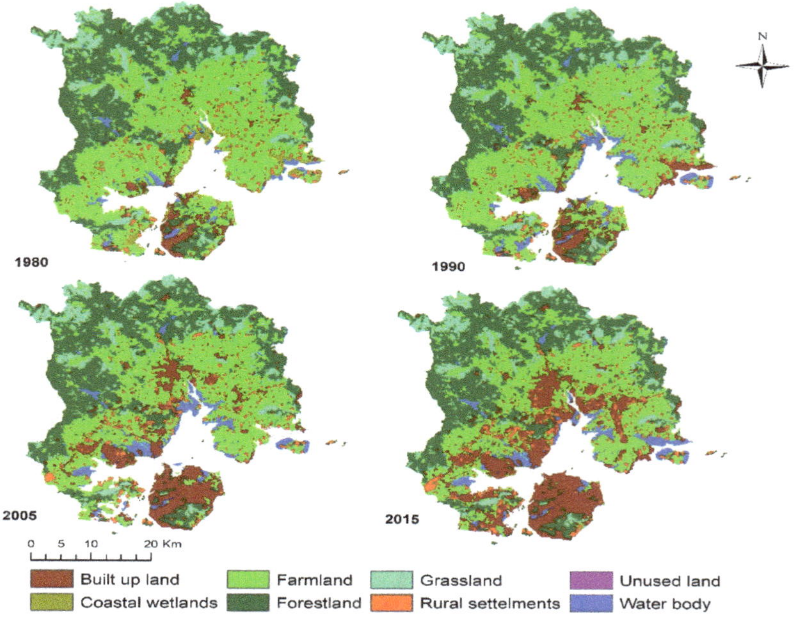

Figure 4. Land use maps of Xiamen in 1980, 1990, 2005, and 2015.

Similarly, grassland and farmland were replaced by built-up areas in the south and south-eastern parts of the mainland and major parts of the island. It can be observed that most coastal wetlands decreased from 1980 to 1990, and built-up areas occupied most areas. The land area of Xiamen City expanded outward by reclamation and construction along the coast after the national survey of 1985, which contributed to the expansion of land area and reduced wetlands in 1990 [42]. A considerable increase in water bodies indicate some farmlands being converted to reservoirs, aqua farms, and other constructed wetlands. Table A1 shows the conversion of the land use in the form of a change matrix for the period of 1980 to 2015. There was a major conversion from farmland to constructed land. The trend of land use change from 1980 to 2015 indicates that urban development has dominated the island city, resulting in vast areas of impervious surfaces.

In Table A2, the percent and area change of each land use type is depicted from 1980 to 2015. It can be observed that farmlands have been reduced by 14.01% and built-up area has increased by 15.7%. The transfer process of land use types in between the study years can be seen from Figure 5. It can be noticed that major portion of change in farmlands in 1990 and 2005 have been replaced by built-up land. Overall, constructed land increased from 9.12% in 1980 to 26.1% in 2015. This indicates that urban impervious areas have increased considerably in the last few decades of the study period.

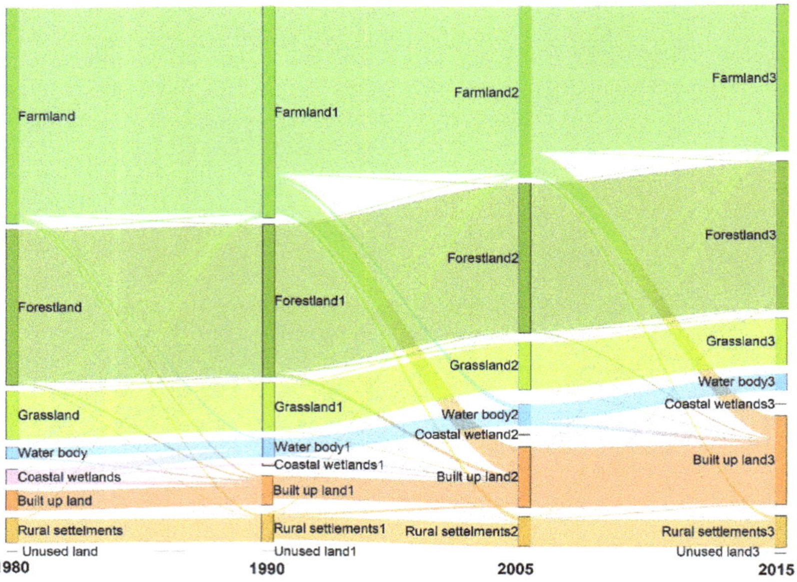

Figure 5. Transfer process of land use type during 1980–2015.

3.2. Spatial Distribution of Runoff in Different Years

The spatial distribution of runoff depth for the highest rainfall return period of 50 years is shown in Figure 6. Surface runoff depth ranges from 176.28 mm to 329.16 mm. The area covered by built up land with higher CN depicts a higher amount of runoff in the period from 1980 to 2015. Xiamen Island, particularly dominated by constructed area, shows larger areas of high runoff. The urban areas with high CN are often the areas with high runoff value.

Under the land use conditions of 1980, 1990, 2005, and 2015, the average surface runoff depth and runoff coefficient show an increasing trend as shown in Figure 7. The average surface depth of the area under four different periods differ from 117.2 to 271.6 mm, and runoff coefficient fluctuated from 0.6 to 0.8. The calculated value of average surface runoff and runoff coefficient increases as the rainfall return period increase from 5 years to 50 years. It can be observed that surface runoff and runoff coefficient significantly increase from the year 1990 to 2015. The amount of runoff percent for different land use types can be noticed in Figure 8. A major portion of runoff is contributed by built-up land and rural settlements which are the major constructed land in the study area with the rise from 38.2% to 48.4%. Runoff contributed by built-up land alone in the study area increased from 14.2% to 27.9% from 1980 to 2015. Similarly, farmlands and coastal wetlands also contribute to the significant portion of surface runoff in the study area.

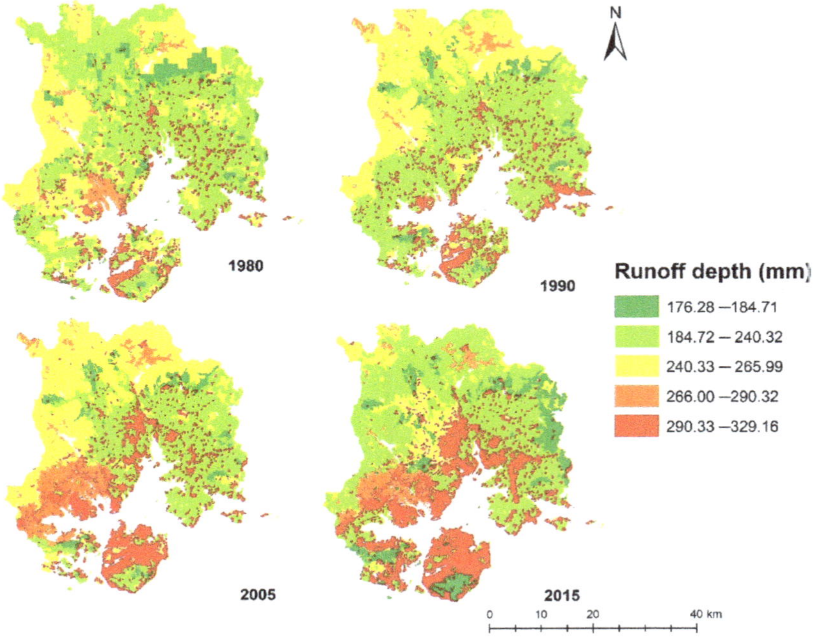

Figure 6. Surface runoff depth in Xiamen from 1980 to 2015.

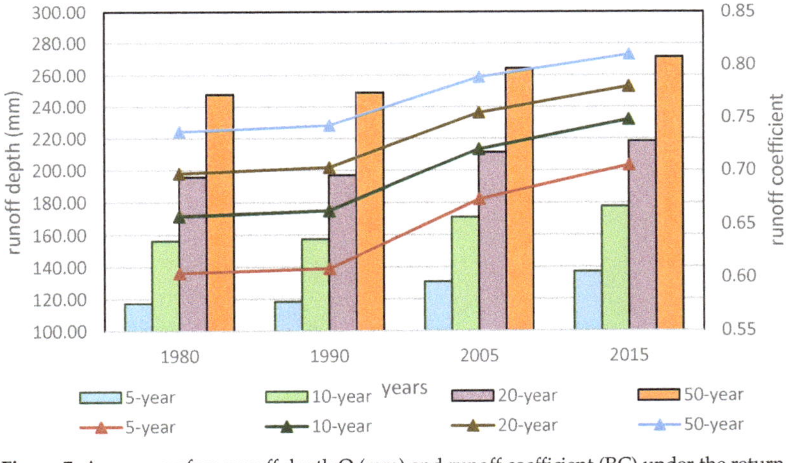

Figure 7. Average surface runoff depth Q (mm) and runoff coefficient (RC) under the return period of 5, 10, 20, and 50 years under the land use conditions of 1980, 1990, 2005, and 2015.

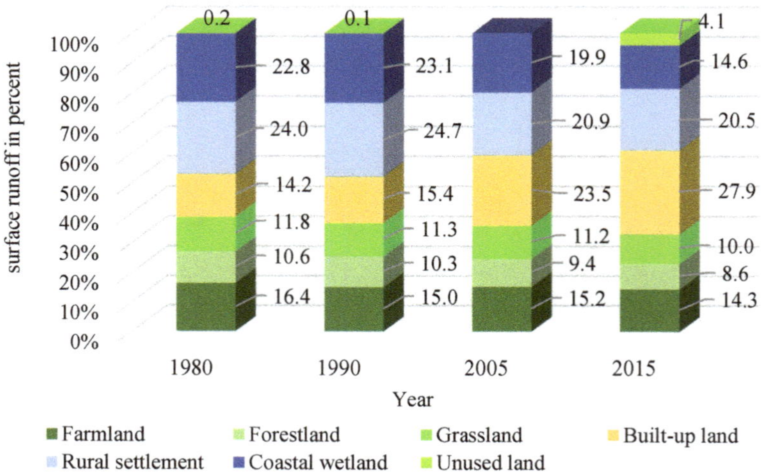

Figure 8. Runoff depth across different land use types.

3.3. *Change in Surface Runoff in Different Land Use Conditions*

Increase in urbanised area gives rise to impervious surfaces, resulting in increased surface runoff. Table 5 depicts the change in surface runoff at different stages of land use conditions for the rainfall return periods of 5, 10, 20, and 50 years. Increase in amount of runoff (ΔQ) is highest during 1990–2005 and lowest during 1980–1990. The surface runoff coefficient change (Δα) during the periods is influenced by the respective change in average surface runoff under return periods of 5, 10, 20, and 50 years.

Table 5. Changes in surface runoff at different stages under rainfall return periods of 5, 10, 20, and 50 years.

Return Period	Time Period	Amount of Runoff Change (ΔQ) mm	Percent Change in Runoff (ΔC) %	Δα
5 years	1980–1990	1.01	0.86	0.005
	1990–2005	12.57	10.63	0.065
	2005–2015	6.15	4.70	0.032
10 years	1980–1990	1.13	0.72	0.005
	1990–2005	13.67	8.68	0.058
	2005–2015	6.62	3.87	0.028
20 years	1980–1990	1.22	0.63	0.004
	1990–2005	14.50	7.37	0.052
	2005–2015	6.97	3.30	0.025
50 years	1980–1990	1.32	0.53	0.004
	1990–2005	15.33	6.16	0.046
	2005–2015	7.32	2.77	0.022

3.4. *Relationship between Surface Runoff and Land Use*

Land use exhibits a high relationship with surface runoff with a statistically significant correlation ($p > 0.05$). As shown in Table 6, farmlands (−0.97), forestland (−0.96), and grassland (−0.97) contribute negatively to the surface runoff, which indicates that increase in these land use types contribute to a decrease in surface runoff. Increase in urban built-up land (0.98) and rural settlements (0.99) corresponded to an increase in average surface runoff, while the decrease in farmland, forestland, and grassland contribute to higher runoff. Furthermore, increase in coastal wetlands contribute to a decrease in runoff, and unused land contributed positively to average runoff. The results are consistent with the common knowledge that increase in urban constructed land causes an increase in surface runoff; however, increases in farmland, grassland, forestland, and wetlands lead to a reduction in surface runoff. Water bodies are considered to have less effect on runoff depth, hence the relationship is less significant.

Table 6. Relationship between land use and surface runoff under rainfall return period of 50 years. t—t-statistic, *p*—*p*-value, and R^2—Pearson's correlation coefficient. Areas in square kilometres.

	Farmland	Forestland	Grassland	Water Body	Coastal Wetlands	Built-Up Land	Rural Settlements	Unused Land	Runoff (Q50)
1980	681.14	494.87	153.78	37.84	48.56	62.85	79.38	0	136.96
1990	667.2	486.48	153.48	62.02	7.7	93.51	88.2	0	177.77
2005	540.02	474.01	151.82	67.94	2.46	191.83	93.22	0	218.28
2015	462.94	470.94	150.91	64.99	1.91	307.54	99.46	0.18	271.62
t	−5.3587	−4.9696	−5.7872	1.77715	−1.9197	7.6905	8.734	2.0111	-
p	0.03311	0.03819	0.02858	0.2185	0.1949	0.01649	0.01286	0.182	-
R^2	−0.97	−0.96	−0.97	0.78	−0.81	0.98	0.99	0.82	-

3.5. Rainfall-Runoff Correlation Analysis

The correlation analysis shown in Figure 9 indicates a strong linear relationship between the SCS–CN runoff and maximum daily rainfall, with a correlation coefficient of 0.99. The study results are comparable with the findings of Rawat and Singh [43] who found a good coefficient of determination (0.91) in a small study area using the SCS–CN model. The slope of the line determines the runoff coefficient, i.e., 0.84. The resulting findings are similar to Al. Ghobari et al. [44], who came to the conclusion that the SCS–CN model has a better simulation effect on study areas with a coefficient of runoff greater than 0.5 than those with a coefficient of runoff less than 0.5. This coefficient may provide valuable information on the extent of the basin response to runoff generation.

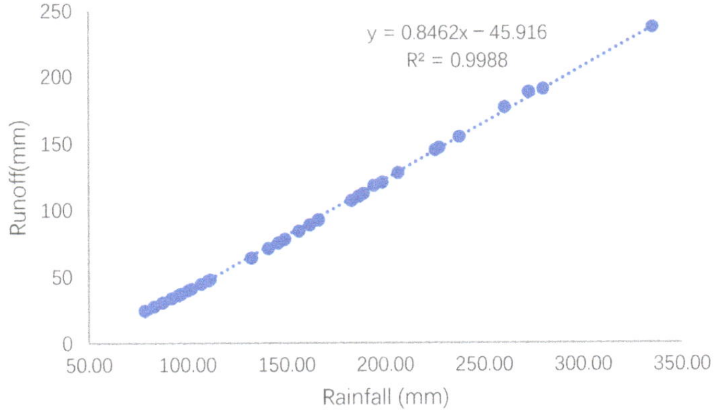

Figure 9. Relationship between rainfall and SCS–CN runoff.

3.6. Validation of SCS–CN

The SCS–CN model was validated using historical observations and simulated flow from 1980 to 2015, as shown in Figure 10. Some statistical efficiency criteria are used to perform evaluation of the validation results between simulated output and observed data which are percent bias (PBIAS), correlation coefficient (r), Nash–Sutcliffe efficiency (NSE) and volumetric efficiency (VE). These statistical indices indicate the goodness of fit between simulated and observed data. The model successfully predicts the annual flow with the high accuracy as depicted by the indices. The PBIAS, r, NSE, and VE were −5.7, 0.82, 0.64, and 0.86, respectively. Although there is an underestimation of streamflow due to static land use, annual flow statistics indicate that there is a good relationship between observed and simulated streamflow. Hence, the model performance was satisfactory and responded well in simulation of runoff.

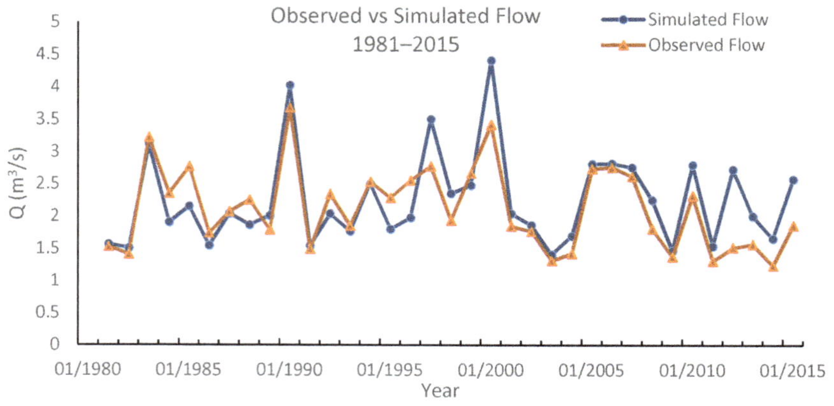

Figure 10. Observed and simulated annual flow from 1981 to 2015.

4. Discussion

The study area has undergone significant land use change from 1980 to 2015. A significant loss of farmland was observed in the period between 1980 and 2015. During this period, most of the farmlands are replaced by built-up lands in the south-eastern part of the study area. After the national survey was carried out in 1985, the land area of Xiamen City expanded outward through reclamation [42]. The major land use change in Xiamen is attributable to land reclamation and urban development in the past years. The process of urban construction in reclaimed land and building new residential areas started to rise after 1985, which is reflected in the runoff increase after 1990 [45,46]. In 1980, after the city was declared as specific economic zone, most of the farmlands and forestlands were converted into urban areas between 1985 and 2005. After 2005, there was major sea reclamation which increased the urban area. New initiatives for industrial and economic development occurred in the period of 1990 to 2005, and an urban renewal program took place between 2003 and 2012 [31]. The change in land use significantly affected the runoff hydrology of the city as a result of urban development. The high values of runoff gradually expanded outside Xiamen Island and were mainly distributed in the areas with increased urban construction [40]. Results show that particular increase in constructed land with higher CN value contributes to higher runoff, whereas farmland, forestland, and grassland with lower CN contribute to a lower percent of runoff [30]. Therefore, change in land use, particularly an increase in urban areas, corresponds to an increase in runoff. As a result of rapid urbanisation, significant increase in surface runoff is observed in the island and outer areas of mainland, which poses higher risks for urban as well as coastal floods [47]. Similar results depicting the strong impact of urbanisation on surface runoff has been obtained by previous studies [2,7,17,48]. Relevant studies have shown that the extent to which urbanisation affects hydrological response depends on spatial and temporal scale, physical geography, landscape composition, and physical and climatic characteristics [49]. The study area, being a lowland coastal area, is relatively flat, and a major part of the area has a slope of less than 15%. Therefore, the slope of the area had a less significant effect on the surface runoff. Additionally, the main objective of the study is to analyse the impact of land use change on the surface runoff, especially due to rapid urbanisation. The impact of slope on the runoff characteristics (peak flow and runoff velocity) are not studied in detail

However, future studies could help better understand the effect of slope on peak flow and runoff velocity.

This study demonstrated the use of a GIS-based SCS–CN method to assess the effects due to land use change on surface runoff by integrating spatial data and hydrological parameters. A GIS-based approach proved to be a reliable tool for quantifying the impact of land use change on runoff with respect to change in CN, which is a function of soil, land use, and moisture conditions [50]. The input data of the soil map and rainfall were based on actual field data, and the CN values of each land use type were obtained from the USDA standard table. The simulation of runoff was validated by comparing the observed and simulated annual flow in the study area, which shows that the design runoff simulated by the model is well accepted. Therefore, the model produced accurate and reliable results of runoff incorporating different spatial aspects. The results were consistent with the results obtained by previous studies of runoff simulations in Xiamen [40,42,51].

To fully comprehend the impacts of spatiotemporal land use change on abrupt or gradual flood peaks and runoff flow, various geographical parameters that influence the runoff should be considered in future studies. In the study, other human and environmental factors affecting runoff, such as the construction of dams, reservoirs and underground extraction, drainage systems, temperature, canopy cover, and soil loss, etc. are given little attention. Therefore, it is necessary to conduct extensive studies including these factors in the future.

5. Conclusions

With the gradual increase in urban areas from 1980 to 2015, a major fraction of land was converted to areas with poor infiltration and low potential storage, which significantly influenced surface runoff. In order to analyse the effect of land use change due to urbanisation on surface runoff, we used GIS-based SCS–CN model. The change in average runoff during the study period was determined, and relation between land use and runoff was obtained. A relation between rainfall and SCS runoff showed that the performance of SCS–CN model is suitable for runoff estimation. The GIS-based approach seemed to be an efficient tool for assessing the land use change and surface runoff through spatial analysis. Hence, the conclusions derived are as follows:

1. The major changes in land use were observed at the expense of conversion of farmland to built-up land. Farmland decreased by 14.02%, and built-up land increased by 15.7%, from 1980 to 2015. Another significant change can be observed in the reduction in coastal wetlands by 2.99% which is attributed to land reclamation and conversion of reclaimed land to constructed land. Overall, the constructed land in the study area increased from 9.12% in 1980 to 26.1% in 2015;
2. Spatial change in surface runoff was noticed from 1990 to 2015 in the south-eastern part of the study area, in which there are areas with higher urban built-up land. Therefore, the increase in runoff in the study area indicates the positive impact of urbanisation. The amount of runoff contributed by land use type shows that, with the increase in total constructed land, the amount of runoff significantly increased from 38.2 to 48.4%. The amount of surface runoff is noticed to be increased from 1990, which is consistent with the rise in urban development that occurred since 1990;
3. The average surface runoff was positively correlated with the built-up and rural settlements, but negatively correlated with the areas of farmland, forestland, grassland, and coastal wetlands. The urbanised land use was determined as a dominant factor for surface runoff increase during the period from 1980 to 2015.

The areas with higher runoff, especially in Xiamen Island, dominated by urban built up land, should be given more attention during land use planning. Forestland, grassland, and farmland in the area have a higher significance for storing runoff. Therefore, these natural green infrastructures should be considered as potential areas for runoff storage. Further research can be focused in exploring the effectiveness of natural infrastructure and nature-based solutions for runoff mitigation and reducing urban flood risks.

Author Contributions: Conceptualization, S.S. and S.C.; Methodology, S.S.; Software, S.S., L.W. and S.D.; Validation, S.S. and L.W.; Formal Analysis, S.S.; Investigation, S.S. and B.M.; Resources, L.W.; Data Curation, S.S. and S.D.; Writing—Original Draft Preparation, S.S.; Writing—Review and Editing, S.S, S.C., L.X. and B.M.; Visualization, S.S.; Supervision, S.C. and L.X.; Project Administration, S.C. and L.X.; Funding Acquisition, S.C. All authors have read and agreed to the published version of the manuscript.

Funding: This research was funded by the National Natural Science Foundation of China (41661144032, 42007418), the National Key Research and Development Program of China (2017YFC0506603), the FJIRSM&IUE Joint Research Fund (Y8L0971), and the Fujian Provincial Natural Science Foundation (2019J05160), China.

Institutional Review Board Statement: Not applicable.

Informed Consent Statement: Not applicable.

Data Availability Statement: The data presented in this study are available on request from the corresponding author.

Acknowledgments: Authors would like to extend their sincere gratitude to the USGS and Geographical Information Monitoring Cloud Platform is acknowledged for making the Land use data product available for the study. We also immensely thank the UCAS International Scholarship for the international Masters Student program. We thank the anonymous reviewers, editor for their constructive comments on the manuscript.

Conflicts of Interest: The authors declare no conflict of interest.

Appendix A

Table A1. Land use and land cover change from 1980 to 2015.

Land Use (km^2)	1980		1990		2005		2015		Land Cover Change (1980–2015)	
	km^2	%	km^2	%	km^2	%	km^2	%	km^2	%
Farmland	681.14	43.71	667.20	42.81	540.02	35.50	462.94	29.70	−218.2	−14.01
Forestland	494.87	31.75	486.48	31.21	474.01	31.16	470.94	30.21	−23.93	−1.54
Grassland	153.78	9.87	153.48	9.85	151.82	9.98	150.91	9.68	−2.87	−0.19
Water body	37.84	2.43	62.02	3.98	67.94	4.47	64.99	4.17	27.15	1.74
Coastal wetlands	48.56	3.12	7.70	0.49	2.46	0.16	1.91	0.12	−46.65	−2.99
Built-up land	62.85	4.03	93.51	6.00	191.83	12.61	307.54	19.73	244.69	15.70
Rural settlements	79.38	5.09	88.20	5.66	93.22	6.13	99.46	6.38	20.08	1.29
Unused land	0.00	0.00	0.00	0.00	0.00	0.00	0.18	0.00	0.18	0.01

Table A2. Land use change transfer matrix between 1980 and 2015.

		Land Use 2015 (In km^2)								
Land Use 1980 (In km^2)	**Land Use Class**	**Built-Up Land**	**Coastal Wetlands**	**Farmland**	**Forestland**	**Grassland**	**Rural Settlements**	**Unused Land**	**Water Body**	**Grand Total**
	Built-up land	**61.82**	0.01	0.58	0.15	0.03	0.02	0.00	0.18	62.78
	Coastal wetlands	22.92	**1.76**	0.51	0.11	0.00	0.84	0.15	22.07	48.37
	Farmland	179.68	0.01	**456.75**	3.60	0.95	25.28	0.00	15.13	681.42
	Forestland	21.36	0.00	1.79	**463.23**	5.19	3.18	0.00	0.18	494.95
	Grassland	5.05	0.00	0.88	2.27	**144.54**	0.69	0.00	0.20	153.62
	Rural settlements	7.41	0.01	1.56	0.88	0.03	**69.04**	0.00	0.07	79.01
	Unused land	0.00	0.00	0.00	0.35	0.00	0.00	**0.00**	0.00	0.35
	Water body	9.10	0.01	1.06	0.48	0.01	0.03	0.00	**27.08**	37.77
	Grand Total	307.350	1.810	463.134	471.062	150.754	99.096	0.159	64.913	1558.277

Note: The bold letters indicate that there is no change in the land use change over the time period.

References

1. United Nations, Department of Economic and Social Affairs. *World Urbanization Prospects: The 2018 Revision*; United Nations: New York, NY, USA, 2019.
2. Hu, S.; Fan, Y.; Zhang, T. Assessing the effect of land use change on surface runoff in a rapidly urbanized city: A case study of the central area of Beijing. *Land* **2020**, *9*, 17. [CrossRef]
3. Ansari, T.A.; Katpatal, Y.B.; Vasudeo, A.D. Spatial evaluation of impacts of increase in impervious surface area on SCS–CN and runoff in Nagpur urban watersheds, India. *Arab. J. Geosci.* **2016**, *9*, 702. [CrossRef]
4. Walsh, C.J.; Fletcher, T.D.; Burns, M.J. Urban stormwater runoff: A new class of environmental flow problem. *PLoS ONE* **2012**, *7*, e45814. [CrossRef]
5. Gill, S.E.; Handley, J.F.; Ennos, A.R.; Pauleit, S. Adapting cities for climate change: The role of the green infrastructure. *Built Environ.* **2007**, *33*, 115–133. [CrossRef]
6. Marelle, L.; Myhre, G.; Steensen, B.M.; Hodnebrog, O.; Alterskjaer, K.; Sillmann, J. Urbanization in megacities increases the frequency of extreme precipitation events far more than their intensity. *Environ. Res. Lett.* **2020**, *15*, 124072.
7. Astuti, I.S.; Sahoo, K.; Milewski, A.; Mishra, D.R. Impact of land use land cover (LULC) change on surface runoff in an increasingly urbanized tropical watershed. *Water Resour. Manag.* **2019**, *33*, 4087–4103. [CrossRef]
8. Shi, P.-J.; Yuan, Y.; Zheng, J.; Wang, J.-A.; Ge, Y.; Qiu, G.-Y. The effect of land use/cover change on surface runoff in Shenzhen region, China. *Catena* **2007**, *69*, 31–35. [CrossRef]
9. Fletcher, T.D.; Andrieu, H.; Hamel, P. Understanding, management and modelling of urban hydrology and its consequences for receiving waters: A state of the art. *Adv. Water Resour.* **2013**, *51*, 261–279. [CrossRef]
10. McGrane, S.J. Impacts of urbanisation on hydrological and water quality dynamics, and urban water management: A review. *Hydrol. Sci. J.* **2016**, *61*, 2295–2311. [CrossRef]
11. Oudin, L.; Salavati, B.; Furusho-Percot, C.; Ribstein, P.; Saadi, M. Hydrological impacts of urbanization at the catchment scale. *J. Hydrol.* **2018**, *559*, 774–786. [CrossRef]
12. Weng, Q. Modeling urban growth effects on surface runoff with the integration of remote sensing and GIS. *Environ. Manag.* **2001**, *28*, 737–748. [CrossRef]
13. Suriya, S.; Mudgal, B.V. Impact of urbanization on flooding: The Thirusoolam sub watershed—A case study. *J. Hydrol.* **2012**, *412*, 210–219. [CrossRef]
14. Vojtek, M.; Vojteková, J. Land use change and its impact on surface runoff from small basins: A case of Radiša basin. *Folia Geogr.* **2019**, *61*, 104.
15. Ruisong, Q.; Min, L.; Lijun, H.; Min, L.; Lijia, Z.; Dongni, O.; Shiyuan, X.; Lizhong, Y. Impact of Land Use Dynamic Change on Surface Runoff: A Case Study on Shanghai Pudong New District. *J. Catastrophol.* **2009**, *1*, 44–49.
16. Sajikumar, N.; Remya, R.S. Impact of land cover and land use change on runoff characteristics. *J. Environ. Manag.* **2015**, *161*, 460–468. [CrossRef]
17. Chen, J.; Theller, L.; Gitau, M.W.; Engel, B.A.; Harbor, J.M. Urbanization impacts on surface runoff of the contiguous United States. *J. Environ. Manag.* **2017**, *187*, 470–481. [CrossRef]
18. Sjöman, J.D.; Gill, S.E. Residential runoff–The role of spatial density and surface cover, with a case study in the Höjeå river catchment, southern Sweden. *Urban For. Urban Green.* **2014**, *13*, 304–314. [CrossRef]
19. Ebrahimian, M. Application of NRCS-curve number method for runoff estimation in a mountainous watershed. *Casp. J. Environ. Sci.* **2012**, *10*, 103–114.
20. Xiao, B.; Wang, Q.-H.; Fan, J.; Han, F.-P.; Dai, Q.-H. Application of the SCS–CN Model to Runoff Estimation in a Small Watershed with High Spatial Heterogeneity. *Pedosphere* **2011**, *21*, 738–749. [CrossRef]
21. Vojtek, M.; Vojteková, J. GIS-based approach to estimate surface runoff in small catchments: A case study. *Quaest. Geogr.* **2016**, *35*, 97–116. [CrossRef]
22. Pandey, A.; Sahu, A. Generation of Curve Number Using Remote Sensing and Geographic Information System. Geospatial World. 2009. Available online: https://www.geospatialworld.net/article/generation-of-curve-number-using-remote-sensing-and-geographic-information-system/ (accessed on 9 August 2021).
23. Zhan, X.; Huang, M.-L. ArcCN-Runoff: An ArcGIS tool for generating curve number and runoff maps. *Environ. Model. Softw.* **2004**, *19*, 875–879. [CrossRef]
24. Shadeed, S.; Almasri, M. Application of GIS-based SCS–CN method in West Bank catchments, Palestine. *Water Sci. Eng.* **2010**, *3*, 1–13.
25. Liu, X.; Li, J. Application of SCS model in estimation of runoff from small watershed in Loess Plateau of China. *Chin. Geogr. Sci.* **2008**, *18*, 235. [CrossRef]
26. Ozdemir, H.; Elbaşı, E. Benchmarking land use change impacts on direct runoff in ungauged urban watersheds. *Phys. Chem. Earth Parts A/B/C* **2015**, *79*, 100–107. [CrossRef]
27. Jahan, K.; Pradhanang, S.M.; Bhuiyan, M.A.E. Surface Runoff Responses to Suburban Growth: An Integration of Remote Sensing, GIS, and Curve Number. *Land* **2021**, *10*, 452. [CrossRef]
28. Chatterjee, C.; Jha, R.; Lohani, A.K.; Kumar, R.; Singh, R. Runoff curve number estimation for a basin using remote sensing and GIS. *Asia-Pac. Remote Sens. GIS J.* **2001**, *14*, 1–7.

29. NRCS. *Urban Hydrology for Small Watersheds-Technical Release 55*; US Department of Agriculture Natural Resources Conservation: Washington, DC, USA, 1986.
30. Al-Juaidi, A.E. A simplified GIS-based SCS–CN method for the assessment of land-use change on runoff. *Arab. J. Geosci.* **2018**, *11*, 269. [CrossRef]
31. Tang, L.; Zhao, Y.; Yin, K.; Zhao, J. Xiamen. *Cities* **2013**, *31*, 615–624. [CrossRef]
32. Bureau, X.S. *Xiamen Statistical Yearbook*; China Statistic Press: Beijing, China, 2020.
33. *Analysis of Xiamen's Population Data in 2019*; China Commercial Industry Research Institute: Xiamen, China, 2020.
34. Geographical Information Monitoring Cloud Platform. Available online: www.dsac.cn (accessed on 20 January 2020).
35. Liu, J.; Liu, M.; Tian, H.; Zhuang, D.; Zhang, Z.; Zhang, W.; Tang, X.; Deng, X. Spatial and temporal patterns of China's cropland during 1990–2000: An analysis based on Landsat TM data. *Remote Sens. Environ.* **2005**, *98*, 442–456. [CrossRef]
36. Nanjing Institute of Soil Science, Chinese Academy of Sciences. Soil Science Database. Available online: vdb3.soil.csdb.cn (accessed on 23 March 2020).
37. Chenyao, X. Hydrological Response Mechanism and Simulations of a Sponge City Residential Block. Master's Thesis, Chinese Institute of Water Resources and Hydropower Research, Beijing, China, April 2018.
38. Sharma, K.; Singh, S. Runoff estimation using Landsat Thematic Mapper data and the SCS model. *Hydrol. Sci. J.* **1992**, *37*, 39–52. [CrossRef]
39. Krajewski, A.; Sikorska-Senoner, A.E.; Hejduk, A.; Hejduk, L. Variability of the Initial Abstraction Ratio in an urban and an agroforested catchment. *Water* **2020**, *12*, 415. [CrossRef]
40. Zhu, W.; Sun, Q.; Li, F.; Gao, Y.; Zhang, L. Assessment of the Effect of Urban Green Space Landscape on Reduction of Storm Water Runoff in Xiamen City. *Res. Environ. Sci.* **2019**, *32*, 74–84.
41. McCuen, R.H. *A Guide to Hydrologic Analysis Using SCS Methods*; Prentice-Hall, Inc.: Hoboken, NJ, USA, 1982.
42. Meixia, L. Study on Rainstorm Water Logging Disaster Risk Base on the Construction of Sponge City: A Case Study of Xiamen. Master's Thesis, Xi'an University of Science and Technology, Xi'an, China, June 2017.
43. Rawat, K.S.; Singh, S.K. Estimation of Surface Runoff from Semi-arid Ungauged Agricultural Watershed Using SCS–CN Method and Earth Observation Data Sets. *Water Conserv. Sci. Eng.* **2017**, *1*, 233–247. [CrossRef]
44. Al-Ghobari, H.; Dewidar, A.; Alataway, A. Estimation of Surface Water Runoff for a Semi-Arid Area Using RS and GIS-Based SCS–CN Method. *Water* **2020**, *12*, 1924. [CrossRef]
45. Quan, B.; Chen, J.-F.; Qiu, H.-L.; RÖMkens, M.J.M.; Yang, X.-Q.; Jiang, S.-F.; Li, B.-C. Spatial-Temporal Pattern and Driving Forces of Land Use Changes in Xiamen1 1Project supported by the Fujian Provincial Natural Science Foundation of China (No. D0210010). *Pedosphere* **2006**, *16*, 477–488. [CrossRef]
46. Cui, S.; Ding, S.; Li, X.; Tang, J.-X.; Li, F.; Shen, Y. Dynamic analysis of land exploitation in Maluan bay during 1957–2017. *J. Oceanogr.* **2019**, *30*, 131–139.
47. Ali, M.; Khan, S.J.; Aslam, I.; Khan, Z. Simulation of the impacts of land-use change on surface runoff of Lai Nullah Basin in Islamabad, Pakistan. *Landsc. Urban Plan.* **2011**, *102*, 271–279. [CrossRef]
48. Gajbhiye, S.; Mishra, S. Application of NRSC-SCS curve number model in runoff estimation using RS & GIS. In Proceedings of the IEEE-International Conference on Advances in Engineering, Science and Management (ICAESM-2012), Nagapattinam, India, 30–31 March 2012; pp. 346–352.
49. Zhou, F.; Xu, Y.; Chen, Y.; Xu, C.-Y.; Gao, Y.; Du, J. Hydrological response to urbanization at different spatio-temporal scales simulated by coupling of CLUE-S and the SWAT model in the Yangtze River Delta region. *J. Hydrol.* **2013**, *485*, 113–125. [CrossRef]
50. Shrestha, M.N. Spatially distributed hydrological modelling considering land-use changes using remote sensing and GIS. In Proceedings of the Map Asia Conference, Singapore, 18–20 August 2009; pp. 1–8.
51. Sheng, L. Study on the Information Extraction of the Urban Heat Island, Run-Off and Impervious Surface in Xiamen City of SE China with Remote Sensing Technology. Master's Thesis, Fujian University, Fuzhou, China, December 2005.

land

MDPI

Article

Flash Flood Events along the West Mediterranean Coasts: Inundations of Urbanized Areas Conditioned by Anthropic Impacts

Francesco Faccini [1,2], Fabio Luino [2,*], Guido Paliaga [2], Anna Roccati [2] and Laura Turconi [2]

1 Department of Earth, Environmental and Life Sciences, University of Genoa, Corso Europa 26, 16132 Genoa, Italy; faccini@unige.it
2 National Research Council, Research Institute for Geo-Hydrological Protection, Strada delle Cacce 73, 10135 Turin, Italy; guido.paliaga@irpi.cnr.it (G.P.); anna.roccati@irpi.cnr.it (A.R.); laura.turconi@irpi.cnr.it (L.T.)
* Correspondence: fabio.luino@irpi.cnr.it; Tel.: +39-011-397-7823

Abstract: Flash floods represent one of the natural hazards that causes the greatest number of victims in the Mediterranean area. These processes occur by short and intense rainfall affecting limited areas of a few square kilometers, with rapid hydrological responses. Among the causes of the flood frequency increase in the last decades are the effects of the urban expansion in areas of fluvial pertinence and climatic change, namely the interaction between anthropogenic landforms and hydro-geomorphological dynamics. In this paper the authors show a comparison between flood events with very similar weather-hydrological characteristics and the ground effects occurred in coastal areas of three regions located at the top of a triangle in the Ligurian Sea, namely Liguria, Tuscany and Sardinia. With respect to the meteorological-hydrological hazard, it should be noted that the events analyzed occurred during autumn, in the conditions of a storm system triggered by cyclogenesis on the Genoa Gulf or by the extra-tropical cyclone Cleopatra. The "flash floods" damage recorded in the inhabited areas is due to the vulnerability of the elements at risk in the fluvio-coastal plains examined. There are numerous anthropogenic forcings that have influenced the hydro-geomorphological dynamics and that have led to an increase in risk conditions.

Keywords: flash floods; intense rainfall; urbanized areas; damage; anthropic impacts; West Mediterranean; Italy

Citation: Faccini, F.; Luino, F.; Paliaga, G.; Roccati, A.; Turconi, L. Flash Flood Events along the West Mediterranean Coasts: Inundations of Urbanized Areas Conditioned by Anthropic Impacts. *Land* **2021**, *10*, 620. https://doi.org/10.3390/land10060620

Academic Editors: Andrea Petroselli, Raffaele Pelorosso and Matej Vojtek

Received: 9 May 2021
Accepted: 7 June 2021
Published: 9 June 2021

1. Introduction

Climatic changes, including the increase in extreme temperatures and number of heavy rainfall events, have been recorded since around 1950 in different areas of the world [1]. Floods are already the most frequent in European coastal areas [2–4] and among the costliest and deadliest natural processes in the Mediterranean area [5].

Based on an international disaster database [6], 200 billion Euros in damages related to various calamities since 1900 have occurred in the countries surrounding the Mediterranean Sea out of which 85 billion are related to river flooding [7,8].

The observed variability of flood frequency and discharge magnitude is therefore the result of a complex interaction between rainfall history and the factors that control the response of the river basins, in particular, run-off modes. In the Mediterranean environment, the global warming process manifests itself with an increase in the average annual air temperature and with a variation in the rainfall regime [9,10]. Some regions such as Liguria, Tuscany and Sardinia in Italy; Provence-Alpes-Cote d'Azur and Corse in France; and Catalonia and the Valencia area in Spain are particularly exposed to flash floods [7,11] for which rainfall peaks and the flood peaks of the watercourses are very close in time. This particular pattern is the result of the interplay between the dominant atmospheric low level flow circulation patterns and the relief and orientations of the northern Mediterranean coasts, which forces convergence and triggers convection.

The magnitude and impact of extreme floods vary significantly over the Mediterranean region with a significative difference between the West and East parts [12]. The western part of the region is much more prone to high impact and high magnitude events [13–16]. This is probably due to: (a) The proximity of the Atlantic Ocean and oceanic climatic influences at latitudes where eastward atmospheric flows dominate [7,17]; (b) the reliefs surrounding the West Mediterranean Sea forces the convergence of low-level atmospheric flows and the uplift of warm wet air masses that drift from the sea to the coasts, thereby triggering active convection and consequently short intense bursts of rainfall.

The coasts of the Mediterranean Sea are characterized by short and intense rainfalls [13,17], which in recent decades have shown an increasing frequency [18,19]. Furthermore, for Italy, the statistics seem to indicate an increase in geo-hydrological phenomena in small basins for the last 30–50 years [20]. These rainfalls, especially if concentrated on reliefs not far from the coasts, can generate the phenomenon of flash floods along the short streambeds that have significant slopes [21–23]. Such precipitations are characterized by convective events, typically with 100 mm or more cumulated rainfalls over a few hours. The affected areas are often limited to a few hundred square kilometers, with rapid hydrological responses, e.g., less than 6 h delay between the peak rainfall intensity and peak water discharge downstream [8].

The coastal urban flooding is a complex phenomenon which may occur in various forms such as: Urban flooding due to high intensity rainfall (pluvial flooding); urban flooding due to inadequate drainage; flooding caused by overtopping in the channels or streams/rivers. In coastal urban cities such as Genoa, Olbia and Livorno, severe flood scenarios mostly take place due to combination of surface flooding and stream overtopping. Urban flooding is one of the most severe problem in numerous parts of the world because they affect goods and can cause casualties. Urban flood, being a natural disaster, cannot be avoided; however, the losses incurred due to flooding can be reduced by proper flood mitigation planning. As such, it is necessary to have a proper estimation of flood extent and flood risk for the different flow conditions so that proper flood evacuation and disaster management plan can be prepared in advance.

The flash flood consequences and the ground effects are amplified if the floodwaters spread to densely urbanized areas [24–27]. They are usually crossed by canalized streams that are often culverted for long stretches. These streambeds, which have been narrowed year after year to acquire new urban spaces [28,29], are often surmounted by bridges that are inadequate, with spans that are clearly insufficient for the discharge of flood waters.

Flash floods characterized by severe ground effects are generally triggered by:

1. Short-lived (often less than 3 h) strongly convective intense rainfall events, with total rainfall amounts (200–300 mm). Such violent events have a limited areal extent (<100 km^2) and generate local floods of small headwater streams that usually possess a surface of <40 km^2. A typical example of such flash floods is the catastrophic flash flood that occurred in eastern part of Genoa Metropolitan Area in November 2002 [30].
2. Mesoscale convective systems can produce stationary rainfall amounts exceeding 200–300 mm in a few hours [31]. During two severe cloudbursts that hit the Liguria coasts in the last 10 years, 539.0 mm/24 h were recorded at the Brugnato rain gauge station, during the famous Cinque Terre event, in Eastern Liguria during October 2011 [32] and 556.0 mm/24 h at Quezzi station, Bisagno Valley, Genoa city in November 2011 [33,34]. The areal extent of such events ranges from less than 100 km^2 to greater than hundreds of km^2.
3. On some occasions, heavy and prolonged rainfall may be part of a large-scale perturbation lasting several days. In such cases, extreme rainfall accumulation may be observed locally: 700 mm over 6 days (up to 1800 mm in October and November) caused floods and further loss of life on the border between Liguria and Piedmont

during the events of 21–22 October 2019 and 23–24 November 2019 [35]. These events generally cover a larger area from hundreds to thousands of km^2.

Along the Italian coasts, during the period from September to November, the so-called "Meteorological Fall" is the main season for flash floods that cause severe damage and often casualties due to their suddenness. This is particularly the case of mesoscale convective systems producing long lasting and stationary rainfall events that lead to strong responses by the watersheds concerned (i.e., high runoff rates due to soil saturation) and substantial coincidence between the peak rainfall and flood peak in small hydrographic basins (<250 km^2).

The north-western coasts of Italy are historically subject to flash floods: Ligurian coasts, along with the Tuscan coasts and coasts in Sardinia. There are at least 46 damaging flood events that have been recorded in the last 30 years, practically one every 7.8 months (Table 1): 9 events occurred in Tuscany, while 19 cases occurred in Liguria and Sardinia.

Table 1. Severe meteorological events during the period 2000–2020 that occurred in Sardinia and along the coasts of Liguria and Tuscany with severe consequences (flooding/flash floods) for urban areas and inhabitants. The content of the brackets displays the victim numbers. The three events described in this article are bolded.

Year	Month	Day	Place/Area (Casualties)	Region
2020	Nov	28	Bitti (3)	Sardinia
2020	Oct	2–3	Ventimiglia (10)	Liguria
2017	**Sept**	**9–10**	**Livorno (8)**	**Tuscany**
2015	Jul	22	Cagliari	Sardinia
2014	Oct	14	Grosseto and Orbetello	Tuscany
2014	Nov	5	Carrara (1)	Tuscany
2014	Nov	10	Recco-Chiavari-Camogli (2)	Liguria
2014	Oct	9–10	Genoa (1)	Liguria
2013	**Nov**	**18**	**Olbia (18)**	**Sardinia**
2012	Nov	28	Carrara and Ortonovo	Tuscany
2012	Nov	12	Maremma and Grosseto (7)	Tuscany
2011	Nov	7	Isola d'Elba (1)	Tuscany
2011	Nov	4	Genoa (6)	Liguria
2011	Oct	24–25	Cinque Terre e Lunigiana (13)	Liguria and Tuscany
2010	**Oct**	**4**	**Genoa Sestri Ponente (1)**	**Liguria**
2010	Sept	7	Genoa	Liguria
2010	Jan	25–26	Olbia	Sardinia
2008	Nov	4	Olbia	Sardinia
2008	Oct	22	Capoterra (5)	Sardinia
2008	Jun	16	Genoa	Liguria
2007	Jun	1	Genoa	Liguria
2006	Nov	13	Cagliari	Sardinia
2006	Sept	25–26	Cagliari	Sardinia
2006	Sept	15	Bordighera and Vallecrosia	Liguria
2006	Aug	16–17	Genoa	Liguria
2005	Nov	13	Cagliari	Sardinia
2005	Apr	5–6	Cagliari, Capoterra, Pula	Sardinia
2004	Dec	6	Villanova Strisaili (2)	Sardinia
2003	Sept	23	Massa-Carrara (2)	Tuscany
2002	Nov	23	Imperia and Genoa	Liguria
2002	Nov	11	Cagliari	Sardinia
2002	Oct	9	Cagliari	Sardinia
2000	Nov	6	Imperia and Savona	Liguria

On the one hand we can affirm that flash floods are geo-hydrological processes linked to particular hydro-meteorological conditions and that their behaviors are significantly affected by climate change [36], but on the other hand we cannot omit the fundamental role played by wild urbanization [37], which has affected many towns on the western Mediterranean coast. This uncontrolled expansion occurred after the Second World War and appears to have been particularly significant in the already notoriously hazardous areas that have undergone and imparted important changes to the hydrographic network over time.

There are some fundamental reasons that constitute the basis of the decision to write this paper. First of all, the fact that no author has so far compared the events that took place in these regions, which are located adjacent to one another along the coast around the Upper Tyrrhenian Sea, in the national and international body of literature and research. The second reason is to compare the dynamics of three similar flood events, both in terms of meteorological and hydrological characteristics, and in terms of effects on the ground, with particular attention to the identification of any anthropogenic factors. In detail, three flash flood events were chronologically analyzed (Figure 1): (1) Genoa Sestri Ponente (Liguria), which occurred on 4 October 2010; (2) Olbia flood (Sardinia) on 18 November 2013 triggered by cyclone Cleopatra; (3) Livorno flood (Tuscany) on 9 September 2017 triggered by the cyclone Genoa Low.

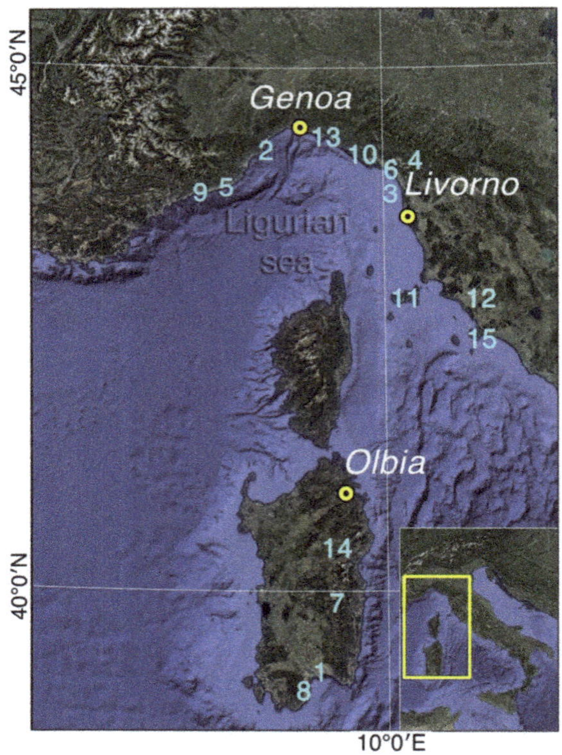

Figure 1. Geographic location of the case-studies: Genoa Sestri Ponente (Liguria), Livorno (Tuscany) and Olbia (Sassari, Sardinia). Other places named in the text: (1) Cagliari; (2) Savona; (3) Versilia; (4) Garfagnana; (5) Imperia; (6) Massa and Carrara; (7) Villanova Strisaili; (8) Capoterra; (9) Bordighera, Vallecrosia, Ventimiglia; (10) Cinque Terre; (11) Elba Island; (12) Grosseto, Maremma; (13) Recco, Camogli, Chiavari; (14) Bitti; (15) Orbetello.

2. Study Area

2.1. Genoa Sestri Ponente City

Sestri Ponente is a district of the Genoa Metropolitan City, the regional chief town of Liguria (northern Italy) (Figures 1 and 2A); it extends over an area of about 8 km^2, with a population of 45,000 inhabitants. Until 1926, Sestri Ponente was an autonomous municipality until it was incorporated, together with other municipalities of the Genoa neighborhood, called "genovesato", into the "Big Genova" City. The historic core of Sestri Ponente is represented by a narrow coastal strip, about 1 km long, running parallel to the shoreline and included in the floodplain between the Chiaravagna stream to the east and the Cantarena stream to the west. The industrial expansion in the 1920s and 1930s developed in the alluvial plain around the historic core, while the subsequent urban sprawl from the 1960s to date has continued upstream and occupies the hill slopes behind (Figure 2A). Now the city is developed for 2.6 km in length along the coastline.

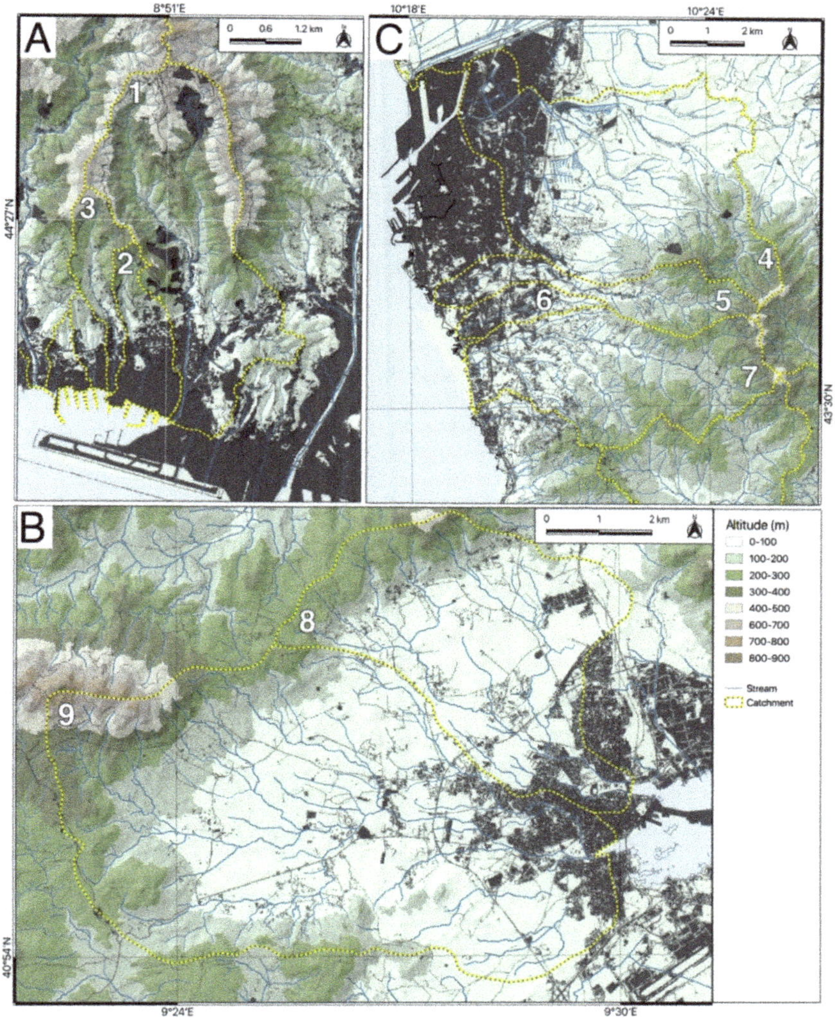

Figure 2. Elevation maps and catchments of the case studies: Genoa Sestri Ponente (**A**), Olbia (**B**) and Livorno (**C**). Numbers refer to catchments in Table 2.

From the geomorphological point of view, Sestri Ponente stands on a narrow alluvial plain that is 2 km length and less than 500 m wide and is genetically linked to the action of several watercourses and significantly modified by anthropogenic landforms. Drainage networks are well developed with a torrential hydrological regime. Catchments are very small in size (generally <10 km^2) and characterized by high energy relief (Table 2) due to elevation exceeding 700 m a.s.l. at a few kilometers from the coast and high slope gradient locally higher than 60% (Figure 3A): Rio Chiaravagna Stream (11 km^2, with an estimated maximum full-flow rate of 213 m^3/s for a 50-years return period), Rio Cantarena Stream (1.58 km^2, 52 m^3/s), Rio Molinassi Stream (2.00 km^2, 66 m^3/s) and Rio Marotto Creek (0.67 km^2, 22 m^3/s) [38]. The final stretches of these waterways are generally canalized and drained.

Table 2. Morphometric parameters of the basins analyzed (the numbers refer to Figure 2). H Max, maximum altitude; H Med, mean altitude; G med, mean gradient; Sup, surface; Ss, soil sealing; L rn, river network total length; Dd, Drainage density.

Area	Catchment	H Max (m)	H Med (m)	G Med (%)	Sup (km^2)	Ss (%)	L Rn (km)	Dd (km^{-1})
Genoa	1	658	262	47.2	11	23.4	60.5	5.7
	2	438	89	31.8	1.9	49.4	5.6	3.0
	3	545	230	49.4	1.8	20.4	9.2	5.1
Livorno	4	430	71	13.4	37.4	23.9	126.0	3.4
	5	456	132	19	8.9	22.9	26.0	2.9
	6	107	26	3	3.0	57.2	5.4	1.8
	7	456	148	20.6	21.9	13.1	71.1	3.3
Olbia	8	457	89	10	20.9	14.1	37.0	1.8
	9	700	115	12.8	48.2	10.7	104.6	2.2

Land use in the studied catchments mainly consists of artificial surfaces (32.5%) in the lower part and forests and seminatural areas (45.0%) in the upper part of the basins; subordinately are agricultural areas (7.0%) (Supplementary Materials Table S1)

The climate of Sestri Ponente is typically Mediterranean, with hot summers and relatively mild winters: the average annual temperature is about 15 °C with a significant upward trend. The mean annual rainfall during the period from 1960 to 2019 is about 1100 mm and the annual rainy days are 71 with an average rainfall rate of 14 mm/d, which is also increasing [21]. The episodes of intense and short-lived rain are frequent, especially in the autumn months (October and November), when typically small-sized and quasi-stationary V-shaped convective systems are generated [31]. These systems result in extreme flood-causing rainfalls, for which ground effects are often disastrous. In the post-war period alone, flood events occurred 10 times in 69 years, which on average amounts to 1 event every 6.9 years (Supplementary Materials Table S2).

2.2. Olbia City

The Olbia city is one of the major cities of Sardinia and the most important link with the mainland and other Mediterranean countries due to its tourist and commercial port (Figures 1 and 2B). It extends for about 40 km^2 (including the more peripheral districts) with a population of about 60,000 inhabitants, while during the summer it can reach 100,000 inhabitants.

The whole coastal sector has undergone important changes since the early 1900s as a result of management and sanitation. In the first decades of the 20th century, large swampy and brackish areas were reclaimed. Then, beginning from the 1960s Olbia experienced a demographic increase related to the tourist development in the neighboring Costa Smeralda, which is world-famous for its sea landscape and resorts. In 50 years, the town has tripled its population, transforming the socio-economic fabric and the population of inhabitants increased from 17,800 inhabitants (1961) to 60,000 today.

The plain of Olbia occupies a structural depression resulting from the kinematics of the Sardinian-Mediterranean block related to the formation of the Western Tyrrhenian Sea [39]. Due to the irregular and indented coastal morphology, with parallel hollow rias separated by prominent ridges featured by typical plateau and sierras, drainage patterns are mainly dendritic and slightly developed; the upper part of catchments locally exceeds 700 m a.s.l. and the watercourses that result are generally steep (Figure 3B). As a consequence of the progressive urbanization of the plain, streams crossing the urban area are strongly modified due to canalization and flow regulation. The main watercourses (Table 2) are the Rio Seligheddu Stream (38.4 km^2, with a maximum capacity of 330 m^3/s for a 200 year return period), the Rio San Nicola Stream (30 km^2, 170 m^3/s) and the Rio Gadduresu Stream (7 km^2, 55 m^3/s) as well as other minor ones (<5 km^2), resulting in an overall flow rate exceeding 600 m^3/s [40,41].

Land use in the studied catchments consists mainly in agricultural use (63.9%); artificial surfaces are present in suborder (12.9%) and forests and seminatural areas (22.7%) (Supplementary Materials Table S1).

The climate is typically Mediterranean, with mild and humid winters and hot and dry summers. The average annual temperature is about 16 °C. Sardinia is located almost at the center of a low-pressure area that determines the convergence of different air masses and the formation of self-regenerating convective systems, especially during the winter period, resulting in strong "V-shaped" marine thunderstorms. The mean annual precipitation ranges between 600 and 900 mm and the annual rainy days are 60 (rainfall rate 10–15 mm/d): rainfall is more abundant in the autumn and winter months (from October to December), while a minimum rainfall peak occurs in summer [41]. The urban development of Olbia is relatively recent. As a consequence, only documents of floods post Second World War have been discovered and considered: at least 16 flash floods in the last 69 years (on average 1 event every 4.3 years) (Supplementary Materials Table S2).

2.3. *Livorno City*

Livorno city extends for just over 100 km^2 of surface along the Tyrrhenian coast and represents one of the most important trade and industrial centers in Tuscany (central Italy) (Figures 1 and 2C): the urban area is subdivided into numerous districts and a total of approximately 158,000 inhabitants are counted. The city has already developed in historical times around the port area, with progressively more important enlargements since the beginning of 1900 linked to the development of communications and industrial activities. Urbanization has resulted in the growing occupation of natural drainage areas and floodplains [29,42]. Most expansions relate to the period after Second World War.

Livorno town stands on a flat sector (in the north-side and along the coast) which corresponds to a polycyclic marine terrace characterized by the presence of three orders of sea terraces that are, at least, aged between the middle Pleistocene and the upper Pleistocene. They are crossed by a hydrographic network that consists of several streams. The alluvial plain is, in fact, a low coastal terrace located north of the city and is shaped by the course of the River Arno in the mouth area.

Catchments are small (Table 2), with a size of generally <30 km^2; they do not reach particularly high elevation (about 400 m a.s.l.) in their upper sectors; they possess mildly steep terrain and gentle hilly slopes (Figure 3C). Drainage networks are well developed; they possess watercourses of limited length; and a well-defined grid of small tributaries: the major ones are the Rio Ugione Stream (33.2 km^2 with a maximum evaluated discharge of 137 m^3/s for a 200-years return period), the Rio Ardenza Stream (21.2 km^2, 284 m^3/s) and the Rio Maggiore Stream (8 km^2, 100 m^3/s) [43].

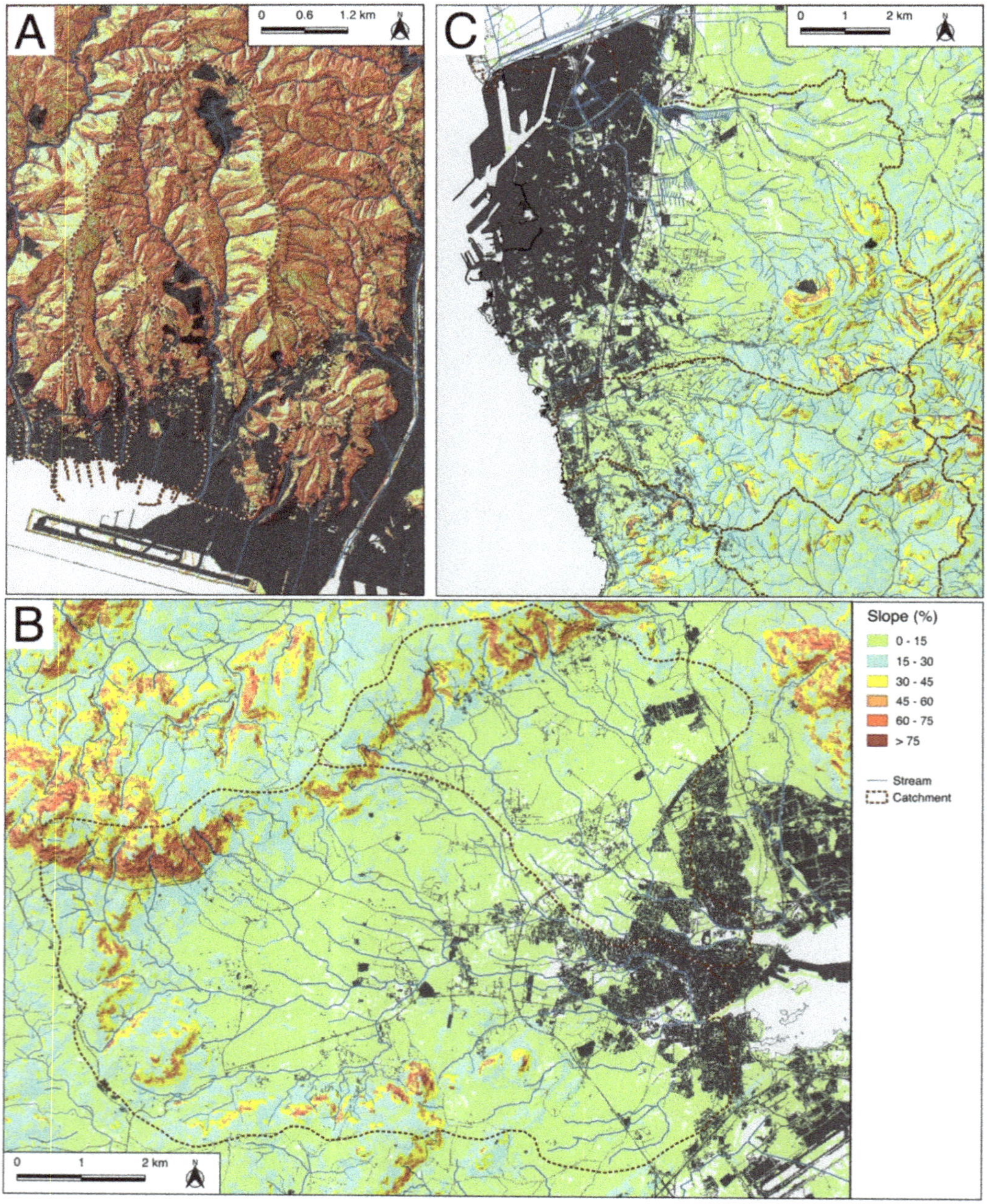

Figure 3. Slope maps and catchments of the case-studies: Genoa Sestri Ponente (**A**), Olbia (**B**) and Livorno (**C**).

Land use in the studied catchments mainly consists of forests and seminatural ar eas (50.0%); agricultural are present in suborder (28.5%) and artificial surfaces (20.4% (Supplementary Materials Table S1).

During the period 1969–2018, the mean annual rainfall recorded in Livorno is abou 800 mm that is mainly concentrated during the autumn, with an average of 74 rain days (rainfall rate 10.7 mm/d) [44]. The climate is Mediterranean, characterized by warm summers, mitigated by the presence of sea breezes and not particularly cold winters. The average annual temperature is about 15.8 °C.

Damaging effects on the ground in Livorno are associated with both flooding of water streams and pluvial floods due to intense rainfall events: before the last event, they occurred at least 16 times in 71 years. On average there is 1 event every 4.4 years (Supplementary Materials Table S2).

3. Materials and Methods

3.1. Research Methodology

The applied methodology has been developed in different phases. First of all, a close bibliographic and cartographic research has been carried out, with the aim of cataloguing recent and historical floods that occurred in the three study areas (Genoa, Olbia and Livorno) and correlate regions. Flood and damage information have been derived from different sources: (i) newspapers articles and chronicles notes from local media, (ii) inedited documents and technical and event reports collected in archives of local municipalities, (iii) books and scientific papers gathered in the libraries and (iv) interviews with local inhabitants.

We obtained rainfall and hydrological data about recent and past events from Hydrological Annals edited by SIMN ("National hydrographic and Tidal Service"), as well as technical and weather-hydrological reports compiled by territorial agencies and regional databases.

For Liguria, information from "Reports of the weather-hydrological events (2003–2019)" by ARPAL and from "Pluviometrical regional database" by OMIRL was mainly used. For Sardinia, information from "Report of the weather-hydrological events (2013)" by ARPAS was mainly used. For Tuscany, information from "Report of the weather-hydrological events (2009–2019)" by CFR and LaMMA Association was mainly used.

Subsequently, a closer pluviometrical examination has been performed for the most recent and damaging flood events: (i) 4 October 2010 in Sestri Ponente [45], (ii) 18 November 2013 in Olbia [46] and (iii) 9 September 2017 in Livorno [47,48].

Flooded areas in the Sestri Ponente, Olbia and Livorno cities during the three considered events have been surveyed and provided by the regional cartographic online database of Liguria, Sardinia, Tuscany and also by the Livorno Municipality. Culverted streams and canals in the Olbia urban area have been surveyed from aerial photointerpretation, technical surveys and the analysis of technical reports for the construction of the drainage canals.

A multi-temporal comparison has been performed using historical and current topographical and cartographical maps in order to identify anthropogenic landforms and to reconstruct the urban evolution for each case study. In addition, field observations have been carried out to evaluate geomorphological and hydrological aspects of the urbanized areas involved in the flood events. All maps and cartographic data used in the research are listed in Table 3.

We integrated all georeferenced data in a Geographical Information System: using QGIS, we derived new original thematic maps which are useful to provide the immediate identification of both flooded areas in relation to urbanization and the main anthropogenic landforms within each study area.

Table 3. Vector and raster data used in the research. Name: DTM, Digital Terrain Model; CORINE, Coordination of information on the environment (European Environment Agency, 1995); TMI, Topographical Map of Italy. Source: GE, Google Earth; IGM, Italian Military Geographical Institute; ISPRA, Higher Institute for Environmental Research and Protection; LR, Liguria Region; PGRA, Flood risk management plan, SR, Sardinia Region; TR, Tuscany Region. Type: R, raster; V, vector.

Name	Source	Type	Scale/Pixel	Date
Catchment	LR	V	1:10,000	2019
DTM	LR	R	5 m	2016
	TR	R	10 m	2017
	SR	R	10 m	2012
Flooded areas—PGRA	LR	V	1:10,000	2015
	TR	V	1:10,000	2016
	SR	V	1:10,000	2017
High-resolution satellite image	GE	R	-	2020
Hydrographical network	LR	V	1:5000	2019
	TR	V	1:10,000	2018
	SR	R	1:10,000	2019
Land use—CORINE Land cover	ISPRA	V	25 m	2012
Soil sealing	ISPRA	R	10 m	2018
TMI—Series 25 "Foglio 082—Tavoletta II-NE (Rivarolo Ligure/Sestri Ponente)"	IGM	R	1:25,000	1878, 1907, 1923, 1930, 1934, 1940
TMI—Series 25 "Foglio 082—Tavoletta II-SE (Genova)"	IGM	R	1:25,000	1899, 1907, 1923, 1930, 1934, 1939
TMI—Series 25 "Foglio 111—Tavoletta I-NO (Tombolo/Tirrenia)"	IGM	R	1:25,000	1881, 1939
TMI—Series 25 "Foglio 111—Tavoletta I-NE (Guasticce)"	IGM	R	1:25,000	1881, 1939
TMI—Series 25 "Foglio 111—Tavoletta I-SE (Salviano)"	IGM	R	1:25,000	1881, 1939
TMI—Series 25 "Foglio 111—Tavoletta I-SO (Livorno)"	IGM	R	1:25,000	1881, 1939
TMI—Series 25 "Foglio 182—Tavoletta I-NE (Muddizza Piana)"	IGM	R	1:25,000	1958
TMI—Series 25 "Foglio 182—Tavoletta IV-SO (Loiri)"	IGM	R	1:25,000	1896, 1931, 1958
TMI—Series 25 "Foglio 182—Tavoletta IV-NO (Terranova Pausania/Olbia"	IGM	R	1:25,000	1896, 1931, 1958

3.2. Hydro-Meteorological Data of the Last Flood Events

Liguria, Sardinia and Tuscany overlook the Ligurian-Tyrrhenian Sea and are three regions that are very prone to violent atmospheric phenomena (see Table 1). Genoa Sestri Ponente, Olbia and Livorno, which are coastal cities that arose at the mouth of rivers, naturally have a long and troubled history of floods that are usually flash floods: their degree of damage has increased year after year in proportion to the degree of urbanization reached by the cities. The areas of river pertinence, that is, those closest to the riverbeds which were once occupied by fields and pastures, have gradually been invaded by buildings; consequently, every time a watercourse overflows during present times, damages result.

This finding stems precisely from the review of the latest cases in the three cities examined. For Genova Sestri Ponente, the last flood of 4 October 2010 was undoubtedly the most serious in terms of damage. The storm cell formed on the first morning of 4 October 2010 stabilized in the neighborhoods of the western city between Pegli, Sestri Ponente and Val Polcevera where it unleashed all its power and caused a true flash flood. In about five hours, between 8 a.m. and 1 p.m., over 400 mm of water fell on the hills behind Sestri Ponente [45]. All the streams reached rapidly exceptional discharges: the waters violently flooded shops, garages, basements, squares, streets, washed parked cars away and also resulted in a victim.

Olbia has also suffered many flood events in the past: 14 events in the period 1946–2010. However, the severest flood was the last flash flood that occurred on 18 November 2013. Six provinces out of the eight existing on the island were affected: the total damage amounted to about 660 million EUR [46] and 18 were casualties; some were drowned at home and others were dragged by the fury of the streamflows while driving their cars. The Cyclone Cleopatra hit the interior of Sardinia with cumulative rainfall greater than 400 mm: the Olbia rain gauge station recorded a value of 117.6 mm, while the Putzolu rain gauge station, in a village close to Olbia, recorded 175.2 mm in 24 h [47]. The waters of the canals and streams crossing the town overflowed with heights greater than two meters on the countryside level. In the town if Olbia, 11 were the casualties and 40 people were hospitalized for symptoms of asphyxia and hypothermia after having been at the mercy of freezing water for hours. Over 2000 displaced people resulted and some hundreds of millions of EURO in damages were incurred.

Livorno also has a long list of flood events; there is at least 14 between 1946 and 2004, but the most recent event was certainly the most serious of its history. On the evening of 9 September 2017 a violent cloudburst hit the Livorno area: 242 mm of rainfall (74.8 mm in 30 min and 210.2 mm in 2 h) was recorded on the hilly areas just east of the city, on the upper basins of the Rio Maggiore and Rio Ardenza streams [48]. The Rio Maggiore stream, culverted in the 1980s, overflooded at the beginning of the culvert and invaded a large area of the city. In total, over 4.3 km^2 were flooded and a large part of this was an urbanized area. The damage to structures and infrastructures was very serious (6.6 million euros), with 8 casualties and a dozen injured in the city alone.

4. Results

4.1. Rainfall Events

With respect to the event of 4 October 2010 on the Ligurian coast, from the analysis of rainfall data recorded in the basins considered (Table 1) it can be highlighted that for the Genoa Sestri Ponente area precipitation resulted from the formation of intense self-regenerating systems (MCS) due to a configuration favorable to a strong convergence between South and South-East, which insisted on the center of the region and in particular on the border between the provinces of Genoa and Savona. Around midnight on 4 October 2010, a stormy event of strong intensity occurred in the area of the Ligurian coast, enhanced by the orographic barrier of the Alpine–Apennine chain and favored by the high sea temperature due to the concomitant presence of an anticyclonic front in the Mediterranean Sea.

After about six hours, a violent weather-pluviometric system reached the town of Varazze with rainfall of about 100 mm/1 h and 220 mm in 3 h. Between 9 p.m. and 12 p.m., the storm cells that had hit the Riviera di Ponente moved towards Genova Sestri Ponente, which was about 20 km to the east. Here, the rainfall recorded at the Mt. Gazzo station (OMIRL-ARPAL hydrological Service) reached 124 mm/1 h, 243 mm/3 h, 360 mm/6 h and 411 mm/12 h (Figures 4–6), compared to annual averages of about 1100 mm. In the areas surrounding Sestri Ponente, high intensities of rain were recorded: 98 mm/1 h peak and a cumulative rainfall of 377 mm/12 h in Pegli (west), while at the Bolzaneto station (north) cumulative rainfall was recorded at 73 mm/1 h and 295 mm/12 h. In Genoa and east of Sestri Ponente, the peak intensity was 40 mm/1 h and the cumulative was 100 mm/12 h (Figure 7).

From 1945 to present day, in which at least ten events with damaging effects on the ground occurred, the event of 4 October 2010 in Liguria the sixth highest for rainfall intensity within a time period of 12 h.

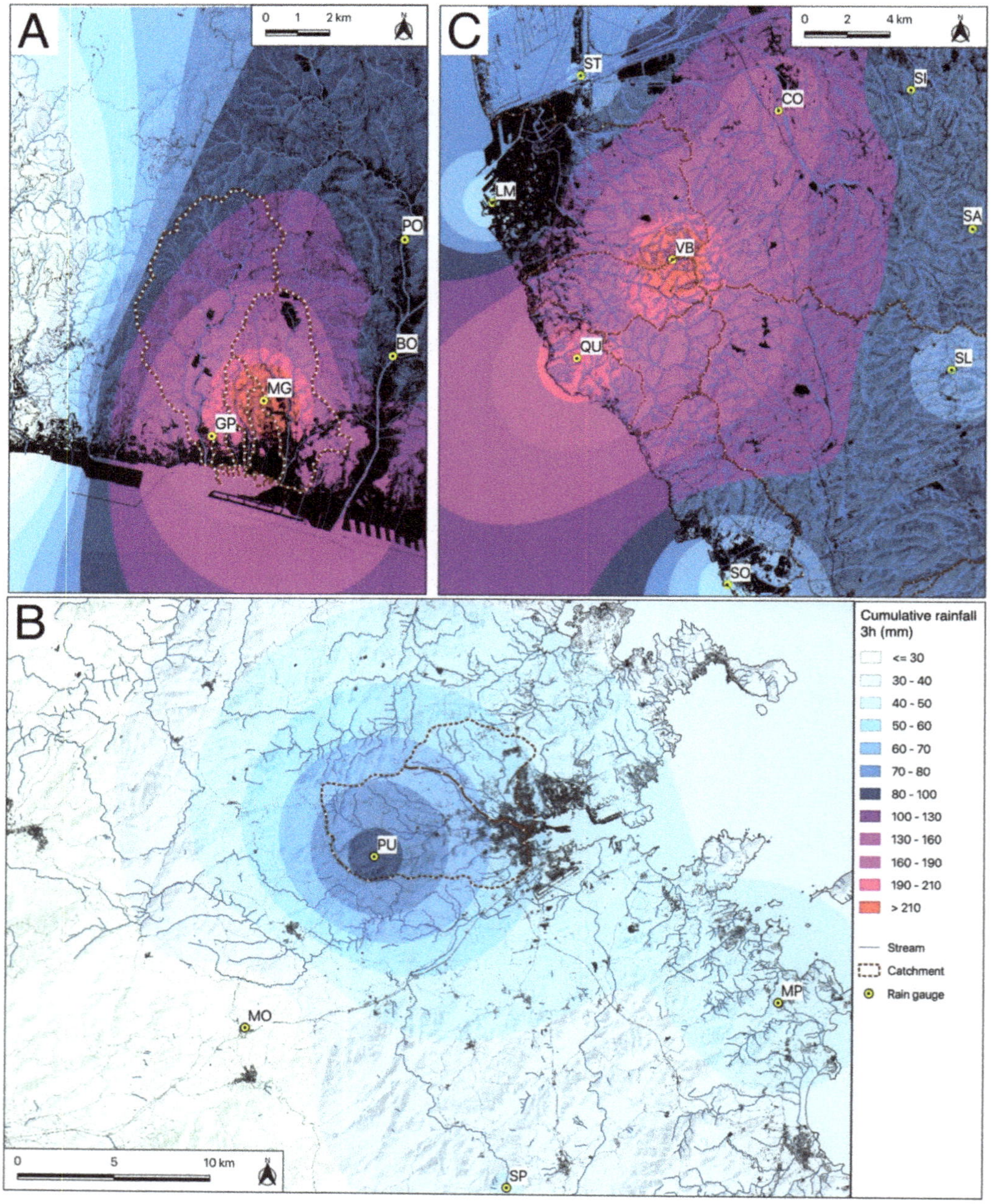

Figure 4. Cumulative rainfall map for a time period of 3 h: (**A**) Genova Sestri Ponente, 4 October 2010; (**B**) Olbia, 18 November 2013; (**C**) Livorno, 9 September 2017. Rain gauge stations: (GP) Genova Pegli; (MG) Monte Gazzo; (PO) Pontedecimo; (BO) Bolzaneto; (LM) Livorno Mareografo; (ST) Stagno; (QU) Quercianella; (VB) Valle Benedetta; (CO) Collesalvetti; (SI) Siberia; (SA) Santermo; (SL) Santa Luce; (SO) Solvay; (PU) Putzolu; (MO) Monti; (MP) Monte Petrosu; (SP) Sa Pianedda.

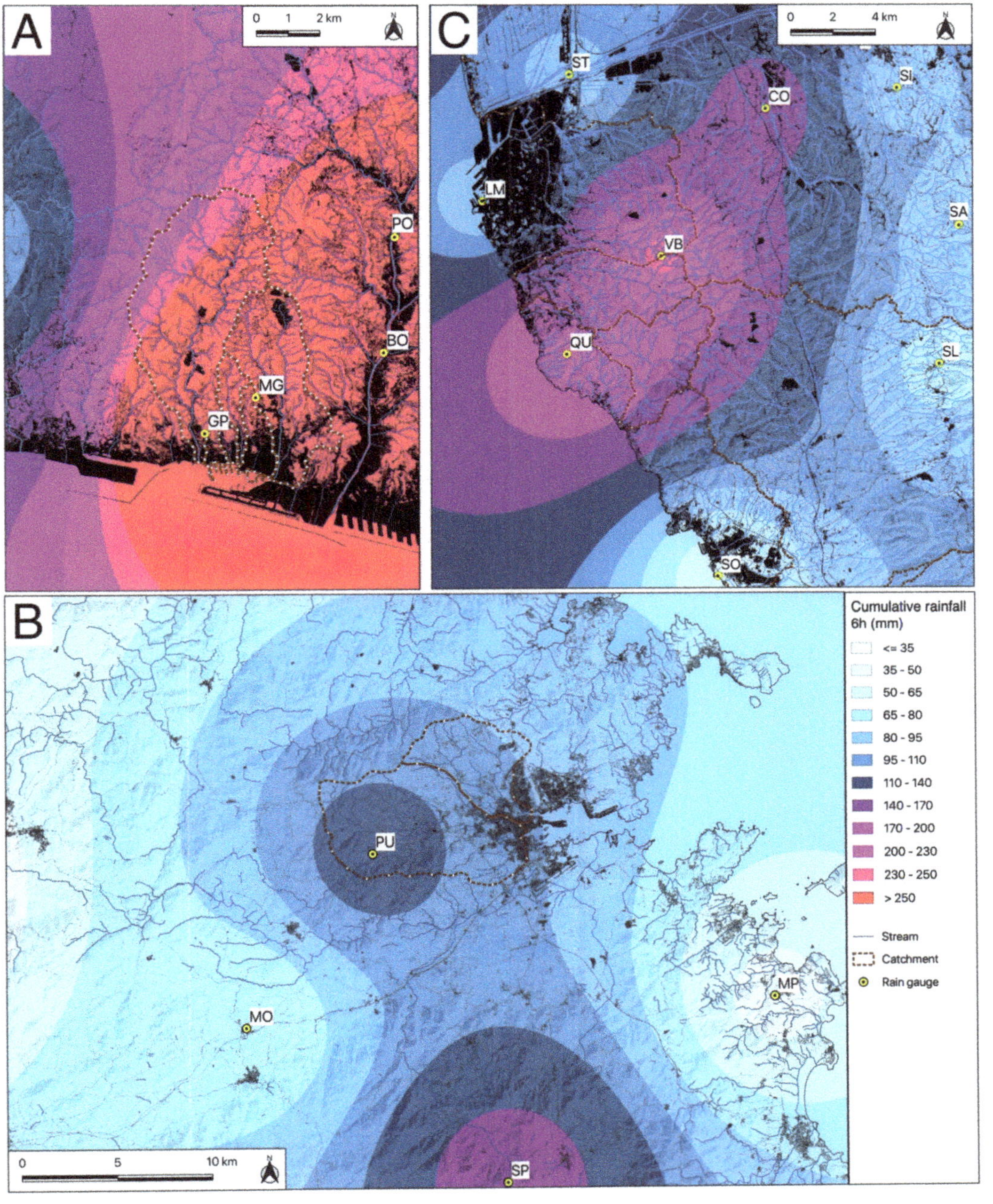

Figure 5. Cumulative rainfall map for a time period of 6 h: (**A**) Genova Sestri Ponente, 4 October 2010; (**B**) Olbia, 18 November 2013; (**C**) Livorno, 9 September 2017. Rain gauge stations: (GP) Genova Pegli; (MG) Monte Gazzo; (PO) Pontedecimo; (BO) Bolzaneto; (LM) Livorno Mareografo; (ST) Stagno; (QU) Quercianella; (VB) Valle Benedetta; (CO) Collesalvetti; (SI) Siberia; (SA) Santermo; (SL) Santa Luce; (SO) Solvay; (PU) Putzolu; (MO) Monti; (MP) Monte Petrosu; (SP) Sa Pianedda.

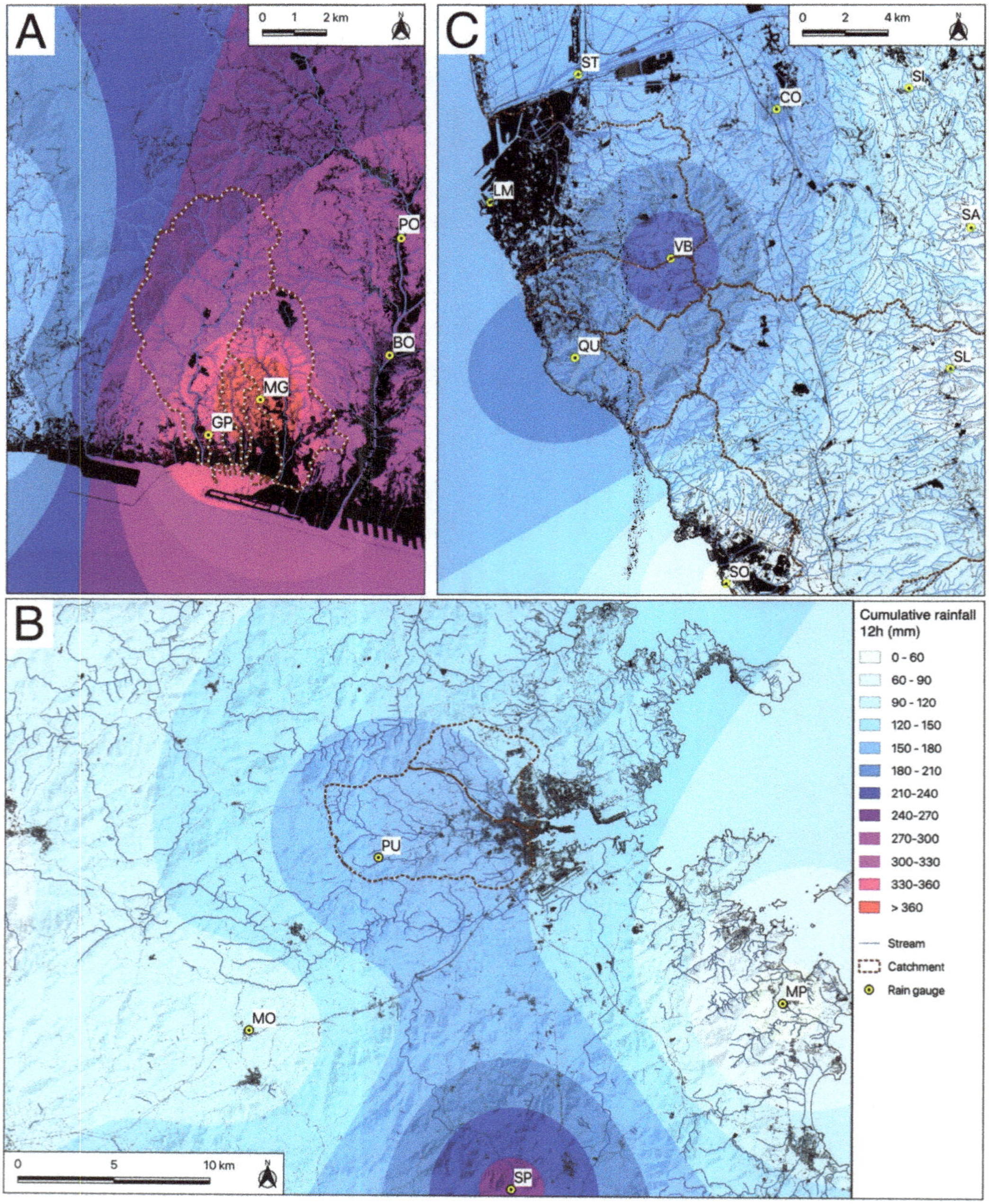

Figure 6. Cumulative rainfall map for a time period of 12 h: (**A**) Genova Sestri Ponente, 4 October 2010; (**B**) Olbia, 18 November 2013; (**C**) Livorno, 9 September 2017. Rain gauge stations: (GP) Genova Pegli; (MG) Monte Gazzo; (PO) Pontedecimo; (BO) Bolzaneto; (LM) Livorno Mareografo; (ST) Stagno; (QU) Quercianella; (VB) Valle Benedetta; (CO) Collesalvetti; (SI) Siberia; (SA) Santermo; (SL) Santa Luce; (SO) Solvay; (PU) Putzolu; (MO) Monti; (MP) Monte Petrosu; (SP) Sa Pianedda.

The meteorological event that affected the Olbia area on 18 November 2013 is among the most serious in recent years that have extensively affected the entire region of Sardinia in a paroxysmal manner. The set of meteoclimatic conditions resulted in the formation of strong "V-shaped" self-healing marine storms on the eastern sector [28] and is linked to the Cleopatra perturbation with consequent effects on the ground, such as flash floods.

The violent thunderstorms, produced by the convergence line of winds from the west and south-east close to the internal reliefs in the upper part of the river basins, have given rise to cumulative rainfall exceeding 230 mm in the area; an exception to this is Olbia, which hovered around the 120 mm mark with a maximum hourly intensity of 61 mm (Sa Pianedda station) [49]. With respect to the rainfall trend, reference was made above all the rain gauges of Olbia and Putzolu (ARDIS hydrological network): the latter, less than 10 km as the crow flies west of Olbia, is more representative of the hydrological conditions persisting in the interior of the hydrographic basins along which the most damaging effects occurred. The set of data available, however, allowed a better spatial reconstruction of the meteoric event.

At the Olbia rain gauge, the maximum intensity was 28 mm/h (between 6 p.m. and 7 p.m.) and it was preceded by at least another 8 h of precipitation with variable intensity up to 18 mm/h. At the Putzolu rain gauge, higher rainfall heights were recorded, with a daily cumulative of 175.2 mm compared to that of Olbia with 117.6 mm. The maximum hourly intensities in Putzolu reached 45 mm/h between 3 p.m. and 4 p.m., while at the Monte Petrosu rain gauge, the intensity over 15 min was 26.4 mm along the coast. The station that recorded the highest cumulative precipitation was that of Sa Pianedda, which is close to the first hills south of Olbia, with values equal to 150 mm/3 h, 167.2 mm/6 h and 247.4 mm/12 h. The estimated return periods are about 200 years (up to 12 h) for the Putzolu station and about 50 years (up to 6 h on the data) for Olbia [49] (Figures 4–6). Over 80% of the recorded rainfall was concentrated in just over six hours, but persisted for over 12 h (Figure 7); due to their continuity, they were sufficient to cause maximum flows not only in the smallest basins but also in the terminal sections of the largest hydrographic basins.

During the night between 9 and 10 September 2017, the flood event that affected the town of Livorno was anticipated by several storms. During the first, which mainly affected the coastal areas between the territories of Livorno city and Marina di Pisa, maximum cumulative rains of 63.4 mm/1 h was recorded over Livorno (between 8:45 p.m. and 9:45 p.m.) and 65.6 mm/1 h in Marina di Pisa (Bocca d'Arno station). In latter area the rainfall continued to intensify, but the rains practically stopped in Livorno after 9.45 p.m.

Starting from 2:00–2:30 a.m. on Sunday, a new and strong thunderstorm, which then turned out to be the most violent, mainly affected the areas between the southern area of Livorno city and Rosignano town [43,50–52]. In these areas, the values of rainfall reached, which on short durations are really extreme, had peaks higher than 42.4 mm/15′, 121.8 mm/1 h (Rp > 200 y), 210 mm/2 h and 230 mm/3 h [44,50] (Figure 7). There is a clear difference between the maximum data recorded in these hours in the different time intervals of the durations 1, 2 and 3 h by the stations of Quercianella and Valle Benedetta compared to the stations located slightly further south or more inland (such as Castellina Marittima and Santa Luce) (Table 4) or further north (as Livorno Mareografo); this difference highlights the strong localization of the thunderstorm phenomenon that locally discharged over 200 mm of rainfall in 2 h (Figures 4–6). Estimated return periods for 1 h and 3 h of rainfall that were recorded during this event are more than 200 years.

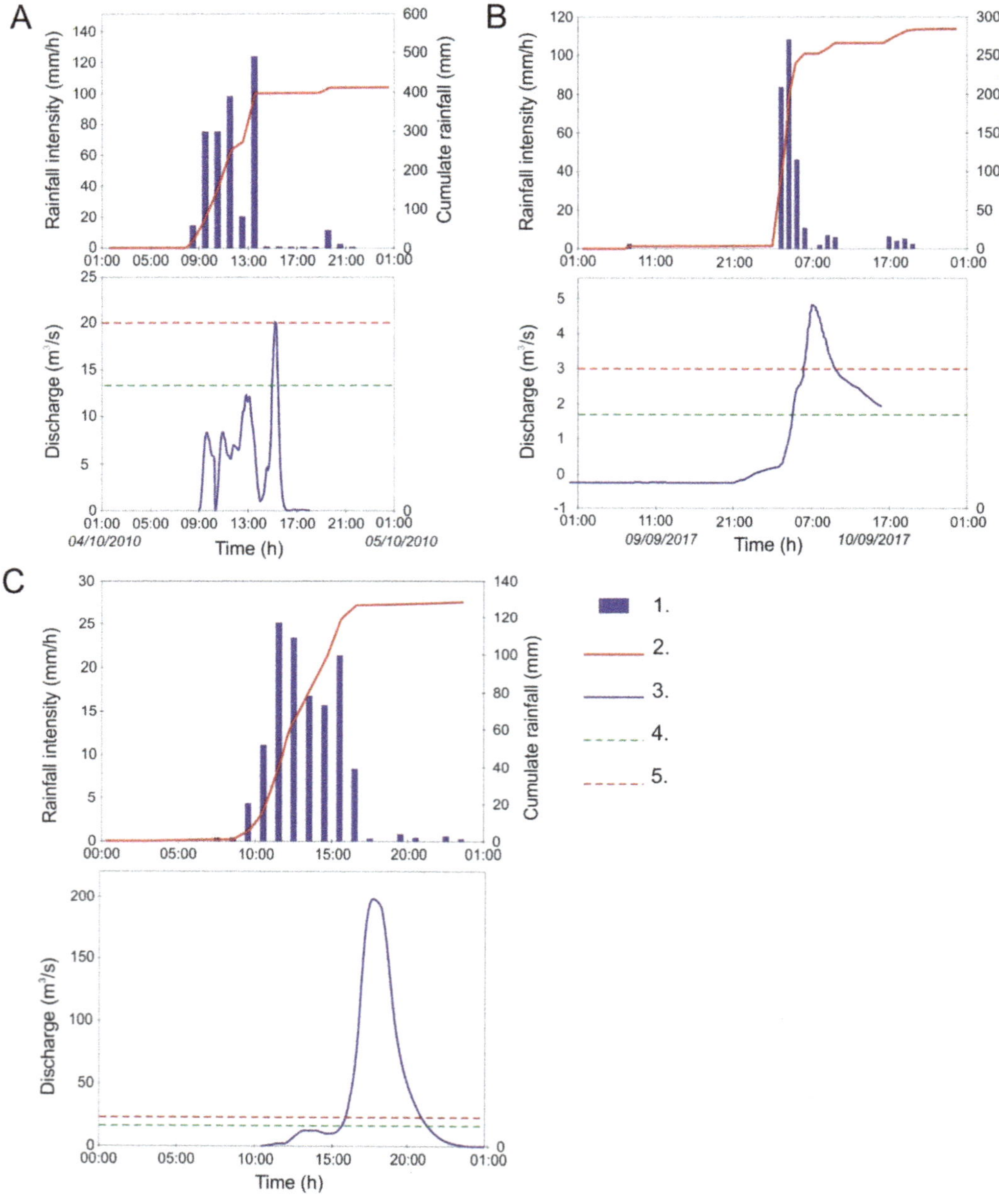

Figure 7. Hyetograms and hydrograms for the events of Genoa (**A**), Olbia (**B**) and Livorno (**C**). Legend: 1. Hourly intensity (mm/h), 2. Cumulated rainfall (mm), 3. Discharges (m^3/s); 4. Discharges for return period = 50 years (m^3/s); 5. Discharges for return period = 200 years (m^3/s).

Table 4. Maximum cumulative rainfall values over short durations (from 15 min to 3 h) recorded for the stations of Quercianella, Valle Benedetta, Santa Luce and Castellina Marittima [48].

Rain Gauge Station	Max Rainfall 15′ (mm)	Max Rainfall 1 h (mm)	Max Rainfall 2 h (mm)	Max Rainfall 3 h (mm)
Quercianella	42.4	121.8	188.6	206.2
Valle Benedetta	38.4	120.8	210.2	235
Santa Luce	23	66.4	98.2	105.8
Castellina Mar.	40.2	89.6	109.6	122.2

4.2. Ground Effects of the Flash Floods

With respect to the 2010 event in Genoa, the effects on the ground were determined by rainfalls produced by the convergence between the northern currents coming from the colder Po basin and the warm humid current coming from the sea. In the face of the quantities of rain received from the Sestri Ponente area, the water level in the streams increased in a short time: the Varenna Stream in Pegli reached a peak of 2.62 m with a rapid increase of 2.08 m in one hour (Pegli hydrometer). The Molinassi, Chiaravagna, Ruscarolo and Cantarena streams flooded the areas close to their beds and the Genoese quarter of Sestri Ponente (Figure 8A). Along the streets, the water level reached variable heights from 20 cm to over 150 cm; the estimated discharge rates for the Chiaravagna and Cantarena streams were comparable to those calculated for a Return time (Rt) of 50 years, while for the flood of the Molinassi Stream it was estimated at Rt = 200 years.

The response times of the meteorological-hydrological event were extremely short: Testimonies and amateur images document that after less than half an hour from the peak of flow, the flood of the Chiaravagna stream occurred and it poured along the roads as the water leaking from the streambed. The rainfall caused a rapid increase in flooding along with the transport of suspended materials and floating shrubs and trees that were eroded along unprotected embankments. The speed of the process made it impossible to implement interventions, unless the event was almost concluded. People were taken by surprise along the streets: water spread into residential areas and businesses and caused extensive damage (Figure 8B).

Along the slopes of the Chiaravagna (11 km^2) and Molinassi (2 km^2) basins, many shallow landslides have been triggered: they interrupted the access roads to the small inhabited areas placed on the hills.

Many streets were flooded and the settlements on the adjacent hills were isolated. The flow of mobilized debris was quickly channeled along the river beds and lower areas which caused critical hydraulic conditions in the secondary hydrographic network and also because the canals that pass culverted under the roads and the inhabited area were not able to dispose the relevant discharges and were quickly clogged.

With respect to the 2013 event, the area affected by the event was estimated at around 1500 km^2 and includes three main basins: the Cedrino and Posada basins and the catchment upstream of Olbia [53]. The city of Olbia was the most affected city, with eleven victims; much of the downtown area was inundated by the flood waters of the San Nicola and Seligheddu streams in the mouth area. Witnesses claim to have seen the hydrometric levels increase by about 3 m and this would be confirmed by the simulations carried out by [54] and associated with flow velocities higher than 3.2 m/s in the upstream sectors along the hydrographic lines. The railway embankment and the various bridges upon the arrival of the flood wave along the aforementioned canals had a dam effect and caused the flooding of the streets and the first floors of the houses. The most acute phase of the flood event was observed between 5:00 p.m. and 7:00 p.m., with more evident manifestations at around 6:00 p.m. on the Rio Seligheddu Stream and at around 6:30 p.m. on the urbanized stretch of Rio Gadduresu Stream which is its left tributary. Around 9:00 p.m. the flood had subsided with evident manifestations in the sectors surrounding the watercourses of Seligheddu, San Nicola, Zozò, Paule Longa and to a lesser extent the Pasana. In some sectors (former Artillery area), with variable tie rods up to about 2 m, the effects of the flood of Rio Gadduresu Stream from the north and east overlapped with those of Rio Seligheddu Stream from the south, which is a condition favored by the existing artifacts which represented temporary structures that are damming to the outflow of flood waters (Figure 9A–C) [55].

Figure 8. (**A**) Chiaravagna Stream flooded the ground floors, where there are many business shops and in some cases the water and mud depending on the preferential outlet flow found that entered the shops from the rear part, as can be seen in a sports shop in the Aprosio Square [56]. (**B**) Image taken from a movie. Chiaravagna Stream during the paroxistic phase of flooding: the black line indicates the submerged left bank wall; the red asterisk represents the bridge of Chiaravagna Street. The building from which the photograph was taken was built just on the riverbed in the 1960s. The October 2010 event (the last one of a long list) was recognized as one of the causes of the flooding of the river: it was demolished a few years later [56]).

Figure 9. (**A**) Olbia: Although about 15 h had passed since the paroxysmal phase, in some morphologically depressed areas of the city the water remained well above 1.3–1.4 m in height. Some cars were totally submerged in one of the streets most affected by the flood in the Baratta quarter (courtesy of private citizen). (**B**) Olbia, left bank of the Rio Seligheddu Stream, canalized stretch (foto Luino). Water level left a marker on the wall, which is evidenced by white line. The height of the waters is evidenced by the bed dragged and placed over the roof of the ground floor (white arrow). (**C**) Olbia: very common situation along the areas close to the streambeds (foto Luino). In this image, the Rio San Nicola (red arrow) waters exceeded the natural bank that was devoid of embankments or retaining walls and easily reached the houses located on the left bank, which was a short distance away. In many houses, the underground garages have been foolishly built and had been totally flooded (black arrow indicates the level), with serious damage inducted to parked cars. The extracted cars are so saturated with fine material (silt and clay) that they had to be demolished (image in the left corner).

During the 2017 event, watercourses flooded the surrounding areas in the hilly sector, where some bridges were damaged and many residents remained isolated. Towards the valley, in the area of the Rio Ardenza and Rio Maggiore mouths, the effects were even more serious with extensive flooding and entire neighborhoods invaded by water and mud. The major effects on the territory (floods, overflows and transport of debris material) were caused by the minor hydrographic network that originates from the Livorno hinterland and flows directly into the sea, as in the case of the Rio Maggiore and Rio Ardenza streams. The culverted stretches, often having insufficient section and occluded by detrital, vegetal and urban material carried by the waters, were bypassed by the floodwaters that retraced the ancient surface river paths. In particular, the waters of the Rio Maggiore Stream, despite the presence of retention basins, managed to flow freely in the area of the Stadio Ardenza District, in the neighboring streets and in Barriera Margherita. The Rio Maggiore Stream escaped from the culvert and poured with high speed into a fenced courtyard that was morphologically depressed compared to the nearby streets. The ground-floor flat in Sauro Street (Figure 10A,B) was flooded within minutes and four people drowned in it. Due to the overflow of the Ardenza Stream and its tributary Forcone stream, four other people lost their lives. The flood event caused in total eight victims in the Livorno area.

In addition to the Rio Ardenza and Rio Maggiore streams, the Ugione, Quercianella and the Chioma streams inundated large areas (Figure 10C). A bridge adjacent to a provincial road collapsed along the Ardenza. The total damage was estimated at 180 million EUR [57].

Figure 10. Livorno. (**A**) aerial photograph of the Ardenza Stadium district. In the foreground isthe house where the four victims drowned (white asterisk). The yellow line highlights the culverted streambed of the Rio Maggiore: its flood waters were the cause of the rapid flooding of the courtyard (evidenced by red lines) and of the ground-floor flat [58]. (**B**) Livorno: the house where four people from the same family drowned (courtesy of Il Tirreno). Their ground-floor flat is located at the end of a large courtyard, below street level on Rodocanacchi Street (area of the Ardenza football stadium). The red arrow indicates the level reached by the floodwaters in a few minutes. (**C**) Livorno: the Stagno district that was largely flooded by the Rio Ugione floodwaters with the Via Aurelia (yellow line) and the refineries in the background [58].

4.3. Urban Geomorphology

The case studies examined are characterized by a similar geomorphological structure and recent evolution, which has profoundly changed almost all Mediterranean urban areas [22,29]. Sestri Ponente, Olbia and Livorno are in fact three cities built on a coastal floodplain at the mouth of small hydrographic basins (Figure 11A–C).

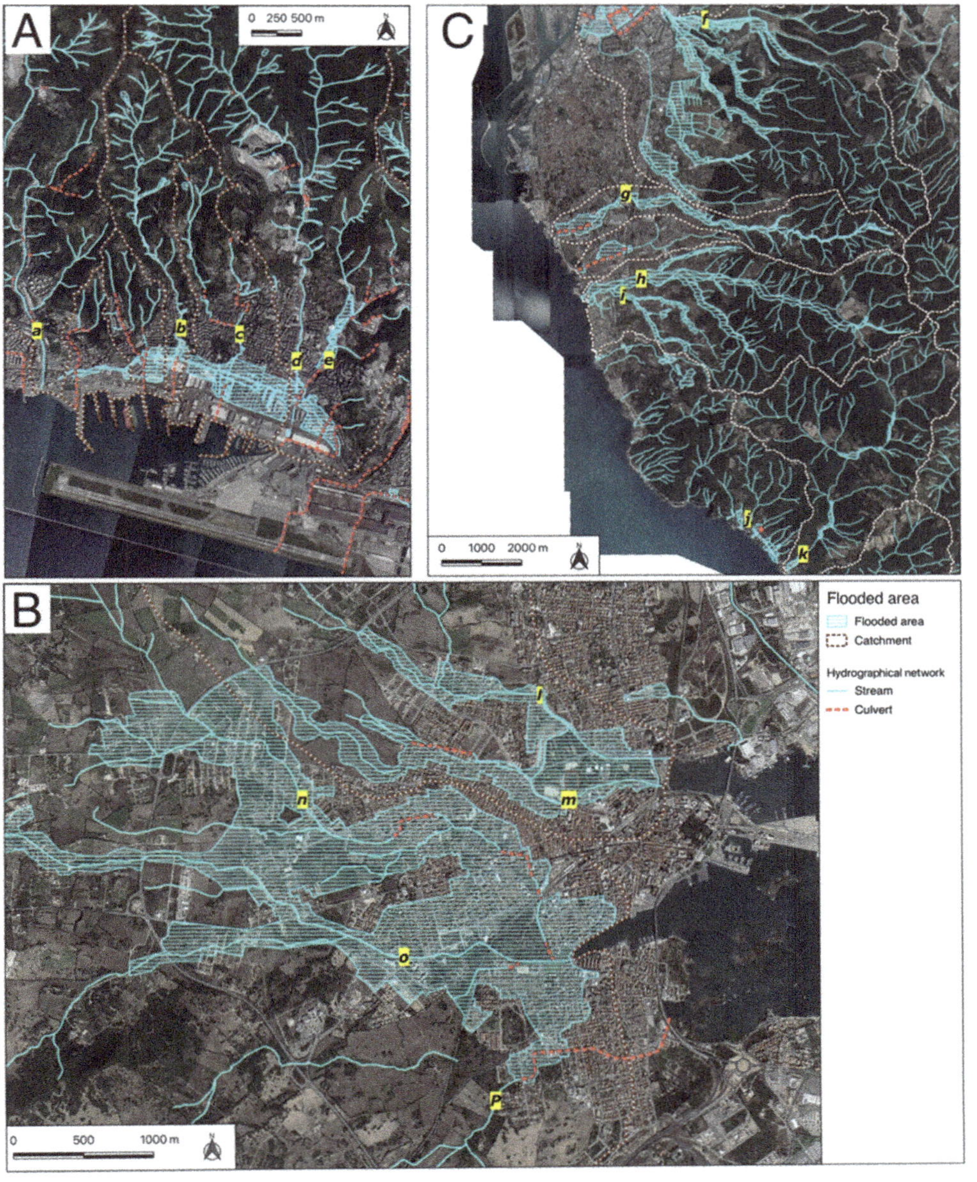

Figure 11. Hydrographic networks: culverted stretches are indicated with red lines. (**A**) Genoa Sestri Ponente; (**B**) Olbia; (**C**) Livorno. Streams: (a) Varenna; (b) Molinassi; (c) Cantarena; (d) Chiaravagna; (e) Ruscarolo; (f) Ugione; (g) Maggiore; (h) Ardenza; (i) Forcone; (j) Quercianella; (k) Chioma; (l) San Nicola; (m) Canale Zazà; (n) Gadduresu; (o) Seligheddu; (p) Paule Longa.

In this morphological situation, it is therefore possible to identify the processes related to the shapes of the landscape: the sea wave action along the coastal strip and the river dynamics in the rear belt, up to the hill base. The footprint of the anthropic landforms dominates the current urban landscape. If natural landforms were almost exclusive until the 18th–19th centuries, then in the last 150 years there has been a gradual increase in anthropogenic activities in the area with consequent major changes [59]. Two historical moments demonstrate great changes in the urban landscape due to anthropic impact: the first around the middle of the nineteenth century (industrial revolution) and a second in the second half of the twentieth century [60], after the conclusion of the Second World War (greater well-being, population growth and finally tourism). At present time, in all the cases examined it is therefore possible to identify modified natural landforms, anthropic landforms and disappearing (or vanished) landforms. The most common modified natural form is the main and secondary riverbed: in all the cases analyzed, the riverbed in the coastal floodplains was narrowed, channeled, rectified, often culverted and sometimes diverted.

In Sestri Ponente, the Chiaravagna Stream is channeled in its 2 km terminal, while the Ruscarolo Stream, originally an autonomous basin, flows entirely into the Chiaravagna. The other four streams of the plain are invisible, that is, culverted practically throughout the terminal stretch and flow within the urban fabric. A multitemporal cartographic comparison allows us to evaluate a narrowing of the riverbed in the last 200 years between 25% and 50% [34].

In Olbia, the Seligheddu Stream to the south and the San Nicola Stream to the north are represented today by artificial canals cemented with an inverted trapezoidal section, with many deviated and culverted sections, and affects the town with a multi-kilometric development [49].

The coastal stretch between Livorno and Antignano is characterized by three watercourses: to the south, the Rio Ardenza Stream is channeled in its final stretch for at least 2 km, with conspicuous narrowing and rectification of the section of the riverbed; on the other hand, the Rio Maggiore Stream and the Fosso Botro canal have been transformed into culverts [43].

The anthropogenic shapes dominate the urban landscape and can be traced back to forms of accumulation: in Sestri Ponente, Olbia and Livorno, among the most significant and common are the embankments, the reclaimed land at sea and the defenses along the coast, while urbanization has involved remodeling of the existing topographic surface with diffuse fillings.

Sestri Ponente is dominated by the sea fill on which the shipyards and the Genoese airport are built on: the railway line runs on an embankment and, further upstream, there is the motorway embankment. The fluvio-coastal plain is practically fully urbanized and the most significant phases of expansion were those of the second half of the 19th century and in the second half of the 20th century.

Olbia has a deeply modified coastal strip for the construction of the maritime station by the filling of the sea; the road and railway embankments emerge in the strip behind the historic core. Particularly significant is the construction of artificial canals built in the 1920s, while urbanization appears significant from the second half of the 20th century, but continued into the third millennium.

The urban coastal stretch south of Livorno has sea fills that are almost continuous, but they are less deep than the ones in Sestri Ponente and are protected by sea defenses. In the strip immediately behind it, there are the railway and motorway embankments; urbanization is continuous, with the exception of some sports facilities, and can be traced back to the second half of the 20th century. The waterways have been channeled and retention basins have been built along the final stretch of the Maggiore Stream.

Lastly, the disappeared forms, i.e., those dismantled or covered by anthropic activities, deserve a mention: In the case of Sestri Ponente, in addition to the consumption of land on the entire coastal floodplain there is the disappearance of the beach that occupied the entire stretch of sea in proximity to the coast for a length of about 2.5 km. In the Livorno coastal strip, the small cliffs modeled in the cemented sandstones alternating with small beaches have been incorporated into the sea fillings built from the second half of the 19th century, while in the coastal plain behind them marshes and coastal dunes have disappeared. In the case of Olbia, even if the imprint of the rias coast remains evident, the disappearance of the marshes has been noted, which characterized much of the coastal strip north and south of Olbia (Terranova Pausania) and into which the hydrographic network used to flow, as well as the salt flats near the historic center and small beaches bordered by cliffs modeled in granite between the Roman Port in the north and the Lepre Island in the south.

5. Discussion

The case-studies presented in this article show some geomorphological similarities: the cities arose on alluvial plains, typical of the coastal strip [22,29,56,61] of the Ligurian-Tyrrhenian Sea. They extend for several kilometers along the coastline (2.6–11.5), occupying variable areas (8–30 km^2) with high hills behind (330–580 m) and some kilometers away (3.8–6). Many streams cross the cities and some of them possess relevant areas (up to 38 km^2): they can reach relevant discharges during the violent rainfall events and in proportion to the basin areas (up to 8.6 m^3/s/km^2) (Table 5).

With respect to the weather-hydrological aspect, it should be emphasized that all the considered events occurred during the autumn season (from September to November), in conditions of a storm system triggered by cyclogenesis at the Gulf of Genoa (Liguria and Tuscany) or by the extra-tropical cyclone Cleopatra (Sardinia).

Table 5. Geomorphological features of the three cases analyzed. City Develop, development of the city along the coastline; Hill Height, height of the hills located behind the city; Distance, distance between the coastline and closest hills; Basin Area, area of the largest hydrographic basin behind the city; Max Discharge, maximum discharge for a 200-years return period.

City	City Area (km^2)	City Develop (km)	Hill Height (m)	Distance (km)	Basin Area (km^2)	Max Discharge (m^3/s)
Genova Sestri Ponente	8	2.6	580	3.8	11 (Chiaravagna)	213 (50 years)
Olbia	27	11.5	460	6	38.4 (Seligheddu)	330
Livorno	30.2	7.8	330	6	33.2 (Ugione)	137

Despite the trend in the number of rainy days being negative and the progressively decreasing annual cumulative rainfall, perturbations capable of generating intense rainfall are increasingly frequent in the Mediterranean area [10], with a growth corresponding to intense geo-hydrological events. A sentence that summarizes this concept is currently widely used: "it rains less, but worse".

If we add the progressive increase in temperatures to this [62], it is possible to confirm the data on climate change underway [63,64] with recent evidence of events studied in the geo-hydrological field: For example, the event of Lavagna-Genoa in 2002 [30], autumn 2011 in Liguria [65] and event of October 2014 in Liguria [33,66]. These data are also supported by studies in other disciplines [67,68].

The events considered had rainfall that reached values between 10.4–15.2% of the annual total in 1 h; between 22% and 25.8% in 3h (minimum difference between the minimum and the maximum); between 26.5% and 32.7% in 6 h; finally, values between 27.2% and 42% in 12 h (Table 6).

Table 6. Rainfall of the three events (for 1,3,6 and 12 h) compared with the mean annual precipitation, MAP (%).

	MAP	1 h	3 h	6 h	12 h
Genoa Sestri Ponente	1100	124 (11.3%)	243 (22%)	360 (32.7%)	411 (37.4%)
Olbia-Sa Pianedda	588	61 (10.4%)	150 (25.5%)	167.2 (28.4%)	247.4 (42%)
Livorno-Quercianella	800	121.8 (15.2%)	206.2 (25.8%)	212.4 (26.5%)	217.6 (27.2%)

The percentages obtained actually indicate the rainfall characteristics of each individual event: the Livorno event was more concentrated on 1–3 h, the Sestri Ponente on 6 h and the Olbia event on 12 h. The intense precipitations occurred in correspondence with strong winds and storm surges, showing hourly cumulative rainfall values comparable to half-yearly or annual averages. The precipitations sent the hydrographic system into crisis: the levels of the streams grew rapidly, reaching and exceeding the alert levels.

The responses of river basins to intense and short rainfalls depended on several factors: (i) land use, (ii) bedrock permeability, (iii) thickness of the eluvium-colluvial cover and (iv) initial content of soil moisture. If, in normal conditions, the Mediterranean catchment areas possess a runoff coefficient of 0.4–0.6, in the conditions of saturated or impermeable soils, runoff coefficients close to 1 can be achieved [69,70].

The streambeds in the cities, over the years, have been gradually narrowed to conquer more urban areas and often the beds have been culverted for long stretches and flow under roads and buildings. It is therefore natural that the streambeds were not able to dispose of the huge discharge of the streams in which water shrubs, large trees (uprooted upstream or in the riverbed), garbage cans and vehicles are often transported.

The increase in urbanized areas has amplified this problem, leading to an increase in waterproofed surfaces; this has caused the irreversible loss of soil and consequent impact on the flow of water, which when unable to infiltrate the soil, is dispersed by surface flow. The latter, increasing in terms of drained volumes and transit speed, is responsible for problems in the control of surface waters, in particular during particularly intense rainfall phenomena. The growth in waterproofed surfaces, in fact, involves an increase in the runoff coefficients and a reduction in the run-off times, making it necessary to construct structures for containment and disposal (bypasses) of exceptional flood events.

In the last 150 years, the Sestri Ponente, Olbia and Livorno cities have suffered remarkable transformations, especially in the coastal areas (Figure 12). This has led to the growth of urbanized areas along the plain, with partial or total impairment of the areas of fluvial pertinence. The waterways have undergone riverbed narrowing, containment and lateral and bottom constraints, which has resulted in the unsuitability for the disposal of significant flood flows.

The serious damage recorded for the cases treated, primarily the 27 overall casualties of the three events described, are linked to the vulnerability of the numerous elements present in the fluvio-coastal plains of Genoa, Olbia and Livorno. Among the anthropogenic forcings that have influenced the hydro-geomorphological dynamics and which have determined the increase in risk conditions, the following are counted [71,72] (Figures 13 and 14):

- Modification of land use in general from agricultural to compact urban cover, with a consequent high decrease in the run-off time;
- Progradation of the coastline by the method of sea filling (e.g., sea filling for the construction of Cristoforo Colombo platform airport and the marine port of Sestri Ponente);
- Modified river forms: the watercourse active beds show marked evidences of canalization and narrowing of the outflow sections, sometimes deviations and often culverts (Figure 15);
- Construction of buildings and infrastructures (especially since the 1960s) in areas known to be hazardous from a hydraulic point of view.

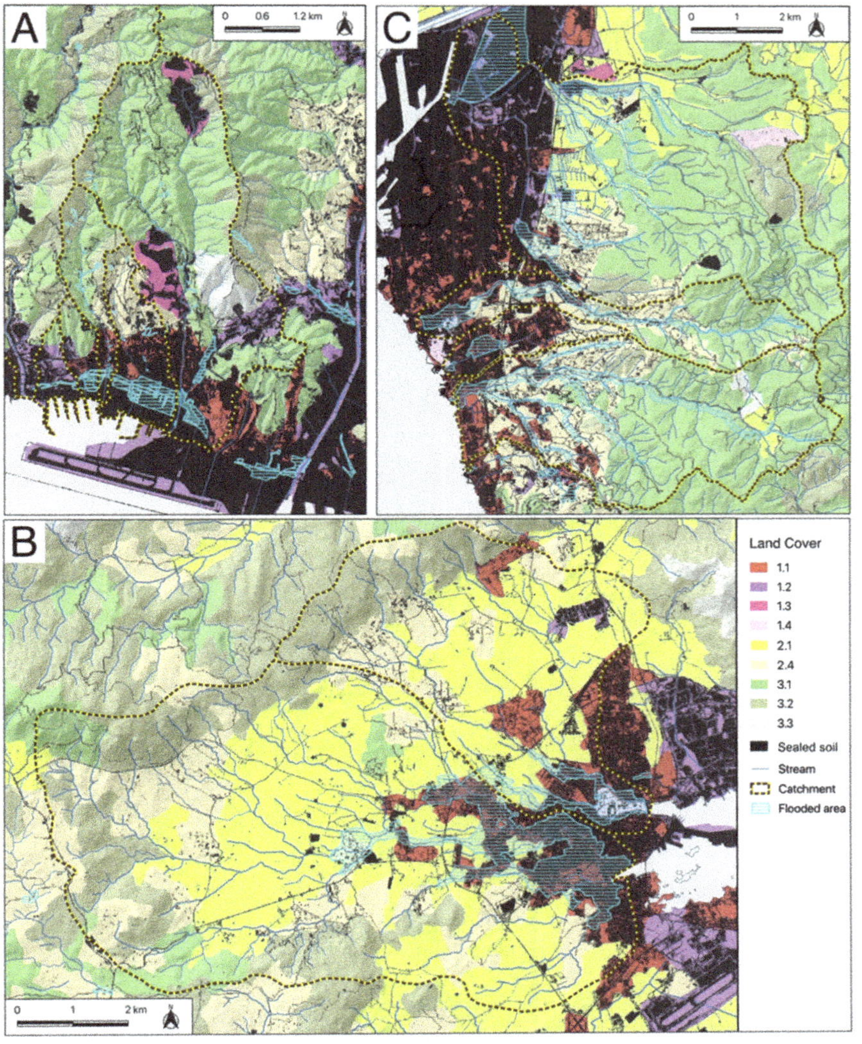

Figure 12. Hydrographic networks, urbanized areas and land use of the basins of Genoa Sestri Ponente (**A**), Olbia (**B**) and Livorno (**C**). Land cover: (1.1) urban fabric; (1.2) industrial; (1.3) mine, dump and construction sites; (1.4) artificial non-agricultural vegetated areas; (2.1) arable land; (2.4) agricultural areas; (3.1) forests; (3.2) shrub and/or herbaceous vegetation associations; (3.3) open spaces with little or no vegetation.

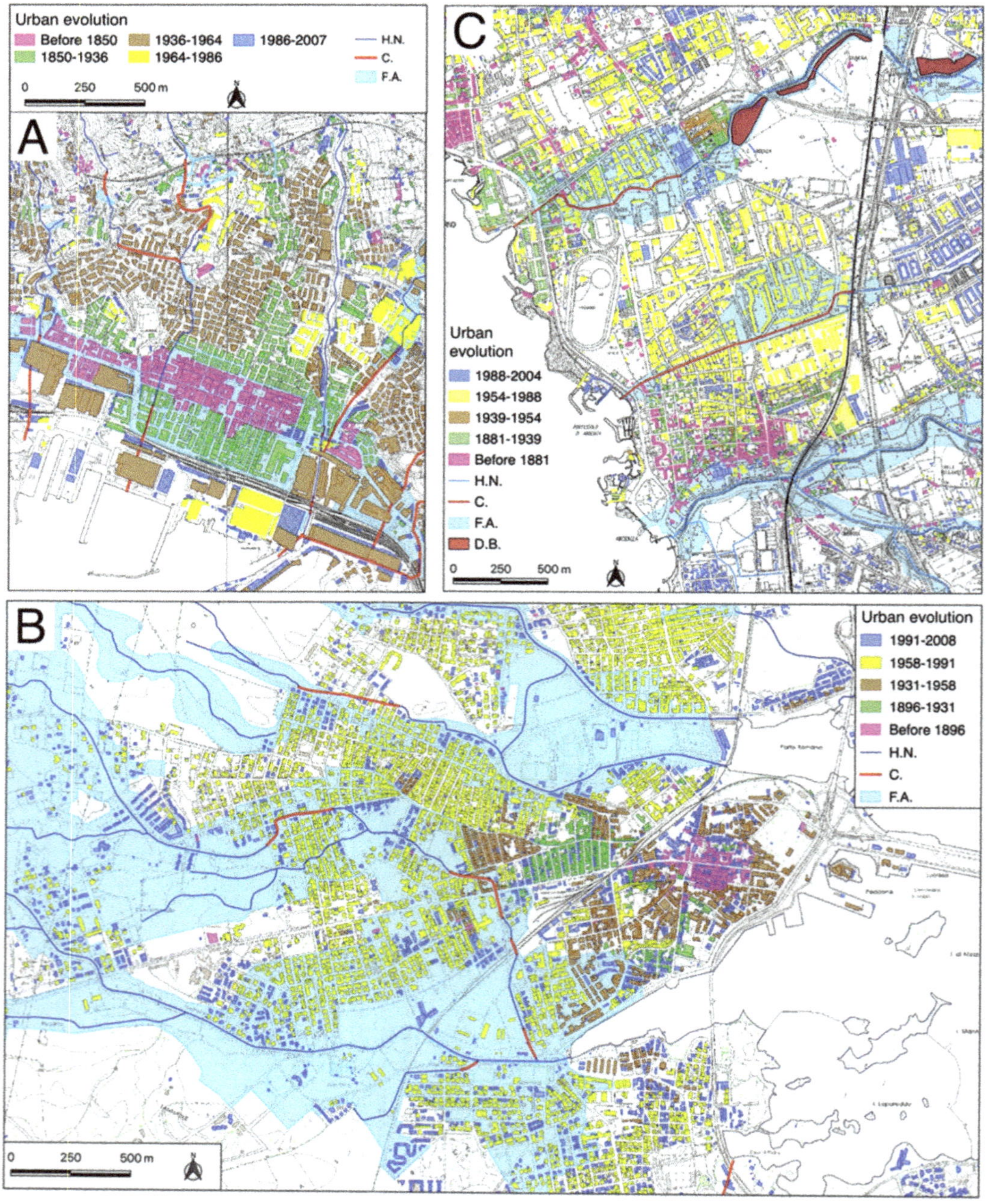

Figure 13. Urban evolution, evidenced by different colors for Genoa Sestri Ponente (**A**), Olbia (**B**) and Livorno (**C**) since the end of the 19th century. In the legend: (H.N.) Hydrographical network; (C) Culvert; (F.A.) Flooded area.

A very important aspect that characterized all three events was the lack of communication with the population before and during the culminating phase of the event [73–75]. In all three cities, the inhabitants were surprised by the flooding of the streams and the rapid growth of water in the city: this made it impossible, in some cases, to escape to safety (15 drowned victims) or to bring their own goods (especially vehicles).

A last and very important topic concerns the historical sources. Moreover, in this study, a careful historical reconstruction has made it possible to detect how much the anthropic changes may have influenced the bed of the watercourses year after year, decisively conditioning the floods during the paroxysmal phase of the event and creating casualties and a lot of damage. A correct utilization of the historical sources could save lives and goods.

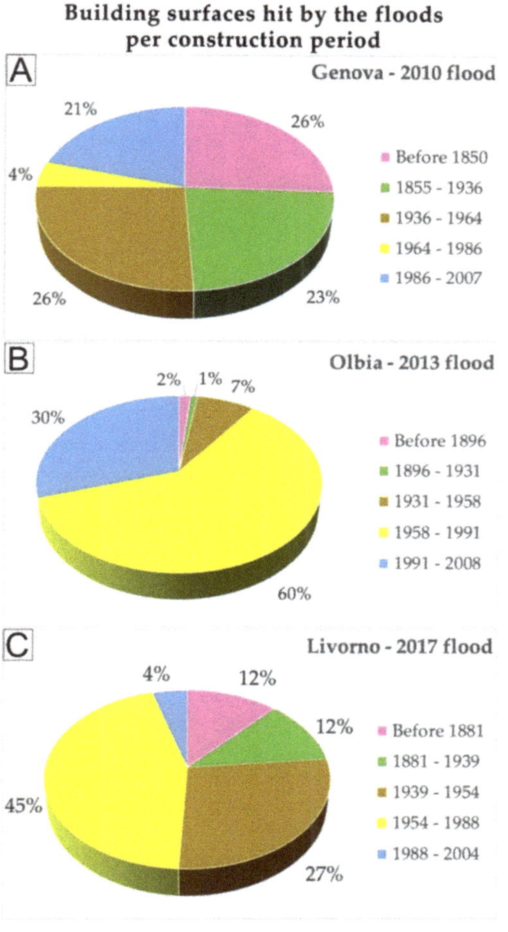

Figure 14. Urban development in the coastal plain of: (**A**) Genoa Sestri Ponente, compared with the percentage of urbanization involved over time by a flood of comparable size in terms of volume and area to that occurred 10 April 2010; (**B**) Olbia, compared with the percentage of urban areas involved over time by a flood of a comparable extent in terms of volume and area to that of 18 November 2013; (**C**) Livorno, compared with the percentage of urbanization involved over time by a flood of comparable magnitude in terms of volume and area to that of 10 September 2017.

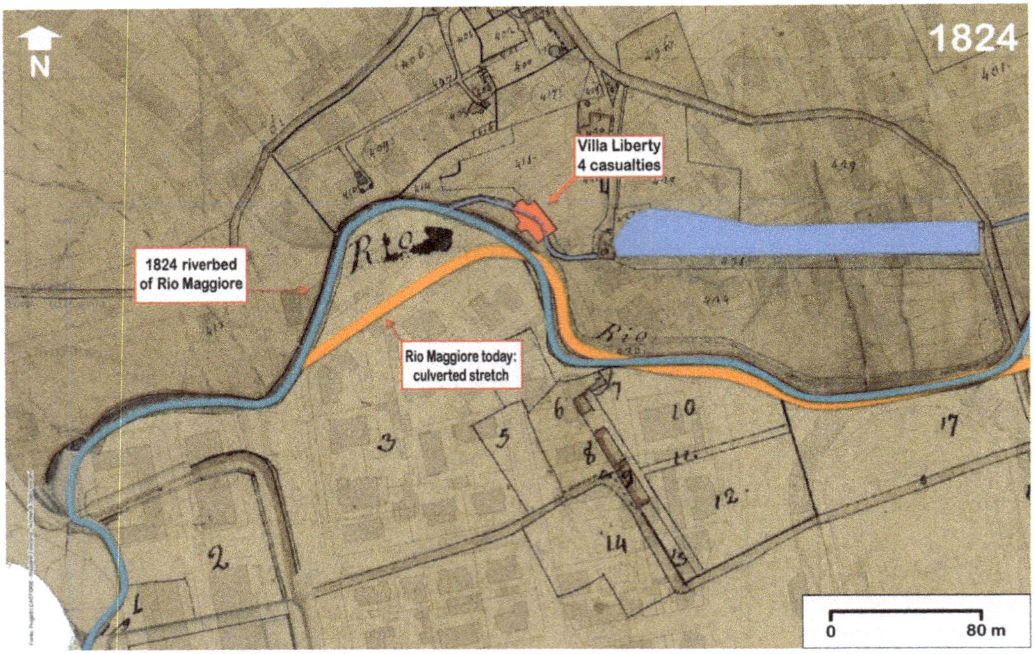

Figure 15. The importance of the historical sources is underlined by this 1824 cadastral map of Livorno. The Rio Maggiore Stream flowed naturally and was free from anthropogenic influences (blue stroke on the map). Its course has been diverted over the years and, above all, the stream has been culverted: it now flows invisibly between the houses of the city (orange stroke). The house where the victims drowned in September 2017, which will be built afterwards in the early 20th century, is highlighted in red ([76], modified).

6. Conclusions

Two factors, which are very different from each other, are at the basis of the geo-hydrological processes that have affected long stretches of the Mediterranean coast in recent decades. The first, which is of anthropogenic origin, is the urban expansion that began after the end of the Second World War. The second, which is of "apparently" natural origin, is climate change (for which the role played by man is now very clear to everyone). The combination of these two factors, which are in some manner very distant from each other, is now causing a situation that is difficult to manage and that places us at the forefront of real emergencies for which it will be necessary to utilize precise analytical tools, sensible land use planning and rapid and resolving interventions.

The floods that occurred in Genoa, Olbia and Livorno revealed the state of vulnerability of the city towards hydraulic processes in spite of many accurate mitigation interventions carried out in the last decades and above all the inadequacy of the current urban fabric in relation to the hydrographic network, both on the hydraulic and geomorphological level.

The comparison between the events described highlights the need to plan flood risk mitigation activities beyond the essential structural interventions on the hydrographic network, whether it be main and minor and at basin scale (extraordinary maintenance) which involves inevitable financial programs that are costly and delayed over time. It appears essential that we prepare non-structural measures which includes both active (routine maintenance, thickening of the weather-hydrological monitoring and construction of expansion tanks) and passive (land use rules with intensive use of urban drainage systems sustainability, foreclosure of fluvial pertinence areas, historical investigation of past events, population trainings, information to inhabitants and insurance coverage) measures

With regard to the actions to be taken in the short-term and medium-term in order to adapt to the resurgence of flash flood effects on the ground, there are several methodological examples for the Mediterranean area that involve urban, social and economic choices. For example, we want to mention the TRIGEAU project [77] and the ADAPT project [78], which suggests strategies to mitigate the effects of climate change. In particular, the ADAPT project suggests: (1) Actions to improve geo-hydrological conditions, such as increasing knowledge on the processes that contribute to the occurrence of geo-hydrological criticalities, adapting existing mitigation works, carrying out interventions for urban flood mitigation and urban greening interventions; (2) actions to increase the resilience of the population and assets at risk with training activities; (3) actions to improve governance with legislative adjustments, urban interventions and limitations on urbanization and restoration of areas of river pertinence and re-naturalization.

Until the state and every single municipality decides to take serious and decisive action to solve this age-old problem, the flash floods will continue to cause victims and massive damage to private and public property.

Supplementary Materials: The following are available online at https://www.mdpi.com/2073-445X/10/6/620/s1, Table S1: Land use in the studied catchments, Table S2: Dates of post-World War II flood events in the three studied cities.

Author Contributions: Conceptualization, F.L., F.F. and L.T.; methodology, F.L., F.F. and L.T.; software A.R. and G.P.; validation, A.R., G.P., F.L., F.F. and L.T.; investigation, G.P., F.L., F.F. and L.T.; resources, A.R., G.P., F.L., F.F. and L.T.; data curation, A.R., G.P., F.L., F.F. and L.T.; writing—original draft preparation, F.L., F.F. and L.T.; writing—review and editing, F.L., F.F. and L.T.; visualization, A.R., G.P., F.L. and L.T.; supervision, G.P. and L.T.; project administration, F.L. and L.T. All authors have read and agreed to the published version of the manuscript.

Funding: This research received no external funding.

Institutional Review Board Statement: Not applicable.

Informed Consent Statement: Not applicable.

Data Availability Statement: Not applicable.

Conflicts of Interest: The authors declare no conflict of interest.

References

1. IPCC. Climate Change 2014: Impacts, Adaptation, and Vulnerability. In Proceedings of the IPCC Working Group II Contribution to the Fifth Assessment Report of the Intergovernmental Panel on Climate Change, Copenhagen, Denmark, 1 November 2014. Available online: http://www.ipcc.ch/report/ar5/wg2/ (accessed on 9 March 2021).
2. López Díez, A.; Máyer Suárez, P.; Díaz Pacheco, J.; Dorta Antequera, P. Rainfall and Flooding in Coastal Tourist Areas of the Canary Islands (Spain). *Atmosphere* **2019**, *10*, 809. [CrossRef]
3. Génova, M.; Máyer, P.; Ballesteros-Cánovas, J.; Rubiales, J.M.; Saz, M.A.; Díez-Herrero, A. Multidisciplinary study of flash floods in the Caldera de Taburiente National Park (Canary Islands, Spain). *Catena* **2015**, *131*, 22–34. [CrossRef]
4. Fragoso, M.; Trigo, R.M.; Pinto, J.G.; Lopes, S.; Lopes, A.; Ulbrich, S.; Magro, C. The 20 February 2010 Madeira flash-floods: Synoptic analysis and extreme rainfall assessment. *Nat. Hazards Earth Syst. Sci.* **2012**, *12*, 715–730. [CrossRef]
5. Swiss RE Institute. SIGMA Annual Report 2015. Available online: https://www.swissre.com/institute/research/sigma-research/sigma-2015-04.html (accessed on 9 March 2021).
6. CRED—Centre for Research on the Epidemiology of Disasters. The International Disaster Database. Available online: https://www.emdat.be/ (accessed on 9 March 2021).
7. Gaume, E.; Bain, V.; Bernardara, P.; Newinger, O.; Barbuc, M.; Bateman, A. A compilation of data on European flash floods. *J. Hydrol.* **2009**, *367*, 70–78. [CrossRef]
8. Gaume, E.; Borga, M.; Llassat, M.C.; Maouche, S.; Lang, M. Mediterranean extreme floods and flash floods. In *The Mediterranean Region under Climate Change. A Scientific Update*; Collection de Synthèses; IRD Editions: Marseille, France, 2016; pp. 133–144.
9. Barrera-Escoda, A.; Llasat, M.C. The role of climatic factors in evolving flood patterns in a Mediterranean region (1301–2012). *Hydrol. Earth Syst. Sci.* **2015**, *19*, 465–483. [CrossRef]
10. Boudou, M.; Danière, B.; Lang, M. Assessing changes on urban flood vulnerability through mapping land use from historical information. *Hydrol. Earth Syst. Sci. Discuss.* **2015**, *12*, 6151–6177. [CrossRef]

11. Tarolli, P.; Borga, M.; Morin, E.; Delrieu, G. Analysis of flash flood regimes in the North-Western and South-Eastern Mediterranean regions. *Nat. Hazards Earth Syst. Sci.* **2012**, *12*, 1255–1265. [CrossRef]
12. Gohar, A.; Kondolf, G.M. Flash flooding as a threat to settlements even in remote areas. *Environ. Urban.* **2017**, *29*, 503–514. [CrossRef]
13. Ducrocq, V.; Braud, I.; Davolio, S.; Ferretti, R. HyMeX-SOP1, the field campaign dedicated to heavy precipitation and flash flooding in the northwestern Mediterranean. *Bull. Am. Meteorol. Soc.* **2014**, *95*, 1083–1100. [CrossRef]
14. Petrucci, O.; Papagiannaki, K.; Aceto, L.; Boissier, L.; Kotroni, V.; Grimalt, M. MEFF: The database of MEditerranean Flood Fatalities (1980 to 2015). *J. Flood Risk Manag.* **2019**, *12*, e12461. [CrossRef]
15. Del Moral, A.; Llasat, M.C.; Rigo, T. Connecting flash flood events with radar-derived convective storm characteristics on the Northwestern Mediterranean coast: Knowing the present for better future scenarios adaptation. *Atmos. Res.* **2020**, *238*, 104863. [CrossRef]
16. Hafnaoui, M.A.; Madi, M.; Hachemi, A.; Farhi, Y. El Bayadh city against flash floods: Case study. *Urban Water J.* **2020**, *17*, 390–395. [CrossRef]
17. Llasat, M.C.; Llasat-Botija, M.; Prat, M.A.; Porcú, F.; Price, C.; Mugnai, A.; Lagouvardos, K.; Kotroni, V.; Katsanos, D.; Michaelides, S.; et al. High-impact floods and flash floods in Mediterranean countries: The FLASH preliminary database. *Adv. Geosci.* **2010**, *23*, 47–55. [CrossRef]
18. Einfalt, T.; Hatzfeld, F.; Wagner, A.; Seltmann, J.; Castro, D.; Frerichs, S. URBAS: Forecasting and management of flash floods in urban areas. *Urban Water J.* **2009**, *6*, 369–374. [CrossRef]
19. Roccati, A.; Faccini, F.; Luino, F.; Ciampalini, A.; Turconi, L. Heavy rainfall triggering shallow landslides: A susceptibility assessment by a GIS-approach in a Ligurian Apennine catchment (Italy). *Water* **2019**, *11*, 605. [CrossRef]
20. Silvestro, F.; Rebora, N.; Rossi, L.; Dolia, D.; Gabellani, S.; Pignone, F.; Trasforini, E.; Rudari, R.; De Angeli, S.; Masciulli, C. What if the 25 October 2011 event that struck Cinque Terre (Liguria) had happened in Genoa, Italy? Flooding scenarios, hazard mapping and damage estimation. *Nat. Hazards Earth Syst. Sci.* **2016**, *16*, 1737–1753. [CrossRef]
21. Faccini, F.; Luino, F.; Sacchini, A.; Turconi, L. The 4th October 2010 flash flood event in Genoa Sestri Ponente (Liguria, Italy). *Disaster Adv.* **2015**, *8*, 1–14.
22. Roccati, A.; Faccini, F.; Luino, F.; De Graff, J.; Turconi, L. Morphological changes and human impact in the Entella River floodplain (Northern Italy) from the 17th century. *Catena* **2019**, *182*, 104122. [CrossRef]
23. Roccati, A.; Paliaga, G.; Luino, F.; Faccini, F.; Turconi, L. Rainfall threshold for shallow landslides initiation and analysis of long-term rainfall trends in a mediterranean area. *Atmosphere* **2020**, *11*, 1367. [CrossRef]
24. Audisio, C.; Turconi, L. Urban floods: A case study in the Savigliano area (North-Western Italy). *Nat. Hazards Earth Syst.Sci.* **2011**, *11*, 2951–2964. [CrossRef]
25. Vietz, G.J.; Walsh, C.J.; Fletche, T.D. Urban hydrogeomorphology and the urban stream syndrome: Treating the symptoms and causes of geomorphic change. *Prog. Phys. Geogr.* **2016**, *40*, 480–492. Available online: https://journals.sagepub.com/doi/full/10.1177/0309133315605048 (accessed on 9 March 2021). [CrossRef]
26. Cutter, S.L.; Emrich, C.T.; Gall, M.; Reeves, R. Flash Flood Risk and the Paradox of Urban Development. *Nat. Hazards Rev.* **2018**, *19*, 05017005. Available online: https://ascelibrary.org/doi/abs/10.1061/%28ASCE%29NH.1527-6996.0000268 (accessed on 9 March 2021). [CrossRef]
27. Ertan, S.; Çelik, R.N. The Assessment of Urbanization Effect and Sustainable Drainage Solutions on Flood Hazard by GIS. *Sustainability* **2021**, *13*, 2293. [CrossRef]
28. Paliaga, G.; Faccini, F.; Luino, F.; Turconi, L. A spatial multicriteria prioritizing approach for geohydrological risk mitigation planning in small and densely urbanized Mediterranean basins. *Nat. Hazards Earth Syst. Sci.* **2019**, *19*, 53–69. [CrossRef]
29. Roccati, A.; Luino, F.; Turconi, L.; Watkins, C.; Faccini, F. Historical geomorphological research of a Ligurian coastal floodplain (Italy) and its value for management of flood risk and environmental sustainability. *Sustainability* **2018**, *10*, 3727. [CrossRef]
30. Faccini, F.; Brandolini, P.; Robbiano, A.; Perasso, L.; Sola, A. Instability, precipitation phenomena and land planning: The flood of 2002 in lower Lavagna valley (Eastern Liguria, Italy). *Geogr. Fis. Din. Quat.* **2005**, *35* (Suppl. VII), 145–153.
31. Cassola, F.; Ferrari, F.; Mazzino, A.; Miglietta, M.M. The role of the sea on the flash floods events over Liguria (northwestern Italy). *Geophys. Res. Lett.* **2016**, *43*, 3534–3542. [CrossRef]
32. Cevasco, A.; Brandolini, P.; Scopesi, C.; Rellini, I. Relationships between geo-hydrological processes induced by heavy rainfall and land-use: The case of 25 October 2011 in the Vernazza catchment (Cinque Terre, NW Italy). *J. Maps* **2013**, *9*, 289–298. [CrossRef]
33. Faccini, F.; Luino, F.; Paliaga, G.; Sacchini, A.; Turconi, L.; Dejong, C. Role of rainfall intensity and urban sprawl in the 2014 flash flood in Genoa City, Bisagno catchment (Liguria, Italy). *Appl. Geogr.* **2018**, *98*, 224–241. [CrossRef]
34. Faccini, F.; Luino, F.; Sacchini, A.; Turconi, L. Flash flood events and urban development in Genoa (Italy): Lost in translation. In *Engineering Geology for Society and Territory*; Lollino, G., Manconi, A., Guzzetti, F., Culshaw, M., Bobrowsky, P., Luino, F., Eds.; Springer International Publishing: Cham, Switzerland, 2015; Volume 5(II), pp. 797–801.
35. Mandarino, A.; Luino, F.; Faccini, F. Flood-induced ground effects and flood-water dynamics for hydro-geomorphic hazard assessment: The 21–22 October 2019 extreme flood along the lower Orba River (Alessandria, NW Italy). *J. Maps* **2021**, 1–16. [CrossRef]

36. Reynard, N.S.; Kay, A.L.; Anderson, M.; Donovan, B.; Duckworth, C. The evolution of climate change guidance for fluvial flood risk management in England. *Prog. Phys. Geogr.* **2017**, *41*, 222–237. Available online: https://journals.sagepub.com/doi/abs/10.1177/0309133317702566 (accessed on 9 March 2021). [CrossRef]
37. Zhou, Q.; Leng, G.; Su, J.; Ren, Y. Comparison of urbanization and climate change impacts on urban flood volumes: Importance of urban planning and drainage adaptation. *Sci. Total Environ.* **2019**, *658*, 24–33. [CrossRef]
38. Regione Liguria. Piano di Bacino Stralcio per la Tutela dal Rischio Idrogeologico. 2018. Relazione Generale. Available online: http://www.pianidibacino.ambienteinliguria.it/GE/chiaravagna/documenti/GE_Chiaravagna_RelGen_rev02.pdf (accessed on 9 March 2021).
39. Righini, M.; Surian, N.; Wohl, E.; Marchi, L.; Comiti, F.; Amponsah, W.; Borga, M. Geomorphic response to an extreme flood in two Mediterranean rivers (northeastern Sardinia, Italy): Analysis of controlling factors. *Geomorphology* **2017**, *290*, 184–199. [CrossRef]
40. Comune di Olbia. Studio di Variante al Piano Stralcio per l'Assetto Idrogeologico (PAI) e del Quadro delle Opere di Mitigazione del Rischio Idraulico nel Territorio Comunale di Olbia. Analisi Idrologica 2015. Available online: https://www.regione.sardegna.it/documenti/1_617_20160420102320.pdf (accessed on 9 March 2021). (In Italian).
41. Comune di Olbia. Soluzione Progettuale Alternativa al Quadro delle Opere di Mitigazione del Rischio Idraulico ("Progetto Mancini") Approvato dal Comitato Istituzionale Dell'autorità di Bacino con Delibera n 1 del 26.05.2015 CIG 693426E03. Studio di Fattibilità; Relazione Generale. 2018. Available online: http://2.45.149.43/001-PubCED/AttiPianificazione/2018/Proposta_n_013/A.01.01_00%20Relazione%20Generale.pdf (accessed on 21 February 2021). (In Italian).
42. Yin, J.; Yu, D.; Yin, Z.; Liu, M.; He, Q. Evaluating the impact and risk of pluvial flash flood on intra-urban road network: A case study in the city center of Shanghai, China. *J. Hydrol.* **2016**, *537*, 138–145. [CrossRef]
43. Arrighi, C.; Castelli, F. The 2017 flash flood of Livorno (Italy): Lessons learnt from an exceptional hydrologic event. In *Advances in Natural Hazards and Hydrological Risks: Meeting the Challenge, Advances in Science, Technology & Innovation*; Fernandes, F., Malheiro, A., Chaminé, H., Eds.; Springer Nature: Cham, Switzerland, 2020; pp. 117–120. [CrossRef]
44. SIR-Servizio Idrologico e Geologico della Regione Toscana-Centro Funzionale Regionale. Un Secolo di Precipitazioni Estreme in Toscana. 2018. Available online: https://www.sir.toscana.it/supports/pdf/precipitazioni/un_secolo_di_precipitazioni_estreme.pdf (accessed on 9 March 2021). (In Italian).
45. ARPAL-Agenzia Regionale per la Protezione dell'Ambiente della Regione Liguria. Rapporto di Evento Meteo-Idrologico del 04/10/2010. Available online: https://www.arpal.liguria.it/contenuti_statici//pubblicazioni/rapporti_eventi/2010/REM_20101004_allerta2_(versione_01032011).pdf (accessed on 9 March 2021). (In Italian)
46. RAS-Regione Autonoma della Sardegna Assessorato dei Lavori Pubblici. Alluvione, un Anno Dopo. 2014. Available online: http://www.regione.sardegna.it/documenti/1_50_20141117184221.pdf. (accessed on 9 March 2021). (In Italian).
47. RAS-Regione Autonoma Della Sardegna. Analisi Dell'evento Meteorologico del 18 Novembre 2013. Available online: http://www.sar.sardegna.it/pubblicazioni/miscellanea/analisi_evento_18nov2013.pdf. (accessed on 9 March 2021). (In Italian).
48. CFR-Centro Funzionale della Regione Toscana. Report Evento Meteo-Idrologico dei Giorni 9 e 10 Settembre 2017. 13p. Available online: https://www.cfr.toscana.it/supports/download/eventi/report_evento_9-10_settembre_2017.pdf (accessed on 10 February 2021). (In Italian).
49. Mancini, M.; Tilocca, G. Ricostruzione Geomorfologica Dell'evento Alluvionale di Olbia del 18 Novembre 2013. Available online: http://2.45.149.43/001-PubCED/AttiPianificazione/2015/Proposta_n_078/Elaborati_Parte_Geologica/REL_RICOSTRUZIONE_EVENTO_GEOMORFOLOGICO_A08.pdf (accessed on 9 March 2021). (In Italian).
50. Castelli, F. Ricostruzione Idrologica Dell'evento del 9/10 settembre 2017 Nella Provincia di Livorno. Available online: https://www.toscana-notizie.it/documents/10180/16217821/08+Studio+Idrologico_Evento2017+OCD+26_2019.pdf/819ce24f-1bec-4b52-8b56-6dd4c8052d8c (accessed on 9 March 2021). (In Italian).
51. Comune di Livorno. Progetto ADAPT: Piano Locale di Adattamento ai Cambiamenti Climatici per Il Rischio Alluvioni del Comune di Livorno; 2019. Available online: https://www.comune.livorno.it/sites/default/files/index/ambiente/adapt_piano_di_adattamento_p7m_qrcode.pdf (accessed on 9 March 2021). (In Italian).
52. LAMMA. Report Meteorologico, 9–10 settembre 2017. Available online: http://www.lamma.rete.toscana.it/clima/report/eventi/evento_09092017.pdf (accessed on 9 March 2021). (In Italian).
53. Borga, M.; Marchi, L.; Nikolopoulos, E.I.; Marra, F.; Crema, S.; Niedda, M.; Pirastru, M.; Amponsah, W. The flash-flood of November 2013 in NE Sardinia (Italy): Post-event documentation and hydrological modelling. In Proceedings of the 8th International HyMeX Workshop, Valetta, Malta, 15–18 September 2014.
54. Beretta, R.; Ravazzani, G.; Maiorano, C.; Mancini, M. Simulating the Influence of Buildings on Flood Inundation in Urban Areas. *Geosciences* **2018**, *8*, 77. [CrossRef]
55. Niedda, M.; Amponsah, W.; Marchi, L.; Zoccatelli, D.; Marra, F.; Crema, S.; Pirastru, M.; Marrosu, R.; Borga, M. Il ciclone Cleopatra del 18 novembre 2013 in Sardegna: Analisi e modellazione dell'evento di piena. *Quad. Idronomia Mont.* **2015**, 32, 47–58. (In Italian)
56. Luino, F.; Turconi, L. *Eventi Alluvionali e Frane in Italia Settentrionale nel Periodo 2005–2016*; SMS: Moncalieri, Italy, 2017; ISBN 978-88-903023-8-1. (In Italian)
57. Swiss Re Institute. SIGMA Annual Report. 2018. Available online: https://www.swissre.com/dam/jcr:1b3e94c3-ac4e-4585-aa6f-4d482d8f46cc/sigma1_2018_en.pdf (accessed on 9 March 2021).

58. Regione Toscana. Alluvione di Livorno: Le Foto Aeree. 2017. Available online: https://www.toscana-notizie.it/-/alluvione-di-livorno-le-foto-aeree (accessed on 9 March 2021).
59. Stefanidis, S.; Stathis, D. Assessment of flood hazard based on natural and anthropogenic factors using analytic hierarchy process (AHP). *Nat. Hazards* **2013**, *68*, 569–585. [CrossRef]
60. Morris, A.E.J. *History of Urban Form before the Industrial Revolution*; Routledge Taylor & Francis Group: London, UK; New York, NY, USA, 2013; 443p.
61. Luino, F.; Paliaga, G.; Roccati, A.; Sacchini, A.; Turconi, L.; Faccini, F. Anthropogenic changes in the alluvial plains of the Tyrrhenian Ligurian basins. *Rend. Online Soc. Geol. Ital.* **2019**, *48*, 10–16. [CrossRef]
62. Kotroni, V.; Lagouvardos, K. Lightning in the Mediterranean and its relation with sea-surface temperature. *Environ. Res. Lett.* **2016**, *11*, 034006. [CrossRef]
63. Vargas-Yáñez, M.; Moya, F.; García-Martínez, M.C.; Tel, E.; Zunino, P.; Plaza, F.; Salat, J.; Pascual, J.; López-Jurado, J.L.; Serra, M. Climate change in the Western Mediterranean Sea 1900–2008. *J. Mar. Syst.* **2010**, *82*, 171–176. [CrossRef]
64. Esposito, G.; Matano, F.; Scepi, G. Analysis of Increasing Flash Flood Frequency in the Densely Urbanized Coastline of the Campi Flegrei Volcanic Area, Italy. *Front. Earth Sci.* **2018**, *6*, 63. Available online: https://www.frontiersin.org/articles/10.3389/feart.2018.00063/full (accessed on 9 March 2021). [CrossRef]
65. Buzzi, A.; Davolio, S.; Malguzzi, P.; Drofa, O.; Mastrangelo, D. Heavy rainfall episodes over Liguria of autumn 2011: Numerical forecasting experiments. *Nat. Hazards Earth Syst. Sci.* **2014**, *14*, 1325–1340. [CrossRef]
66. Paliaga, G.; Faccini, F.; Luino, F.; Turconi, L.; Bobrowsky, P. Geomorphic processes and risk related to a large landslide dam in a highly urbanized Mediterranean catchment (Genova, Italy). *Geomorphology* **2019**, *327*, 48–61. [CrossRef]
67. Moonena, A.C.; Ercoli, L.; Mariotti, M.; Masoni, A. Climate change in Italy indicated by agrometeorological indices over 122 years. *Agric. For. Meteorol.* **2002**, *111*, 13–27. [CrossRef]
68. Fantappiè, M.; L'Abate, G.; Costantini, E.A.C. The influence of climate change on the soil organic carbon content in Italy from 1961 to 2008. *Geomorphology* **2011**, *135*, 343–352. [CrossRef]
69. Marchi, L.; Borga, M.; Preciso, R.; Gaume, E. Characterisation of selected extreme flash floods in Europe and implications for flood risk management. *J. Hydrol.* **2010**, *394*, 118–133. [CrossRef]
70. Pattison, I.; Lane, S.N. The Link between Land-Use Management and Fluvial Flood Risk: A Chaotic Conception? *Prog. Phys. Geogr.* **2011**, *36*, 72–92. Available online: https://journals.sagepub.com/doi/full/10.1177/0309133311425398 (accessed on 14 February 2021). [CrossRef]
71. Paliaga, G.; Faccini, F.; Luino, F.; Roccati, A.; Turconi, L. A clustering classification of catchment anthropogenic modification and relationships with floods. *Sci. Total Environ.* **2020**, *740*, 139915. [CrossRef]
72. Piana, P.; Faccini, F.; Luino, F.; Paliaga, G.; Sacchini, A.; Watkins, C. Geomorphological landscape research and flood risk management in heavily modified Thyrrenian catchment. *Sustainability* **2019**, *11*, 4594. [CrossRef]
73. Salman, A.M.; Li, Y. Flood Risk Assessment, Future Trend Modeling, and Risk Communication: A Review of Ongoing Research. *Nat. Hazards Rev.* **2018**, *19*, 04018011. Available online: https://ascelibrary.org/doi/pdf/10.1061/%28ASCE%29NH.1527-6996.0000294 (accessed on 28 February 2021). [CrossRef]
74. Bustillos Ardaya, A.; Evers, M.; Ribbe, L. Participatory approaches for disaster risk governance? Exploring participatory mechanisms and mapping to close the communication gap between population living in flood risk areas and authorities in Nova Friburgo Municipality, RJ, Brazil. *Land Use Policy* **2019**, *88*, 104103. [CrossRef]
75. Walkling, B.; Tusker Haworth, B. Flood risk perceptions and coping capacities among the retired population, with implications for risk communication: A study of residents in a north Wales coastal town, UK. *Int. J. Disaster Risk Reduct.* **2020**, *51*, 101793. [CrossRef]
76. Regione Toscana. Progetto Castore. Available online: https://www.regione.toscana.it/-/castore (accessed on 9 March 2021).
77. TRIGEAU Project (2014–2020) Progetto Interreg Marittimo Italia-Francia-Transfrontalierità, Resilienza, Innovazione & Governance per la prevenzione del Rischio Idrogeologico. Available online: http://interreg-maritime.eu/fr/web/t.r.i.g-eau (accessed on 28 February 2021).
78. ADAPT Project (2014–2020) Assistere l'aDAttamento ai Cambiamenti Climatici dei Sistemi Urbani Dello sPazio Transfrontalier Available online: http://interreg-maritime.eu/web/adapt (accessed on 28 February 2021).

land

MDPI

Review

Environmental Forest Fire Danger Rating Systems and Indices around the Globe: A Review

Ioannis Zacharakis and Vassilios A. Tsihrintzis *

Centre for the Assessment of Natural Hazards and Proactive Planning, Laboratory of Reclamation Works and Water Resources Management, School of Rural, Surveying and Geoinformatics Engineering, National Technical University of Athens, 9 Heroon Polytechniou St., Zographou, 15780 Athens, Greece
* Correspondence: tsihrin@otenet.gr or tsihrin@survey.ntua.gr

Abstract: The objective of the present review is to analyze and evaluate the most used and well-performing environmental forest fire danger rating systems and indices globally, aiming to the creation of an integrated forest fire danger system for Greece. The analysis emphasizes the core input parameters that have been associated with forest fire danger (i.e., weather, vegetation, topography, and hydrology) and the computational procedure of each system index as well as the categorization of the output values. Online search engines such as Scopus, Google Scholar, WorldWideScience, ScienceDirect, and ResearchGate were used in the search for relevant literature published in scientific journals, manuals, and reports. The retrieved studies were classified and reviewed. Studies were selected for analytically describing the calculation process related to forest fire danger ignition and not being strictly geographically bound. A total of 210 studies were included in the current review, describing 63 forest fire danger systems and indices. These were analyzed and evaluated based on a scoring system. Overall, the top-rated indices were the: Nesterov's index, Sharples' index, Keetch and Byram's drought index, Telicyn logarithmic, and vapor pressure deficit, and the 3rd and the 4th also proved to be the most accurate for fire-prone regions. Remote sensing indices also proved to be promising in forest fire danger estimation.

Keywords: forest fire; fire danger rating systems; environmental fire danger; fire indices; drought indices; remote sensing fire indices; fire ignition probability; climate change extremes

Citation: Zacharakis, I.; Tsihrintzis, V.A. Environmental Forest Fire Danger Rating Systems and Indices around the Globe: A Review. *Land* **2023**, *12*, 194. https://doi.org/10.3390/land12010194

Academic Editors: Matej Vojtek, Andrea Petroselli and Raffaele Pelorosso

Received: 8 December 2022
Revised: 30 December 2022
Accepted: 31 December 2022
Published: 6 January 2023

1. Introduction

The role of forests—which cover approximately 31% of the global land [1]—is of great importance in ecological-environmental and socio-economic terms [2]. However, internationally, there has been a radical increase in the annual number of forest fire danger days and forest fire incidents, with climate change being one of the major contributors [3–10]. During the past two years, more than 45 million hectares have been burned across regions in Russia, Brazil, Canada, the United States of America, the European Union, and Australia [11–16], where forested areas cover 56% of the global forest land [1,17,18].

Several studies and reports indicate that most of the forest fire incidents are man-driven, in the form of either arson or negligence [19–24]. Nevertheless, wildfires occur mostly during periods of high temperature, intense drought, strong winds, low relative humidity, and inadequate precipitation [25–27].

Fire danger rating systems and indices are the products of systematic research both in theoretical and in empirical terms. Hence, many environmental fire danger rating systems throughout the world focus on the calculation of the condition of dead or alive fuels, such as fuel moisture codes, alongside meteorological parameters that have an impact on the source of heat as well as the ambient oxygen supply [28–31]. However, these systems ignore the human-driven ignition causes, with the latter being covered to a certain extent in the related literature [19–21,32–38]. Furthermore, only a limited number of studies adopts an integrated and/or holistic approach combining natural and human-driven causing factors as well as weather indices and conditions [39–41].

The objective of this review article is to report, analyze, compare, and evaluate the most applied and well-established environmental fire danger rating systems and indices around the world, aiming at the development of an integrated fire danger rating system for Greece.

2. Materials and Methods

Adopting the approach by Chuvieco et al. [34], fire risk assessment consists of two pillars: danger and vulnerability. Fire danger—also reported as fire hazard—is related to the conditions that favor the fire outbreak and its spread, while vulnerability is related to the possible outcome of a fire event as far as effects and value loss are concerned [34,42,43].

The present article focuses mainly on systems and indices that estimate fire danger ignition probability related to environmental factors as proposed by Cardille et al. [33]. The present review follows the Preferred Reporting Items for Systematic Reviews and Meta-Analyses (PRISMA) 2020 guides for systematic reviewing; thus, the following selection criteria for the considered references were set [44]: (1) studies must be papers published in scientific journals, manuals that are operationally in use, or technical documents supporting fire agency policies; (2) studies must contain systems or indices that focus on fire ignition probability; (3) studies must include systems indices that are not strictly geographically bound; and (4) the indices and systems must include environmental input parameters (such as weather, vegetation, hydrology, and others). Although there are already several research and review articles related to fire danger rating systems [45–50], to the best of our knowledge, the present study is the most complete as far as the number, the geographical scale, and the analysis of the computational procedure of systems and indices are concerned.

The research commenced with the examination of fire danger rating systems currently in use in countries with significant fire history and forest land, such as the USA, Canada, Russia, and Brazil, in the official websites of the respective ministries or agencies. Since the original publications, which describe the systems of the mentioned countries, were gathered, further research was held in the cited literature of the above publications. In addition, online search engines such as Scopus, Google Scholar, WorldWideScience, ScienceDirect, and ResearchGate were used, with the use of the following keywords: "fire danger", "fire danger rating systems", "fire danger indices", "fire ignition probability", "fire danger and remote sensing", "fire danger and drought", and "forest fire danger rating systems", among others. The research was conducted during an eight-month period lasting until July 2022, while the consulted sources were scrutinized according to the following steps: (1) the titles of the studies were compared with the above keywords; (2) those that matched were examined by their respective abstract; and (3) those whose abstracts fulfilled the selection criteria mentioned above were included. Moreover, filters such as "natural caused fires", "risk assessment", and "year of publication" were used.

Systems and indices included in the current study were divided into two groups: (1) the ad hoc fire systems indices; and (2) the indirect indicators. The first one consists of all the systems developed exclusively for fire danger estimation gathered based on geographical criteria, while the second one contains indices that have been proven to be to a certain extent related to fire danger estimation and are divided into drought or moisture presence and into remote sensing indices. All systems and indices involve three major procedures: (1) the collection of the input data; (2) the computational procedure; and (3) the outcome categorization in danger classes. These three procedures were identified and extracted from the collected studies. The first two are described in the next sections. The last one, alongside supporting material from the computational procedure in the form of tables, is included in the Supplementary Material (SM) of the present paper. Tables and sections in the Supplementary Material are cited in the manuscript with the indication "S" followed by the number of the respective table or paragraph.

All systems indices were eventually evaluated based on the cited literature. For validating the accuracy of each system index, the values of the latter were calculated for the two-month period from June–July 2022 using an ad hoc calculating package created in the Python programming language. Hence, the computed values were correlated to days with and without fire occurrences. The principles and the material gathered in the current review are expected to contribute positively to the forest fire science. All parameters not described after the presented equations are included in the Supplementary Material as Section S3. Nomenclature.

3. Results

Overall, a total of 210 studies met the inclusion criteria and were considered in the review. The selection process in numbers is presented in Figure 1. From the selected studies, a total of 63 systems indices were gathered—including modified versions—across 16 countries, as presented in Table 1.

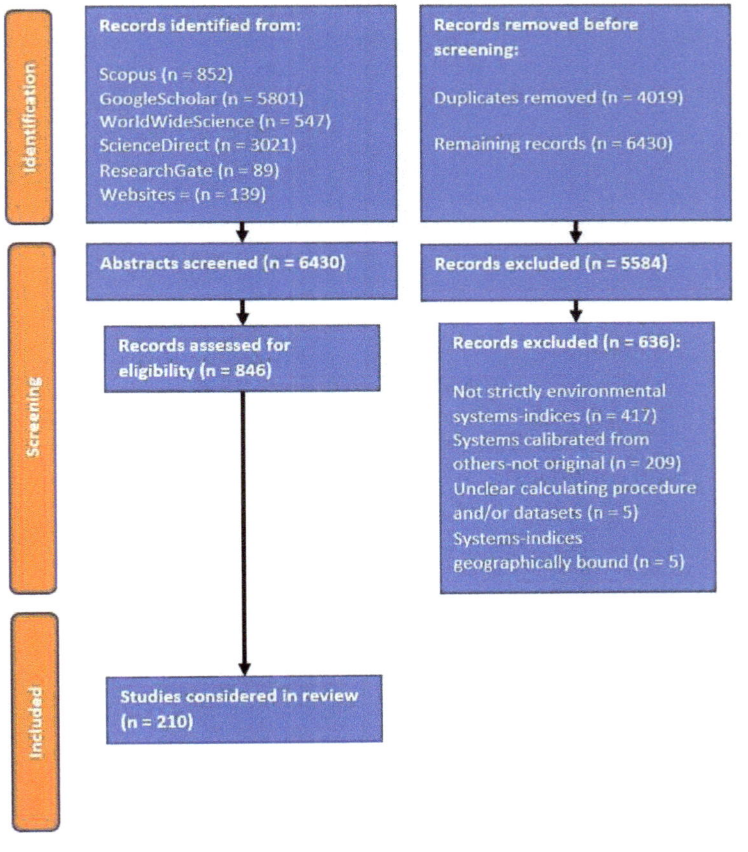

Figure 1. The selection process of consulted studies based on PRISMA 2020 flowchart for systematic reviews.

3.1. *North America Fire Danger Systems and Indices*

3.1.1. Canadian Forest Fire Danger Rate System

The Canadian Forest Fire Danger Rate System (CFFDRS) is a meteorologically based approach in fire danger rating developed in Canada in 1968, consisting of four subsystems: Fire Weather Index (FWI), Fire Behavior Prediction (FBP), Fire Occurrence Prediction (FOP), and Accessory Fuel Moisture (AFM) [30,51,52]. In the current study, though, only the FWI system will be considered, as it is related to fire ignition probability. The FWI system comprises six modules: Fine Fuel Moisture Code (FFMC), Duff Moisture Code (DMC), Drought Code (DC), Initial Spread Index (ISI), Buildup Index (BI), and Fire Weather Index (FWI) [30,53]. The three moisture codes refer to the moisture levels of three different fuel type categories, respectively, depending on fuel weight and fuel layer depth, while the two following indices are intermediate products that are related to fire spreading and total available fuel accordingly and produce the FWI [30,52,54–56]. The final index, FWI, is a general measure of fire danger representing potential fire-line intensity [56–58]. The calculation of all FWI system components is complex, as a total of 30 computation steps lead to the final output [59].

Depending on FWI values, six danger classes were defined in the original publication [30] as displayed in Table S1. However, when FWI is used in other countries, classes are redefined based on local calibrations [47,60–64].

3.1.2. National Fire Danger Rating System

The National Fire Danger Rating System (NFDRS), developed in 1972 [29] and revised in 1978, 1988 [14,65,66], and 2016, is the USA fire danger rating approach. It is designed to be scientifically based, applicable across the USA, adaptable, and inexpensive in operating terms [27]. The NFDRS uses meteorological as well as fuel moisture and topographical inputs, while a series of equations and calculations lead to six hourly output components: Spread Component (SC), Energy Release Component (ERC), Burning Index (BI), Ignition Component (IC), Human- and Lightning-Caused Fire Occurrence Index, and Fire Load Index [65,67,68]. The first three indices are based on combustion physics [69] and correspond to fire behavior characteristics, while the remaining indices provide an estimation on fire danger rating [29,65]. In the computation procedure, in which computer programs such as "AFFIRMS" and "FireFamily" or nomograms can be utilized, land slope is divided into three classes and fuel types into five, while nine fuel models are implemented in the system [29,65,70–73].

The NFDRS has various outputs; therefore, the categorization of fire danger rating can be estimated through the evaluation of different aspects of the system's components. In the current study, fire danger rating (Table S2) was produced based on the fire characteristics charts.

Table 1. Environmental fire danger rating systems indices.

No	Systems Indices	Origin	Publications
Ad hoc Fire Danger Rating Systems			
North American			
1	CFFDRS	Canada	[30,51,55,56,59]
2	NFDRS	USA	[29,65–67]
3	Fosberg	USA	[74,75]
4	Fosberg +	USA	[76]
5	BEHAVE	USA	[69,73,77,78]
6	CBI	USA	[79]
7	HDWI	USA	[80]
8	LASI	USA	[81]
Southern Hemisphere			
9	FFDI	Australia	[28,82,83]
10	GFDI	Australia	[28,82,84]
11	FFBT	Australia	[85,86]
12	SFDI	Australia	[87,88]
13	LFDI	S. Africa	[89,90]
14	FMA	Brazil	[91]
15	FMA+	Brazil	[92]

Table 1. *Cont.*

No	Systems Indices	Origin	Publications
16	IRM	Argentina	[93]
17	RF	Brazil	[94]
18	EPI	Brazil	[95]
19	PEI	Brazil	[95]
Mediterranean			
20	r (Orieux)	France	[96]
21	I87	France	[97]
22	Numerical	France	[98]
23	Lourenco	Portugal	[99]
24	Lourenco_m100	Portugal	[99]
25	Lourenco_f	Portugal	[99]
26	Ifa	Portugal	[100,101]
27	ICONA	Spain	[102]
28	CFS	Italy	[103]
29	IREPI	Italy	[104]
30	IFI	Italy	[105,106]
31	DMRIF	Tunisia	[107]
North Eurasian			
32	AI	Sweden	[108]
33	BI_t	Germany	[109]
34	I_{Br}	Germany	[110]
35	TLI	Russia	[111]
36	NI	Russia	[112]
37	mNI	Russia	[113]
38	Zhdanko	Russia	[114]
39	M68	Germany	[113]
40	mM68	Germany	[113]
41	DW	Finland	[115]
Indirect Indicators			
Drought–Moisture			
42	MDI	USA	[116]
43	KBDI	USA	[117]
44	SDI	Australia	[118]
45	PDSI	USA	[119]
46	RDI	Greece	[120]
47	CWD	USA	[121]
48	VPD	USA	[122]
49	DI	France	[123,124]
Remote Sensing			
50	NDVI	USA	[125]
51	RG	USA	[31]
52	VG	USA	[31]
53	NDWI	USA	[126]
54	NDWI_m	USA	[127]
55	NDII_6	USA	[128]
56	NDII_7	USA	[128]
57	NMDI	USA	[129]
58	SAVI	USA	[130]
59	EVI	USA	[130]
60	VARI	USA	[131]
61	FPI	USA	[70,132]
62	FPI_m1	USA	[133]
63	FPI_m2	USA	[61,134]

3.1.3. Fosberg and Modified Fosberg Indices

Fosberg developed an index supplemental to the NFDRS based on wind speed and equilibrium moisture content [74–76]. The basic equation, in SI units, is given below:

$$\text{FFWISI} = \frac{\eta}{0.3002}\left[1+\left(\frac{w}{1.690344}\right)^2\right]^{0.5} \tag{1}$$

where FFWI_{SI} is Fosberg index (SI units), η is a factor related to equilibrium moisture, and w is wind speed (km/h). The index value categorization is shown in Table S3 [76,135].

Since the original index ignores precipitation, a modified version was proposed by Goodrick [76], which includes a drought index developed by Keetch and Byram [117]:

$$\text{FAF} = 0.000002\ \text{KBDI}^2 + 0.72 \tag{2}$$

where FAF is a correction function, and KBDI is the Keetch and Byram drought index. The improved formula of Fosberg index can be computed as follows:

$$\text{FFWI}_{\text{m}} = \text{FAF} \cdot \text{FFWI}_{\text{SI}} \tag{3}$$

The original form of the Fosberg index is described in paragraph S2.1 in Supplementary Material, including the calculation of the other parameters.

3.1.4. BEHAVE System

BEHAVE is a system for evaluating fire potential that uses the same mathematical model as NFDRS although the equations differ as well as the input parameters, which can vary according to the available information, while the concept of BEHAVE focuses on fire behavior prediction rather than fire danger rating [77,136,137]. It consists of two subsystems, one for fuel modeling—in which 13 fuel types are introduced—and one for fire prediction, in which Rothermel's models are deployed alongside Byram's fire intensity [57,69,78].

3.1.5. Chandler Burning Index

Chandler proposed a simple index as a function of air temperature and relative humidity that estimates fuels ignitability and is calculated as follows [79]:

$$\text{CBI}_{\text{d}} = \left[(104.5 - 1.373\text{RH} + 0.54\text{T})124 \cdot 10^{-0.0142\text{RH}}\right]\frac{1}{60} \tag{4}$$

$$\text{CBI}_{\text{m}} = \left\{[(110 - 1.373\text{RH}) - 0.54(10.2 - \text{T})] \cdot 124 \cdot 10^{-0.0142\text{RH}}\right\}\frac{1}{60} \tag{5}$$

where CBI_{d} and CBI_{m} are daily and monthly, respectively, Chandler Burning Indices. For calculating CBI_{m}, average values of dry bulb air temperature T (in °C) and relative humidity RH (%) over a 30-day period are required. The categorization of the index is presented in Table S4.

3.1.6. Hot-Dry-Windy Index

A simple index combining air temperature, relative humidity, and wind speed was developed in 2018 in the USA and named the Hot-Dry-Windy Index (HDWI). It can be computed with the following equation [80,138]:

$$\text{HDWI}\left(\frac{\text{km}}{\text{h}}\right) = \frac{\text{W}_{max}\ \Delta\text{e}_{max}}{3.6} \tag{6}$$

where W_{max} is the maximum wind speed (m/s), and Δe_{max} is the maximum vapor pressure deficit (hPa) on daily basis, and the factor 3.6 is needed for conversion to SI units. The HDWI has been evaluated for a short number of incidents. A better accuracy for severe fire occurrences than the following LASI index was observed, so it has also been proposed to replace the latter one in the USA; however, further analysis is required, as it is presumed to perform poorly in thunderstorm-caused fires [80,138–140]. The higher the index values, the higher the fire danger.

3.1.7. LASI Index

The Lower Atmosphere Stability Index (LASI) is based on the atmospheric stability conditions and severe fires of a 20-year period [81]. Haines divided the USA into three climatic zones based on average elevation, where the LASI index is applied differently, as shown in Table S5 [81,141–143].

The LASI index is significantly different from all other indices analyzed in the current study, as it uses dry bulb air temperature values from two different atmospheric pressure levels based on elevation as well as the difference between dry bulb air temperature and dew point temperature on a third level, which is either the first or the second one, according to Table S5. LASI has been proven to perform well in diverted climatic regions of the world and has a very simple computational process [143–148]. Nevertheless, the low availability of the input data from a typical meteorological station limits the wider usage of LASI, while the absence of wind speed and drought factors substitutes LASI as a supplemental index for forest fire danger rating.

3.2. *South Hemisphere Fire Danger Systems and Indices*

3.2.1. Australian Systems and Indices

Two major fire danger rating systems were developed in Australia: McArthur Fire Danger Meters and Forest Fire Behavior Tables [28,82]. The first system was designed for Eastern Australia in the 1960s and has undergone since then several revisions, with the final versions being Mark 5 (FFDI) for forest fires and Mark 5 (GFDI) for grassland fires. The second was designed for western Australia in 1980s and is based on tables that predict fire behavior based on fuel characteristics and types of six dominant tree species [85,86]. These systems were produced based on empirical data from experimental fires in the Australian wildland [28,85]. The following equations can be used for the computation of the systems mentioned above [82]:

$$\mathrm{FFDI} = 2e^{(-0.45 + 0.987\ln(\mathrm{DF}) - 0.0345\mathrm{RH} + 0.0338\mathrm{T} + 0.0234\mathrm{W})} \quad (7)$$

where DF is a drought factor calculated as in paragraph S2.2 in Supplementary Material [82]. Accordingly, grassland fire danger indices (Mark 5) can be calculated based on paragraph S2.2 in Supplementary Material [82–84].

Eventually, the forest fire behavior tables (FFBT) system was designed to be deployed in a different manner than the previous systems, based on parameters provided in the aforementioned tables [85]. Nevertheless, a set of 72 equations was developed in the 1990s, from which the basic ones that describe the final index are displayed in paragraph S2.2 in Supplementary Material [99,102]. The categorization of the indices' danger classes analyzed above is based on the fire danger rating and displayed in Table S6; hence, the rest of FFBT output categorization is omitted [84,149,150].

In 2009, due to the complexity of the established systems in Australia, Sharples developed a computationally simple index. Firstly, a simple fuel moisture index was introduced, validated in the Australian eucalypt forests, and given by the following equation [87]:

$$\mathrm{SFMI} = 10 - 0.25(\mathrm{T} - \mathrm{RH}) \quad (8)$$

where SFMI represents Sharples' fuel moisture index.

Secondly, embodying SFMI, Sharples developed a fire danger index taking into account wind speed values, according to the following equation [88]:

$$\mathrm{SFDI} = \frac{\max(\mathrm{Wo}, \mathrm{W})}{\mathrm{SFMI}} \quad (9)$$

where SFDI (Equation (9)) stands for Sharples' fire danger index; Wo is set to 1 km/h in order to avoid zero values. The categorization of the index danger classes is presented in Table S7 [88].

3.2.2. Lowveld Fire Danger Index

Lowveld Fire Danger Index (LFDI) was developed in South Africa based on the Angstrom and the Canadian Forest Fire Danger Rating Systems and has been the official system used in the country, with the computational procedure being the following [89,90,151]:

$$\text{LFDI} = (\text{BI} + \text{WF})\ \text{RCF} \tag{10}$$

where LFDI is the Lowveld Fire Danger Index, BI is the Burning Index, WF is the Wind Factor, and RCF is the Rain Correction Factor. BI is related to the Angstrom index (with $R^2 = 0.99$), while it has been proven to be accurate in Greece—a typical Mediterranean country with long fire seasons [151]. The components of the index are presented in paragraph S2.3 in Supplementary Material, while the fire danger classes of the index are presented in Table S8.

3.2.3. Formulas of Monte Alegre

Megafires occur frequently in the tropical forests in the greater Amazon area; hence, a significant number of indices have been proposed and applied in Latin America Countries, mostly at the regional level [152,153]. Thus, a plethora of indices has been in use currently in Latin America. The first one, FMA, is a simple index that combines the number of days without any precipitation and relative humidity values, as attested by the following equation [91,153–155]:

$$\text{FMA} = \sum_{i=1}^{n} \frac{100}{RHi} \tag{11}$$

where *RH* stands for relative humidity on day *i*; *n* is the total amount of days without rain greater than 12.9 mm. However, in the case of rain between 2.5 and 12.9 mm, the FMA index must be reduced, as displayed in Table S9.

The FMA formula takes into account only two of the core meteorologic parameters; hence, an alteration has been proposed in order to include wind speed in the computational procedure [92,153–155]:

$$\text{FMA}^{+} = \sum_{i=1}^{n} \left(\frac{100}{RHi}\right) \text{e}^{(0.04\text{W})} \tag{12}$$

The categorization of danger classes of both indices is shown in Table S10.

3.2.4. Rodriguez–Moretti Index

The Rodriguez–Moretti Index (IRM), elaborated in the regions of Andean and Patagonia, combines the four basic meteorological components: dry bulb air temperature, wind speed, relative humidity, and days without any rain greater than 2 mm [93,153,156,157] Each of the four components is converted to input values using respective tables, as summarized in Table S11, according to the following equation [93,153,156,157]:

$$\text{IRM} = \text{Ti} + \text{RHi} + \text{Wi} + \text{Ri} \tag{13}$$

where Ti is temperature index, RHi is relative humidity index, W is wind speed index, and Ri is rainless days. The fire danger classes of the index are presented in Table S12 [93,153,156,157].

3.2.5. Risco do Fogo Index

The Risco do Fogo (RF) was developed by the Brazilian "Instituto Nacional de Investigaciones Espaciales" (INPE) based on simple meteorological inputs and vegetation type [94,157–159]. However, RF requires precipitation data for a period of 120 days in advance of the day of interest in order to estimate a series of respective factors [94,157–159

The equations describing the precipitation factor—which is of great importance in fire danger estimation for the current index—the period of drought, as well as the other components of the index are presented in paragraph S2.4 in Supplementary Material and in Table S13.

$$RF = RF_o \cdot FLAT \cdot \tag{14}$$

where FLAT and FELV are latitude and elevation factors, accordingly, and RF is the final fire risk (Risco do Fogo). The categorization of the index danger classes is displayed in Table S14.

3.2.6. Evaporation-Precipitation Indices

Two indices related to evaporation and precipitation are described in the current section. The first one is based on the division of the two parameters, while the second one is based on the respective difference [93,95,160]:

$$EPI = \sum_{i=1}^{t} \left(\frac{E_i}{P_i}\right) \tag{15}$$

where EPI is the evaporation divided by precipitation index, E is evaporation (mm), and t is the number of days since the start of the calculations. Respectively, the second index is computed as follows [93,95,160]:

$$PEI = \sum_{i=1}^{t} (P_i - E_i) \tag{16}$$

where PEI is the precipitation subtracted with evaporation. Both indices are cumulous, and their calculations follow the restrictions shown in Table S15 [95,160]: The higher the EPI index values, the higher the fire danger, while the lower the PEI index values, the higher the fire danger [95,154,155,160,161].

3.3. Mediterranean Indices

3.3.1. Orieux Index

In recent decades, forest fires have been on the rise across the Mediterranean, a region that has been indicated as severely vulnerable to climate change impacts, including increased forest fire season duration [162–164]. Moreover, the largest percentage of human-caused fires (95%) worldwide has been reported in the Mediterranean [165,166]. Hence, a great number of systems have been developed and used although other indices under calibration, such as the FWI from CFFDRS and FFDI, are currently preferred to be used [63,105,166].

The first of the Mediterranean indices presented here is the one developed in France by Orieux and is based on the exponential decrease of soil water reserve as well as wind speed values. The calculation of the index requires the estimation of the potential evapotranspiration through the Thornthwaite equation [96,167,168]. A maximum water reserve value of soil is considered to be 150 mm and represents the starting point of the index calculation, as the following equation suggests [96]:

$$r = R \cdot e^{\left(-\frac{\Sigma ETP}{R}\right)} \tag{17}$$

where r is the daily value of soil water reserve (which represents Orieux index), R is the maximum value of r equal to 150 mm, and ETP is the potential evapotranspiration computed through Thornthwaite's equation as in paragraph S2.5 in Supplementary Material and Table S16 [119,167,169].

The Orieux index is cumulative; thus, the previous day ETP is needed for present-day calculations. If precipitation occurs, the index increases although in the original publication [96], the method of applying the rainfall event effects is not clarified. Index categorization classes can be estimated as presented in Table S17.

3.3.2. Carrega's I_{87} Index

Another index developed in France as well, by Carrega [97,170], embodies the soil water reserve index as proposed by Orieux [96] alongside superficial water reserve, temperature, relative humidity, and wind speed. Index I_{87}, which is an amelioration of I_{85}, was proposed by Carrega [96], and it additionally includes temperature and superficial water reserve and can be calculated as follows [170]:

$$I_{87} = \frac{\max(10, T)W \cdot C}{RH \cdot r_s \sqrt{r}} \tag{18}$$

where T is dry bulb air temperature in °C, W is wind speed in m/s, RH is relative humidity (%), C is a phenological coefficient that corresponds to 200 in summer and winter and 100 in autumn and spring, r is Orieux water reserve saturated at 150 mm, and r_s is superficial water reserve saturated at 10 mm and computed using Thornthwaite's equation, as in paragraph S2.5 in Supplementary Material.

Carrega's index is cumulative as well; hence, the previous-day evapotranspiration is needed in the computation of current-day parameters. In case of a rain event, the values of soil water reserve and superficial reserve must be augmented by the amount of rain, while the second reserve must be further increased by 1 mm if dew occurs during night or 2 mm if the dew is strong [170]. The categorization of Carrega's index is not represented clearly in the original publication, where a very high danger corresponds to values greater than 200.

3.3.3. Numerical Index

The third index, called the Numerical Index and developed by Drouet and Sol in southern France, is based on the product of soil water reserve, wind speed, and false relative humidity, while other meteorological factors, such as temperature and cloud coverage, are used although considered less important [98,171,172]. The estimation of the index undergoes the following procedure:

$$\text{Numerical Index} = 25 - \frac{(FHR \cdot C_{res} \cdot C_{vent})}{15} + A \tag{19}$$

where FHR is the false relative humidity, C_{res} is the coefficient of soil water reserve, C_{vent} is the coefficient of wind, and A is a correction coefficient. These parameters are computed as described in paragraph S2.6 in Supplementary Material [173,174]. The fire danger classes are shown in Table S18.

3.3.4. Portuguese Indices

Another index developed in the Mediterranean is the Portuguese index. It was developed by the Portuguese Meteorological and National Institute by modifying the Nesterov index (presented in a later section). The Portuguese index is an estimation of atmospheric conditions at the fuel layer and consists of three indicators [100,101]: an ignition index, a rain coefficient, and a wind speed coefficient. The final index can be estimated according to the following equations [100,101,175]:

$$Ifa_i = I_i + Ia_{(i-1)} + CW \tag{20}$$

where Ifa_i is the Portuguese Index on day i, I_i is the ignition index, $Ia_{(i-1)}$ is a variant of the Nesterov Index, and CW is a wind coefficient. The calculations can be found in paragraph S2.7 in Supplementary Materialand and Table S19.

Finally, Ifa_i and $Ia_{(i-1)}$ are re-estimated according to tables presented in the original publication as well as the Swiss Federal Institute's fire weather danger wiki [176], while the danger classes are shown in Table S20.

Lourenço [99]—one of the two developers of the Portuguese Fire Danger Index—describes five simple fire danger indices that require only two to three of the basic meteorological parameters. These indices are described in the following equations:

$$\text{LFDRI} = \frac{\text{T}}{\text{RH}} \tag{21}$$

where LFDRI stands for Lourenço's fire danger index, T is dry bulb air temperature, and RH is relative humidity. Three variations of Lourenço's fire danger index (LFDRImax, LFDRIm, and LFDRIm100) have been documented, which use the maximum and minimum values of temperature and the relative humidity, respectively, as well as wind speed values (see paragraph S2.7 in Supplementary Material). The final index provided by Lourenço combines meteorological data from the day of interest plus the sequent day's forecast, as given below:

$$\text{LFDRIf} = \left\{\frac{Ti}{RHi} + \frac{Wi}{100} + \left\{[2(Tj - Ti) + (RHi - RHj) + (Wj - Wi)]\frac{1}{100}\right\}\right\}R \tag{22}$$

where LFDRIf is Lourenço's fire danger index for forecast, i represents current day, j the next day, and R is a risk factor based on each region's historical profile in fires, as shown in Table S21 with the categorization of the LFDRI indices fire danger classes [99].

3.3.5. ICONA Index

Another method for predicting forest fire danger rating, in the Mediterranean, was developed in 1993 in Spain [102] although a calibrated and enhanced version of the Canadian FWI is currently in operation in the country [177]. The ICONA index is based on fine fuel moisture content and wind speed, embodying the parameters for fuel modelling that the BEHAVE system utilizes [78]. The calculation process depends on tables provided by the system [102], and the terms and parameters have been translated as presented in paragraph S2.8 in Supplementary Material. The original publication defines four danger classes, as presented in Table S22 [102].

3.3.6. Italian Indices

There are two major indices used in continental Italy: the first one has been applied mostly in the Mediterranean part of Italy (CFS—Italian Fire Danger Index) and is based on McArthur's meters, while the second one (IREPI) was developed especially for the Alpine regions [103,104]. The Italian Fire Danger Index consists of both equations and tables, where simple meteorological data are inserted. The computational process is presented in paragraph S2.9 in Supplementary Material [103,175]. The Italian Fire Danger Index main equation is shown below:

$$\text{CFS} = 3.9\text{Ar}2^{(0.048\text{T} - 0.051\text{RH} + 0.033\text{W})} \tag{23}$$

where Ar is a parameter estimated as given in paragraph S2.9 in Supplementary Material. The index is cumulative, as the previous day's soil water deficit is needed, while the categorization can be estimated as displayed in Table S23 [178,179].

The second index, designed for the Italian Alps, is based on the relationship of potential and daily evapotranspiration and is called "Indice di Riduzione Evapotranspirazione per il Pericolo d'Incendio" (IREPI). The following equation can be used for the IREPI estimation [104,180]:

$$\text{IREPI} = \left(\frac{(\text{ETP} - \text{ETR})}{\text{ETP}}\right)100 \tag{24}$$

where ETP is the potential evapotranspiration, and ETR the real evapotranspiration in mm per day. There is a plethora of equations and methods in order to compute ETR and ETP, as are presented analytically in Xiang et al. [181] and McMahon et al. [182]. As this difference increases, the fire danger decreases.

A third index designed for Mediterranean vegetation characteristics is operational in Sardinia, named the Integrated Fire Index or Ichnusa Fire Index (IFI), and consists of four components [105,106,183]. Two versions of the index have been documented; thus, both will be presented in the current study:

$$\mathrm{IFI} = \mathrm{DC} + \mathrm{FC} + \mathrm{MC} + \mathrm{TC} \quad (25)$$

where DC is Drought Code indicating the evapotranspiration rate, FC is Fuel Code, MC is Meteo Code, and TC is Topological Code.

$$\mathrm{IFI}' = \mathrm{DC} + \mathrm{FC} + \mathrm{MC} + \mathrm{R} \quad (26)$$

with R being the code for solar radiation, replacing the TC parameter. The calculation of the above codes is analyzed in paragraph S2.9 in Supplementary Material. The categorization of the index is based on five danger classes (Table S24) after the normalization of the index, a process not clearly described in the original publication [106].

3.3.7. Tunisian Index

DMRIF is a simple index used in Tunisia, which requires a limited number of weather parameters [107,184]. According to the number of days since last rainfall event, there are two cases:

$$\mathrm{DMRIF} = -131.7\mathrm{r} + 5.9\mathrm{W} + 26.8\mathrm{N_d} + 1.4\mathrm{T_{max}} - 32.8\mathrm{Q},\ \text{if } \mathrm{N_d} > 6 \quad (27)$$

$$\mathrm{DMRIF} = -26.3\mathrm{r} + 4.6\mathrm{W} + 0.5\mathrm{T_{max}},\ \text{if } \mathrm{N_d} \le 6 \quad (28)$$

where r represents the soil water reserve (mm), which can be estimated according to Thornthwaite equation; W defines wind speed (m/s); N_d is the number of days since the last rainfall event; T_{max} is maximum dry bulb air temperature (°C); and Q is the amount of precipitation (mm). The index categorization differs for each case, as shown in Table S25 [107,185].

3.4. Northern Eurasian Indices

3.4.1. Angstrom Index

Forest fires are rapidly increasing in the temperate and boreal forests of northern Europe and Russia as well due to climate change, among other causes [37,186–188]. Several indices have been developed in the greater area, which have been also deployed in diverted climatic zones [64,68,155,189].

One of the simplest but also highly effective indices in the respective literature is the index developed in Sweden by Angstrom [152,153]. The index is calculated based on the following equation [79,108]:

$$\mathrm{AI} = \frac{\mathrm{RH}}{20} + \frac{27 - \mathrm{T}}{10} \quad (29)$$

where RH is relative humidity (%), and T is temperature (°C), both measured at 13:00 local time. The index categorization is in line with Table S26 [79,108].

3.4.2. Baumgartner Index

The Baumgartner index was developed and destined to be deployed in Bavaria although some findings indicate that its suitability in the area is limited [190]. The index relies directly on precipitation and indirectly on temperature, wind speed, net solar radiation, elevation, and relative humidity of five days in advance of the current one, according to the following equation [68,109]:

$$\mathrm{BIt} = \sum_{i=1}^{5} (\mathrm{ETP_{i-1}} - \mathrm{P_{i-1}}) \quad (30)$$

where BIt is Baumgartner Index for day t, ETP is potential evapotranspiration calculated by Penman equation [191], and P is precipitation in mm. The respective categorization, which excludes winter months fire danger, is shown in Table S27.

3.4.3. Bruschek Index

Another index developed in Germany, called Bruschek Index, that uses simple meteorological data can be computed by the following equation [110,192]:

$$I_{Br} = \frac{\sum_{i=vs}^{ve} sdi}{\sum_{i=vs}^{ve} Pi} \quad (31)$$

where I_{Br} is Bruschek Index, sdi is a parameter that equals 1 whether daily dry bulb air temperature is equal or greater than 25 °C and 0 in all other cases on day i, Pi is the precipitation depth on day i, vs is the starting period of vegetation green-up (the 1st of April), and ve is the end period (the 30th of September). The higher the index values, the higher the fire danger.

3.4.4. Telicyn Logarithmic Index

Another index mostly used in the tropical forests of Latin America—although developed in the Soviet Union [111]—is the Telicyn logarithmic Index (TLI), which requires simple meteorological inputs. The computational procedure is based on the following equation [153–155]:

$$TLI = \sum_{i=1}^{n} \log 10(Ti - Tdew, i) \quad (32)$$

where Ti is the dry bulb air temperature on day I; Tdew,i is dew point temperature on day i; n is the number of days without rain greater than 2.5 mm; and log10 is the logarithm on base 10. When the rainfall depth exceeds 2.5 mm, the index is set to zero. In line with the Monte Alegre formulas, the index is cumulative, requiring values of the previous day. The categorization of the index can be concluded from Table S28 [92,160,193].

3.4.5. Nesterov, Modified Nesterov, and Zhdanko Indices

The index of Nesterov—amongst the most widely used [47,68,101,105,112,155]—was developed in the Soviet Union. The index uses the dry bulb air temperature and dew point temperature as well as the number of days with precipitation depth less than 3 mm, according to the following equation [47,79,112]:

$$NIt = \sum_{i=1}^{t-1} (Ti - Tdew, i)Ti \quad (33)$$

where NI_t is the Nesterov index on day t. The meteorological data must be recorded at 15:00 local time. Two modified versions of Nesterov index were proposed by Käse and Zhdanko, as in paragraph S2.10 in Supplementary Material and in Table S29 [113,114,186]. The categorization of the Nesterov indices is displayed on Table S30 [194,195]. However, no categorization of the Zhdanko index was found in the respective literature although the index is similar to Nesterov but with much lower values.

3.4.6. M68 and Modified M68 Indices

M68 was developed by Käse in east Germany based on the same principle as the Nesterov index; however, three coefficients are implemented representing corrections related to precipitation, snow coverage, and vegetation condition, as presented below [113,192,196]:

$$M68_t = \sum_{15\ Febr}^{30\ Sept} (Tt - 10)\Delta e_t, \text{ (without coefficients)} \quad (34)$$

The calculations of the modified versions of M68 index are presented in paragraph S2.11 in Supplementary Material and Table S31. The categorization of the M68 and the modified M68 are displayed in Tables S32 and S33.

3.4.7. Finnish Fire Index

The Finnish Fire Index (FFI) is based on the calculation of volumetric moisture content changes, and most of the parameters can be computed according to Allen et al. [173] and Monteith [197]. The FFI relies on three components [115,198,199]:

$$\mathrm{DW} = \mathrm{E_{pot}} \cdot \mathrm{DE} + \mathrm{P_i} \tag{35}$$

where DW is the volumetric moisture content change of the total surface layer, Epot is the potential evaporation according to Penman and Monteith equation, DE is the drying efficiency, and P_i the precipitation depth in mm remaining in the surface layer. These three components can be calculated as in paragraph S2.12 in Supplementary Material [199]. The value of DW can be calculated as well as the fire danger class according to Table S34 [64,199].

3.5. Drought–Moisture Indices

3.5.1. Munger Drought Index

The first index, proposed in 1916 by Munger, is based on the number of consecutive days with precipitation height less than 1.27 mm and has been proven to be efficient for short-time drought periods [116,200]:

$$\mathrm{MDI} = 0.5 \cdot \mathrm{d}^2 \tag{36}$$

where MDI represents Munger's Drought Index; d is the number of consecutive days with rain height less than 1.27 mm. The higher the index value, the higher the fire danger is.

3.5.2. Keetch–Byram Drought Index

The drought index proposed by Keetch and Byram (KBDI) is one of the most used in fire danger rating systems. It is based on the next principles: the rate of moisture loss dependent on vegetation density, vegetation and rainfall have an exponential relationship, evapotranspiration determines the rate of soil's moisture loss—which is depleted with time exponentially, and an arbitrary layer depth of 8 in. (~20 cm) of soil capacity is arbitrarily used [47,81,82,117,201]. The following equations can be used for KBDI calculation:

$$\mathrm{KBDI_t} = \mathrm{Q} + \left[(800 - \mathrm{Q})\left(0.968\mathrm{e}^{0.0486(1.8\mathrm{T}+32)} - 8.3\right)\mathrm{dt}\right] \frac{10^{-3}}{1 + 10.88\mathrm{e}^{\left(\frac{-0.0441 P_a}{25}\right)}} \tag{37}$$

The components of the KBDI can be computed according to paragraph S2.13 in Supplementary Material as well as the original form of the KBDI equation [202,203]. The index categorization can be concluded according to Table S35 [117,204]. The KBDI index has been criticized for underestimating soil drying or wetting rates, especially in the critical phase between spring and summer, while it ignores the contribution of wind [47,118,205].

3.5.3. Soil Dryness Index

In order to solve the KBDI's inaccuracies, Mount developed the Soil Dryness Index (SDI), which embodies a different calculation approach for interception and runoff components into the soil–moisture deficit relationship [118,205,206]. The computational procedure is analyzed below:

$$\mathrm{SDI_t} = \mathrm{SDI_{t-1}} - \mathrm{P_{net}} + \mathrm{ET} \tag{38}$$

where SDI represents Soil Dryness Index on day t, P_{net} is the net precipitation, and ET is the evapotranspiration. The latter two components can be estimated based on Table S36 and paragraph S2.14 in Supplementary Material [118]. The SDI categorization for fire danger is described in Table S37 [205].

3.5.4. Palmer Drought Severity Index

One of the most-used drought indices in the USA is the Palmer Drought Severity Index (PDSI) developed in 1965; this index has also been associated with forest fire danger rating estimation [207–209]. The index relies on the hydrologic balance of water supply and loss—using historical drought data—dividing soil into layers with different water storage capacity, according to the following formulas [207–209]:

$$\mathrm{PDSI_i} = 0.897\mathrm{PDSI_{i-1}} + \frac{\mathrm{z_i}}{3} \tag{39}$$

where $\mathrm{PDSI_i}$ is Palmer's drought severity index on month i, and Z is the current moisture anomaly on the same month, as described in detail in paragraph S2.15 in Supplementary Material and Table S38 [173,207–212].

Although PDSI is the most broadly used drought index in the USA, there are some skeptical reviews considering the evapotranspiration calculation approach, the simplification method used for potential runoff and recharge delay, the ignorance of freezing conditions, the monthly basis estimation of the index, as well as the arbitrary subdivision of drought classes [208,213,214]. The index categorization is presented in Table S39 [207].

3.5.5. Reconnaissance Drought Index

A drought index embodying cumulative precipitation and potential evapotranspiration is the Reconnaissance Drought Index (RDI), and its three versions are as described below [120,215]:

$$\mathrm{RDIk} = \frac{\sum_{j=1}^{k} P_j}{\sum_{j=1}^{k} ETP_j} \tag{40}$$

where $\mathrm{RDI_k}$ is the index value for month k, P_j is the precipitation on month j, and ETP_j is the potential evapotranspiration on month j. The two other versions of the index are presented in paragraph S16 in Supplementary Material. The categorization of the last version of the index is presented in Table S40.

3.5.6. Climatic Water Deficit and Vapor Pressure Deficit

Another two indicators of water presence in the vegetation and in the air are climatic water deficit and vapor pressure deficit, respectively, with the latter already being mentioned. These indices have been directly correlated to some extent with fire danger, while they are included in other systems as well [3,121,122]:

$$\mathrm{CWD} = \mathrm{ETP} - \mathrm{ETR} \tag{41}$$

$$\Delta \mathrm{e} = \mathrm{E_{sat}} - \mathrm{Eact} = 0.6108\mathrm{e}^{\frac{17.27\mathrm{T}}{\mathrm{T}+237.3}}\left(1 - \frac{\mathrm{RH}}{100}\right) \tag{42}$$

where CWD is Climatic Water Deficit; ETP and ETR represent potential and real evapotranspiration, respectively.

3.5.7. Darcy's Law

Although Darcy's law describes the flow of a fluid through a porous medium and was basically developed to describe the flow of groundwater, a hydraulic corollary has been developed based on this law for the estimation of the possibility of a certain plant surviving harmful conditions, such as drought, wildfires, and pest attacks [123,124]. The equation is strongly related to plants physiology as well as meteorological parameters, according to the subsequent formula [124]:

$$\mathrm{DBI} = \frac{\mathrm{A_s} \cdot \mathrm{K_s}(\psi_\mathrm{s} - \psi_\mathrm{L})}{\mathrm{h} \cdot \eta \cdot \mathrm{A_L} \cdot \Delta\mathrm{e}} \tag{43}$$

where DBI is the Darcy-law-based index; A_s is the cross-sectional area of the conducting sapwood in cm^2 that has to be measured on field; K_s is the specific conductivity of the sapwood that has to be measured on field; ψ_S, ψ_L are soil and leaf water potential, respectively; h is the plant height that has to be estimated by field measurements or through remote sensing techniques; η is water viscosity that can be estimated through tables related to fluid mechanics [216]; A_L is the average leaf area that can be estimated or measured; Δe is vapor pressure deficit. The ψ_s and ψ_L can be calculated as in paragraph S2.17 in Supplementary Material [217,218]. Darcy's law requires a significant amount of field measurements, while it refers to a single tree, meaning that average values have to be inserted for vegetated areas—requiring similar plant types.

3.6. *Remote Sensing Indices*

3.6.1. Normalized Difference Vegetation Index

The advance in satellite technologies alongside with the remote sensing techniques has been proven significant in the development of fire danger rating systems, as the spatial and temporal resolution of the input data as well as the output indices has followed the improvement in computational systems [219]. In the current section, some of the most used and documented indices of remote sensing related to fire danger estimation will be described.

The Normalized Difference Vegetation Index (NDVI) is one of the most used concerning this category [34,220]. The index is based on the high reflectance of vegetation's chlorophyll in the near-infrared spectrum of radiation and the respective low reflectance on red, which differentiate the healthy plants containing an important amount of chlorophyll from the water-stressed ones [125,221]. The calculation of NDVI is given below [221,222]:

$$\mathrm{NDVI} = \frac{\mathrm{NIR} - \mathrm{R}}{\mathrm{NIR} + \mathrm{R}} \tag{44}$$

where NIR is the near infrared value of a pixel band (with wavelength from 0.80 to 0.90 μm), and R is the respective red one band (with wavelength from 0.63 to 0.70 μm). The NDVI has been used as a proxy of fuel moisture content and has been proven to be a reliable option when the required satellite data are available [223–230]. NDVI values are in range of −1 to 1, with positive values close to 1 showing healthy vegetation and negative values close to −1 showing water stressed vegetation.

3.6.2. Relative and Visual Greenness

The Relative and Visual Greenness are indices produced by NDVI values for long and medium periods of observations and have been used in several studies as a long- and medium-term, respectively, fuel moisture index [31,231,232] according to the following equations:

$$\mathrm{RG} = 100\left[\frac{\mathrm{NDVIo} - \mathrm{NDVImin}}{\mathrm{NDVImax} - \mathrm{NDVImin}}\right] \tag{45}$$

$$\mathrm{VG} = \frac{100}{0.66}\mathrm{NDVI_o} \tag{46}$$

where RG and VG are relative and visual greenness, respectively; $\mathrm{NDVI_o}$ is the NDVI index with values over a two-week period of observations; $\mathrm{NDVI_{min}}$ and $\mathrm{NDVI_{max}}$ are the NDVI index minimum and maximum values, respectively, for historical observations. Higher index values show more water presence in the vegetation.

3.6.3. Liquid Water Presence-Based Indices

Another category of indices based on the infrared bands consists of indices enhancing the visibility of plants that contain water in liquid form, such as the normalized difference water index NDWI, the normalized Multi-Band Drought Index (NMDI), and the Normalized Difference Infrared Indices (NDII), calculated as follows [126]:

$$NDWI = \frac{NIR - SWIR}{NIR + SWIR} \tag{47}$$

where NIR as in Equation (44) and SWIR is the shortwave infrared band (with wavelength from 1.00 to 2.50 μm). A second version—less used—is given by the following expression [127]:

$$NDWI_m = \frac{G - NIR}{G + NIR} \tag{48}$$

where G is the green band (with wavelength from 0.53 to 0.60 μm). Higher values of both indices show greater water presence. Two versions also have been utilized for NDII indices concerning the bandwidth of the SWIR wavelength [128]:

$$NDII_6 = \frac{NIR - SWIR_2}{NIR + SWIR_2} \tag{49}$$

$$NDII_7 = \frac{NIR - SWIR_3}{NIR + SWIR_3} \tag{50}$$

where NDII6 is the index using the 6th band of MODIS satellites, which is represented here by SWIR2 with bandwidth 1.628–1.652 μm; NDII7 is the index using the 7th band of the respective satellite represented by SWIR3 with bandwidth 2.105–2.155 μm. In Equations (49) and (50), NIR corresponds to the 2nd band of MODIS with a bandwidth of 0.841–0.876 μm. Finally, the NMDI uses two bands for liquid water absorption, enhancing the sensitivity to drought severity concerning both plants and soil [129]:

$$NMDI = \frac{NIR - (SWIR_2 - SWIR_3)}{NIR + (SWIR_2 + SWIR_3)} \tag{51}$$

where the proposed wavelengths are 0.860 μm, 1.640 μm, and 2.130 μm for NIR, SWIR2, and SWIR3, respectively.

3.6.4. Soil Adjusted Vegetation Index

The Soil-Adjusted Vegetation Index (SAVI) also refers to NDVI, as it was proposed as an amelioration of the latter for soil reflectance correction in sparsely vegetated areas, according to the following relationship [130]:

$$SAVI = \left[\frac{NIR - R}{NIR + R + L}\right](1 + L) \tag{52}$$

where L is a factor representing vegetation density ranging from 0 (very high vegetation density—SAVI equals to NDVI) to 1 (very low vegetation density). The output values are slightly lower than the respective NDVI ones, as for leaf area index (LAI as mentioned above) equal to 0.5 and 1, NDVI ranges from 0.24 to 0.60 and from 0.44 to 0.74, while the respective SAVI values range from 0.21 to 0.24 and 0.38 to 0.40 accordingly [131].

3.6.5. Enhanced Vegetation Index and Visible Atmospheric Resistant Index

These two indices are alternatives to NDVI. The Enhanced Vegetation Index (EVI) considers atmospheric and canopy noise, while the sensitivity to high density vegetation is greater, according to the following equation [233]:

$$EVI = G\left[\frac{NIR - R}{NIR + C_1 \cdot R - C_2 \cdot B + L}\right] \tag{53}$$

where G, C1, and C2 are the gain factor coefficient (equal to 2.5) and the aerosol resistance terms (equal to 6 and 7.5, respectively); B stands for the blue band (with a wavelength range from 0.43 μm to 0.50 μm), while all others are as in previous equations.

The Visible Atmospheric Resistance Index (VARI) considers the atmospheric noise as well; however, it requires data from the visible spectrum, in line with the following relationship [133]:

$$\mathrm{VARI} = \frac{\mathrm{G} - \mathrm{R}}{\mathrm{G} + \mathrm{R} - \mathrm{B}} \tag{54}$$

3.6.5.1. Fire Potential Index Model and Modifications

Fire Potential Index (FPI), which combined satellite and field observations with high correlation to fire incidents. FPI was developed as an alternative to the complicate NFDRS calculations, while its intended accuracy approximates the 1 km [70]. FPI embodies some of the remote sensing indices mentioned in earlier sections as well as the fuel models proposed by the developers of NFDRS. The computational procedure of FPI alongside the two modifications is presented in paragraph S2.18 in Supplementary Material, Tables S41 and S42 [31,61,65,67,133,134]. Finally, the FPI has been proven to be reliable in environmentally diverted areas [133,134] although the ignition source of the fire incidents is ignored.

4. Discussion

The 63 systems indices reported in the current review incorporate environmental parameters for assessing directly or indirectly the fire danger in wildlands. However, there is a variety in parameters used as input for the analyzed systems, while their respective significance, impact, and calculation procedures also differ from case to case. Therefore, inputs, calculations, and outputs are the core parts of the evaluation process. Table 2 displays the relationships between systems and input parameters except for the remote sensing indices, as in these, the input parameters are the reflectance values of satellite images. For assessing the performance of the systems and indices in terms of inputs, calculations, and outputs, thirteen criteria were established according to the cited references of the current review, which were grouped in four categories: (1) computational procedure; (2) fire characteristics; (3) modularity; and (4) credibility. The final grade is the sum of all points derived from these criteria.

4.1. Computational Procedure

Five criteria are incorporated in the present group: (1) calculation complexity, defined by the number of equations and tables needed for the calculation of an index; (2) required data volume, defined by the number and the form of the data; (3) input data complexity, defined by the difficulty to acquire or measure the input data; (4) units which can be in SI or in US customary units; and (5) accumulated index related to previous calculations. The first three are rated on the scale 0–5, with 5 implying the less complex and 0 the most challenging to estimate. If the unit system differs from SI, then a point is subtracted from the sum of the first five criteria, while the same happens in the case that an index is cumulous, as different units and cumulus indices require additional calculations. The final result is divided by two in order to ensure equal weights to the following criteria.

4.2. Fire Characteristics

Two criteria are included in fire characteristics: (1) input variables type, defined by the first line of Table 2, and (2) fire danger aspect, related to fire ignition, spread, or severity. For every variable type and fire danger aspect, a point is added to the sum of the criteria for every index, as indices including more types of parameters and fire characteristics are considered as the most integrated.

Table 2. Systems and indices analyzed in terms of input parameters.

	Meteorology									Vegetation								Topography		Hydrology										
Index	**A**	**B**	**C**	**D**	**E**	**F**	**G**	**H**	**I**	**J**	**K**	**L**	**M**	**N**	**O**	**P**	**Q**	**R**	**S**	**T**	**U**	**V**	**W**	**X**	**Y**	**Z**	**a**	**b**	**c**	**Σ**
CFFDRS	X	X	X		X			X																						5
NFDRS	X	X	X		X		X			X																				6
Fosberg	X	X	X																											3
mFosberg	X	X	X		X																X									5
BEHAVE	X		X	X							X	X	X			X	X		X	X										10
CBI	X	X																												2
HDWI	X	X	X																											3
LASI	X					X												X												3
FFDI	X	X	X		X																X									3
GFDI	X	X	X									X	X																	3
FFBT	X	X	X			X			X		X																			5
SFDI	X	X	X																											3
LFDI	X	X	X		X																X									5
FMA		X																			X									2
FMA+		X	X																		X									3
IRM	X	X	X																		X									4
RF	X	X			X			X			X							X												6
EPI					X																			X						2
PEI					X																			X						2
r (Orieux)	X							X														X								3
I87	X	X	X					X	X													X								6
Numerical	X		X			X	X																							4
Ifa	X		X		X	X																								4
ICONA	X	X	X				X		X	X	X								X	X										9
CFS	X	X	X		X																X									5
IREPI																						X	X							2
IFI	X	X	X		X					X					X	X			X	X					X				X	11
DMRIF	X		X		X																X	X								5
Lourenco	X	X																												2
Lourenco_m100	X	X	X																											3
Lourenco_f	X	X	X																											3
AI	X	X																												2
BIt					X																	X								2
I_{Br}	X				X																									2
TLI	X					X															X									3

Table 2. *Cont.*

	Meteorology									Vegetation								Topography			Hydrology									
Index	**A**	**B**	**C**	**D**	**E**	**F**	**G**	**H**	**I**	**J**	**K**	**L**	**M**	**N**	**O**	**P**	**Q**	**R**	**S**	**T**	**U**	**V**	**W**	**X**	**Y**	**Z**	**a**	**b**	**c**	**Σ**
NI	X					X															X									3
mNI	X					X															X									3
Zhdanko	X					X															X									3
M68	X	X																												2
mM68	X	X																												2
DW	X	X			X					X				X										X	X				X	8
MDI																					X									1
KBDI	X				X																					X	X	X		5
SDI	X				X											X						X				X	X	X		7
PDSI					X			X														X	X			X	X	X		6
RDI					X																	X								2
CWD																						X	X							2
VPD	X	X																												2
DI										X							X													2
Σ	38	27	22	1	20	8	3	5	3	5	4	2	2	1	1	3	2	2	3	3	13	9	3	3	2	3	3	3	2	

Legend (Column heading)

A	Temperature	K	Vegetation type	U	Drought period
B	Relative humidity	L	Curing degree	V	Potential Evapotranspiration
C	Wind speed	M	Fuel quantity	W	Actual Evapotranspiration
D	Wind direction	N	Leaf area index	X	Evaporation
E	Precipitation	O	Leaf area density	Y	Latent Heat of Evaporation
F	Dew point temperature	P	Canopy characteristics	Z	Interception
G	Clouds concentration	Q	Plants physiology	a	Runoff
H	Latitude	R	Elevation	b	Recharge
I	Date and/or Time	S	Slope	c	Solar Radiation
J	Fuel moisture	T	Aspect	Σ	Sum

4.3. Modularity

Two criteria are included as well in this group: (1) useful subcomponents related to intermediate outputs of the fire danger systems and (2) embodiment of other indices. Both are rated in a binary form, adding a point to the overall sum for every criterion that is fulfilled. Modularity is not a necessity; however, it can be considered as a leverage in the calculation process as well as in the integration of the deployed indices.

4.4. Credibility

The final category can be considered as the most essential in operating terms and consists of four criteria: (1) calculation basis, referring to the theoretical background concerning each system's development; (2) output categorization, concerning the clarity and the relevance of the output to the fire danger rating; (3) validation, concerning the estimation of the system index values in real scenarios and based on the selected studies of this review; and (4) adaptability, related to the degree that a system has been successfully tested in different environments. In the first one, six cases can be distinguished: arbitrary, empirical, scientific-based systems, and their combinations per two. Arbitrary systems are considered less credible, especially in different environments; thus, no points are added in the evaluation process. Empirical systems are developed to be more well-suited to local conditions; hence, these systems are rated with two points, while the mixture of arbitrary-empirical basis is rated with one point. Scientific-based systems provide enhanced credibility; therefore, the respective rate equals three, the mixture of scientific-arbitrary basis is rated with two, and lastly, the mixture of scientific-empirical basis is rated with four, as it is the most complete approach. In the second one, the output categorization can be clear and immediately related to fire danger, moderately clear and related to fire danger, and not clear or not immediately related to fire danger, with ratings being 1, 0, and −1, respectively. In the third and the fourth criteria, the validation results and adaptability are rated from 0 concerning systems that have not been yet deployed and from 3 concerning systems that have been used in almost every environment for a long period of time. The results of the evaluation procedure are displayed in Table 3. The results presented in Table 3 use the same weight for all criteria although the first and the last group are considered of higher importance as more criteria and points are included. Other combinations of weights can also be defined. NI, AI, KBD and NDVI are the top-performing indices, while IREPI, I_{Br}, CWD, and DI have the lowest scores.

Table 3. The evaluation of systems and indices for fire danger rating.

	S1					**Σ**	**S2**		**Σ**	**S3**		**Σ**	**S4**				**Σ**	**Grade**
Index	**A**	**B**	**C**	**D**	**E**		**F**	**G**		**H**	**I**		**J**	**K**	**L**	**M**		
CFFDRS	1	2	4	SI	−1	6	m	i	2	1	0	1	se	1	3	3	11	17
NFDRS	0	0	3	O	0	2	m,v,t	i,b,S	6	1	1	2	se	0	3	2	9	18
Fosberg	3	4	5	O	0	11	m	i	2	0	0	0	s	1	2	1	7	14.5
Fosberg +	2	3	5	O	0	9	m	i	2	0	1	1	s	1	1	1	6	13.5
BEHAVE	0	0	3	O	0	2	m,v,t	b,S	5	1	1	2	se	1	2	1	8	17
CBI	5	5	5	SI	0	15	m	i	2	0	0	0	e	1	1	1	5	14.5
HDWI	4	4	5	SI	0	13	m	i	2	0	0	0	s	−1	1	1	4	12.5
LASI	5	4	1	SI	0	10	m	i,S	3	0	0	0	a	1	3	2	6	14
FFDI	2	2	5	SI	−1	8	m	i	2	1	1	2	se	1	3	2	10	18
GFDI	4	3	4	SI	0	11	m,v	i	3	0	0	0	se	1	2	1	8	16.5
FFBT	1	2	3	SI	0	6	m,v	i	3	0	0	0	se	1	1	0	6	12
SFDI	5	5	5	SI	0	15	m	i	2	1	0	1	se	1	1	1	7	17.5
LFDI	4	4	4	SI	0	12	m,h	i	3	0	0	0	se	1	2	1	8	17
FMA	5	5	5	SI	−1	14	m	i	2	0	0	0	ea	1	2	1	5	14
FMA+	5	4	5	SI	−1	13	m	i	2	0	0	0	ea	1	2	1	5	13.5
IRM	5	4	5	SI	0	14	m	i	2	0	0	0	e	1	1	1	5	14
RF	3	2	3	SI	0	8	m,v	i	3	1	0	1	se	1	2	2	9	17

Table 3. *Cont.*

	S1					Σ	S2		Σ	S3		Σ	S4				Σ	Grade
Index	**A**	**B**	**C**	**D**	**E**		**F**	**G**		**H**	**I**		**J**	**K**	**L**	**M**		
EPI	5	5	5	SI	−1	14	m,h	i	3	0	0	0	a	−1	1	1	1	11
PEI	5	5	5	SI	−1	14	m,h	i	3	0	0	0	a	−1	1	1	1	11
r (Orieux)	2	3	4	SI	−1	8	m,h	i	3	1	0	1	ea	1	2	1	5	13
I87	2	3	4	SI	−1	8	m,h	i	3	0	1	1	sa	−1	2	1	4	12
Numerical	2	4	5	SI	−1	10	m	i	2	1	1	2	sa	1	1	1	5	14
Lourenco	5	5	5	SI	0	15	m	i	2	0	0	0	a	1	2	1	4	13.5
$Lourenco_{\_m100}$	5	4	5	SI	0	14	m	i	2	0	0	0	a	1	2	1	4	13
Lourenco_f	5	5	4	SI	0	14	m	i	2	0	0	0	ea	1	2	1	5	14
Ifa	4	4	5	SI	−1	12	m	i	2	0	0	0	se	1	2	1	8	16
ICONA	3	1	3	SI	0	7	m,v,t	i	4	1	0	1	se	1	2	1	8	16.5
CFS	3	3	3	SI	−1	8	m,h	i	3	0	0	0	sa	1	2	1	6	13
IREPI	3	5	5	SI	0	13	h	i	2	0	0	0	a	−1	2	0	1	9.5
IFI	2	0	1	SI	0	3	m,v,h,t	i	5	1	0	1	sa	1	1	0	4	11.5
DMRIF	4	3	4	SI	−1	10	m,h	i	3	0	1	1	sa	1	2	1	6	15
AI	5	5	5	SI	0	15	m	i	2	0	0	0	sa	1	3	3	9	18.5
BIt	4	4	5	SI	−1	12	m,h	i	3	0	0	0	ea	1	1	1	4	13
I_{Br}	5	5	4	SI	−1	13	m	i	2	0	0	0	a	−1	1	1	1	9.5
TLI	5	4	5	SI	−1	13	m	i	2	0	0	0	sa	1	2	2	7	15.5
NI	5	5	4	SI	−1	13	m,h	i	3	0	0	0	s	1	3	3	10	19.5
mNI	4	4	4	SI	−1	11	m,h	i	3	0	0	0	se	1	1	1	7	15.5
Zhdanko	4	4	4	SI	−1	11	m,h	i	3	0	0	0	se	1	1	1	7	15.5
M68	3	4	4	SI	−1	10	m,h	i	3	0	0	0	se	1	1	1	7	15
mM68	3	4	4	SI	−1	10	m,h	i	3	0	0	0	se	1	2	1	8	16
DW	2	0	0	SI	0	2	m,h	i	3	0	0	0	se	1	2	1	8	12
MDI	5	5	5	SI	−1	14	h	i	2	0	0	0	s	−1	1	1	4	13
KBDI	4	3	4	O	−1	9	m,h	i	3	0	0	0	se	1	3	3	11	18.5
SDI	3	3	3	SI	−1	8	m,h	i	3	0	0	0	se	1	3	2	10	17
PDSI	1	2	1	O	−1	2	m,h	i	3	0	0	0	se	0	3	2	9	13
RDI	5	5	4	SI	−1	13	m,h	i	3	0	0	0	s	0	1	0	4	13.5
CWD	4	3	3	SI	0	10	h	i	2	0	0	0	a	−1	2	1	2	9
VPD	5	5	5	SI	0	15	m	i	2	0	0	0	e	−1	2	2	5	14.5
DI	3	1	1	SI	0	5	h	i,S	3	0	0	0	s	−1	0	0	2	7.5
NDVI	5	5	5	SI	0	15	v	i,S	3	0	0	0	s	−1	3	3	8	18.5
RG	5	5	4	SI	−1	13	v	i,S	3	0	1	1	se	−1	2	2	7	17.5
VG	5	5	4	SI	−1	13	v	i,S	3	0	1	1	se	−1	1	1	5	15.5
NDWI	5	5	5	SI	0	15	v	i,S	3	0	0	0	s	−1	1	1	4	14.5
NDWIm	5	5	5	SI	0	15	v	i,S	3	0	0	0	s	−1	0	1	3	13.5
NDII6	5	5	5	SI	0	15	v	i,S	3	0	0	0	s	−1	0	1	3	13.5
NDII7	5	5	5	SI	0	15	v	i,S	3	0	0	0	s	−1	0	1	3	13.5
NMDI	5	5	5	SI	0	15	v	i,S	3	0	0	0	s	−1	0	1	3	13.5
SAVI	5	5	5	SI	0	15	v	i,S	3	0	0	0	s	−1	0	1	3	13.5
EVI	5	5	5	SI	0	15	v	i,S	3	0	0	0	s	−1	0	1	3	13.5
VARI	5	5	5	SI	0	15	v	i,S	3	0	0	0	s	−1	0	1	3	13.5
FPI	2	3	4	SI	−1	8	m,v	i,S	4	0	1	1	se	−1	2	2	7	16
FPI_m1	3	3	4	SI	−1	9	m,v	i,S	4	0	1	1	se	−1	1	1	5	14.5
FPI_m2	3	3	4	SI	−1	9	m,v	i,S	4	0	1	1	se	−1	1	1	5	14.5

Legend (Column heading)

S1	Computational procedure	D	Units	L	Validation	B	Behavior
S2	Fire characteristics	E	Accumulated index	M	Adaptability	S	Severity
S3	Modularity	F	Fire danger variables	N	Accuracy	A	Arbitrary
S4	Credibility	G	Fire danger aspect	M	Meteorology	E	Empirical

Table 3. *Cont.*

	S1					Σ	S2		Σ	S3	Σ	S4					Σ	Grade
Index	**A**	**B**	**C**	**D**	**E**		**F**	**G**		**H**		**I**	**J**	**K**	**L**	**M**		

Σ	Sum	H	Useful subcomponents	V	Vegetation	S	Scientific
A	Calculation complexity	I	Embodiment of other indices	T	Topography	Si	International system
B	Required data volume	J	Development basis	H	Hydrology	O	Other
C	Input data complexity	K	Output interpretation	I	Ignition		

4.5. Accuracy

The last but also the most important step for selecting the best fire danger rating system or index is estimating the respective accuracy. For the evaluation process, four regions within the Greek territory were selected, as depicted in Figure 2: (1) Mt. Penteli region in Attica; (2) the Regional Authority of Evros, northeastern Greece; (3) the Region of Kimi-Aliveri in Evoia; and (4) the Regional Authority of Helia-Achaia. As shown in Figure 2, fire incidents for the period 01/06/2022–31/07/2022 were gathered from satellite images provided by the NASA-FIRMS (https://firms.modaps.eosdis.nasa.gov/map/, accessed on 1 August 2022) [234] as well as meteorological data from the local weather stations, provided by the National Weather Service of Greece and the National Observatory of Athens.

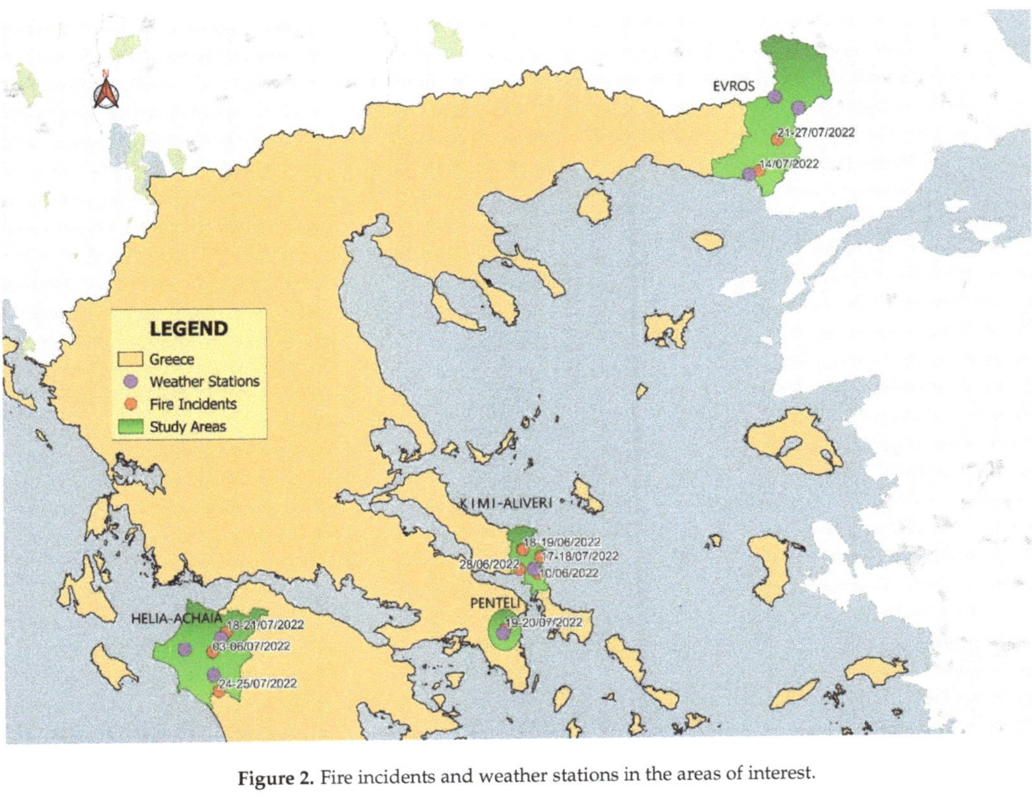

Figure 2. Fire incidents and weather stations in the areas of interest.

For calculating the values of the indices for every region, a computer program in Python was developed according to the equations presented in this article and the respective SM file. The program uses as input the meteorological data, i.e., temperature, relative humidity, wind speed, and precipitation depth, while it calculates other intermediate parameters needed for the computation of the fire indices, such as the number of days of drought, evapotranspiration, etc. The computational procedure concludes with the output values of every index on daily basis from 1 June 2022 to 31 July 2022. Unfortunately, due to lack of some specialized data, not all presented indices were used. The following indices were excluded from this accuracy experiment: BEHAVE, LASI, FFBT, ICONA, IREPI, PDSI, CWD, Darcy, as well as remote sensing indices, as a different approach would have to be adopted; thus, the comparison would be unrepresentative.

According to the index values in relation to fire incident or no fire occurrence, the following four cases were examined: (1a) no fire occurrence and index "hit"; (1b) no fire occurrence and index "miss"; (2a) fire occurrence and index "hit"; and (2b) fire occurrence and index "miss". In case of no fire incident, an index hit is considered as the outcome value out of the range of extreme fire danger class—according to SM tables. Accordingly, in case of a fire event, an index hit is when the respective outcome value is in the range of the extreme fire danger class. Fire danger classes for some of the indices had to be redefined to be realistic in the Greek environment in order to fit with the outcome value ranges. Cases 1a and 1b were marked with 1 and 0 points, respectively, per day of calculations, while cases 2a and 2b, which were considered more important for fire management, were marked with 2 and −2, respectively. For the final score, all index marks were normalized by dividing with 277 (the maximum mark for all days and all regions). The five most-accurate indices for the tested period and regions in Greece were proven to be the following: (1) NI; (2) KBDI; (3) SFDI; (4) FFDI5; and (5) SDI. The five least-accurate indices were as follows: (1) DW; (2) r (Orieux); (3) IBr; (4) IFI; and (5) PEI. The NI has also been successfully applied in mountainous areas in Greece [235–237], corroborating the findings of the present article.

Eventually, the scores of indices (presented in Table 3) and the accuracy marks were summed, using equal weights (divided with the respective maximum grade), to produce the final evaluation of the included indices, as shown in Table 4.

Table 4. Overall performance of environmental fire danger systems and indices based on the five groups of criteria.

Index	Score	Index	Score	Index	Score
NI	0.68	Zhdanko	0.56	pmM68	0.51
KBDI	0.65	M68	0.55	HDWI	0.49
SFDI	0.64	CBI	0.55	FMA	0.49
FFDI5	0.62	RF	0.55	RDI	0.49
SDI	0.61	DMRIF	0.54	CFS	0.48
LFDI	0.60	IRM	0.54	EPI	0.47
GFDI5	0.59	Fosberg	0.53	Numerical	0.46
TLI	0.59	MDI	0.53	FMA+	0.46
CFFDRS	0.59	Lourenco_f	0.53	I87	0.44
Ifa	0.59	Lourenco	0.53	DW	0.44
AI	0.58	BIt	0.52	r (Orieux)	0.43
NFDRS	0.58	Lourenco$_{-m100}$	0.52	I_{Br}	0.43
mNI	0.57	Fosberg+	0.51	IFI	0.42
VPD	0.57	mM68	0.51	PEI	0.41

5. Conclusions

A total of 63 environmental fire danger rating systems from across the globe wer analyzed and compared. The most important parameters were associated with weathe and hydrology although the most accurate indices required only two to five inputs. Some c the most-used systems—also reported in the present review—require complex calculation However, the top-rated indices and the most accurate as well were those with simple formulas and procedures. In addition, indices developed in a specific region have been proven to be more accurate in different environments—as most of the Mediterranean indices included in the current study underperformed in Greece. Additionally, the most complete systems—such as the CFFDRS and the NFDRS—had a fine performance, while the FFDI5 reached near the top, leading to the conclusion that if these systems adapt better to the local conditions, their performance will be greater than the respective one of the simpler indices. Finally, this review corroborated the inadequacy of the existing environmental fire danger rating systems in predicting modern day incidents, as the top-performing systems had an accuracy of 60–66% and a total score of 59–68%, indicating the need for an integrated approach including social and other factors.

Supplementary Materials: The following supporting information can be downloaded at: https://www.mdpi.com/article/10.3390/land12010194/s1, Paragraphs S2.1–S2.18; Tables S1–S42. The SM file presents the calculation procedure and value range for each system and index. Also, the SM includes the nomenclature.

Author Contributions: Conceptualization, I.Z. and V.A.T.; methodology, I.Z. and V.A.T.; software, I.Z.; validation, I.Z. and V.A.T.; formal analysis, I.Z.; investigation, I.Z. and V.A.T.; resources, V.A.T.; data curation, I.Z.; writing—original draft preparation, I.Z.; writing—review and editing, V.A.T.; visualization, I.Z.; supervision, V.A.T.; project administration, V.A.T. All authors have read and agreed to the published version of the manuscript.

Funding: This research received no external funding.

Data Availability Statement: Meteorological open data were downloaded from the National Observatory of Athens, https://www.meteo.gr/Gmap.cfm (accessed on 1 September 2022). Fire incidents data were downloaded from the National Fire Service of Greece, https://www.fireservice.gr/en_US/stoicheia-symbanton (accessed on 1 September 2022). Spatial open data were downloaded from https://geodata.gov.gr/en/ (accessed on 1 September 2022). For meteorological data processing, the Python open-source programming language was used, and for spatial analysis and mapping, the open-source Quantum GIS was deployed. All data are included in the paper.

Conflicts of Interest: The authors declare no conflict of interest.

References

1. FAO; UNEP. The State of the World's Forests 2020. In *Forests, Biodiversity and People*; FAO: Rome, Italy, 2020. [CrossRef]
2. Agrawal, A.; Cashore, B.; Hardin, R.; Shepherd, G.; Benson, C.; Miller, D. Economic Contributions of Forests. In *United Nations Forum on Forests Tenth Session 8–19 April 2013*; United Nations: Istanbul, Turkey; p. 132. Available online: https://www.un.org/esa/forests/wp-content/uploads/2015/12/EcoContrForests.pdf (accessed on 13 March 2022).
3. Abatzoglou, J.T.; Williams, A.P. Impact of anthropogenic climate change on wildfire across western US forests. *Proc. Natl. Acad. Sci. USA* **2016**, *113*, 11770–11775. [CrossRef] [PubMed]
4. Boer, M.M.; Nolan, R.H.; De Dios, V.R.; Clarke, H.; Price, O.F.; Bradstock, R.A. Changing Weather Extremes Call for Early Warning of Potential for Catastrophic Fire. *Earth's Futur.* **2017**, *5*, 1196–1202. [CrossRef]
5. Ponomarev, E.; Shvetsov, E.; Litvintsev, K.; Bezkorovaynaya, I.; Ponomareva, T.; Klimchenko, A.; Ponomarev, O.; Yakimov, N.; Panov, A. Remote Sensing Data for Calibrated Assessment of Wildfire Emissions in Siberian Forests. *Multidiscip. Digit. Publ. Inst. Proc.* **2018**, *2*, 348. [CrossRef]
6. Westerling, A.L. Wildfire Simulations for California's Fourth Climate Change Assessment: Projecting Changes in Extreme Wildfire Vents with a Warming Climate. California's Fourth Climate Change Assessment, California Energy Commission. Publication Number: CCCA4-CEC-2018-014. 2018. Available online: https://www.energy.ca.gov/sites/default/files/2019-11/Projections_CCCA4-CEC-2018-014_ADA.pdf (accessed on 4 May 2022).
7. Bentz, B.J.; Jönsson, A.M.; Schroeder, M.; Weed, A.; Wilcke, R.A.I.; Larsson, K. Ips typographus and Dendroctonus ponderosae Models Project Thermal Suitability for Intra- and Inter-Continental Establishment in a Changing Climate. *Front. For. Glob. Chang.* **2019**, *2*, 1. [CrossRef]
8. Jia, J.; Cao, Z.; Liu, C.; Zhang, Z.; Lin, L.; Wang, Y.; Haghipour, N.; Wacker, L.; Bao, H.; Dittmar, T.; et al. Climate warming alters subsoil but not topsoil carbon dynamics in alpine grassland. *Glob. Chang. Biol.* **2019**, *25*, 4383–4393. [CrossRef]
9. Abram, N.J.; Henley, B.J.; Gupta, A.S.; Lippmann, T.J.R.; Clarke, H.; Dowdy, A.J.; Sharples, J.J.; Nolan, R.H.; Zhang, T.; Wooster, M.J.; et al. Connections of climate change and variability to large and extreme forest fires in southeast Australia. *Commun. Earth Environ.* **2021**, *2*, 1–17. [CrossRef]

10. Fernandez-Anez, N.; Krasovskiy, A.; Müller, M.; Vacik, H.; Baetens, J.; Hukić, E.; Solomun, M.K.; Atanassova, I.; Glushkova, M.; Bogunović, I.; et al. Current Wildland Fire Patterns and Challenges in Europe: A Synthesis of National Perspectives. *Air Soil Water Res.* **2021**, *14*. [CrossRef]
11. EMERCOM—Ministry of Emergency Situations of Russian Federation. Available online: https://www.chs.gov.ru/ (accessed on 13 March 2022).
12. INPE; Brazilian National Institute for Space Research. Available online: https://www.ov.br/inpe/pt-br (accessed on 18 March 2022).
13. Natural Resources, Canada. Available online: https://www.rcan.gc.ca/home (accessed on 14 March 2022).
14. National Interagency Fire Center, USA. Available online: https://www.ifc.gov/ (accessed on 15 March 2022).
15. European Forest Fire Information System. Available online: https://effis.rc.ec.europa.eu/ (accessed on 13 March 2022).
16. Department of Agriculture, Water and the Environment, Australia. Available online: https://www.griculture.gov.au/ (accessed on 15 March 2022).
17. ISFR (India State of Forest Report, Forest Survey of India—Ministry of Environment & Forests), Dehradun, India. 2019. Available online: https://fsi.nic.in/forest-report-2019 (accessed on 13 March 2022).
18. Eurostat. Available online: https://ec.uropa.eu/eurostat/web/main/data/database (accessed on 13 March 2022).
19. Del Hoyo, L.V.; Isabel, M.P.M.; Vega, J.M. Empleo de Técnicas de Regresión Logística Para la Obtención de Modelos de Riesgo Humano de Incendio Forestal a Escala Regional. Boletín de la Asociación de Geógrafos Españoles. 2008. Available online: https://www.bage.age-geografia.es/ojs/index.php/bage/article/download/2027/1940 (accessed on 13 March 2022).
20. Vilar, L.; Woolford, D.G.; Martell, D.L.; Martín, M.P. A model for predicting human-caused wildfire occurrence in the region of Madrid, Spain. *Int. J. Wildland Fire* **2010**, *19*, 325–327. [CrossRef]
21. Srivastava, S.K.; Saran, S.; De By, R.A.; Dadhwal, V.K. A geo-information system approach for forest fire likelihood based on causative and anti-causative factors. *Int. J. Geogr. Inf. Sci.* **2013**, *28*, 427–454. [CrossRef]
22. Forest Survey of India. Available online: https://fsi.ic.in/index.php (accessed on 20 March 2022).
23. Civil Protection—Greece. Available online: https://www.ivilprotection.gr/en/forest-fires (accessed on 2 February 2022).
24. U.S Forest Service Data Archive. Available online: https://www.s.usda.gov/rds/archive/ (accessed on 10 March 2022).
25. Houghton, J.T.; Meira-Filho, L.G.; Callander, B.A.; Harris, N.; Kattenberg, A.; Maskell, K. *Climate Change 1995: The Science of Climate Change*; Cambridge University Press: New York, NY, USA, 1996; Available online: Digitallibrary.un.org/record/223181/files/ipcc_sar_wg_I_full_report.pdf?ln=en (accessed on 13 March 2022).
26. Keane, R.E.; Burgan, R.; van Wagtendonk, J. Mapping wildland fuels for fire management across multiple scales: Integrating remote sensing, GIS, and biophysical modeling. *Int. J. Wildland Fire* **2001**, *10*, 301–309. [CrossRef]
27. Schlobohm, P.; Brain, J. Gaining an Understanding of the National Fire Danger Rating System. National Wildfire Coordinating Group, PMS, 932. 2002. Available online: https://www.nwcg.gov/sites/default/files/products/pms932.pdf (accessed on 30 November 2021).
28. McArthur, A.G. Fire Behaviour in Eucalypt Forests. 1967. Available online: http://vgls.sdp.sirsidynix.net.au/client/search/asset/1299701/0 (accessed on 10 December 2021).
29. Deeming, J.E.; Burgan, R.E.; Cohen, J.D. *The National Fire-Danger Rating System—1978*; General Technical Report INT-39; USDA Forest Service, Intermountain Forest and Range Experiment Station: Ogden, UT, USA, 1977; 63p. Available online: https://www.fs.usda.gov/rm/pubs_series/int/gtr/int_gtr039.pdf (accessed on 25 November 2021).
30. Van Wagner, C.E. *Structure of the Canadian Forest Fire Weather Index*; Environment Canada, Forestry Service: Ottawa, ON, Canada, 1974; Volume 1333, Available online: https://meteo-wagenborgen.nl/wp/wp-content/uploads/2019/08/van-Wagner-1974.pdf (accessed on 25 November 2021).
31. Burgan, R.E.; Hartford, R.A. *Monitoring Vegetation Greenness with Satellite Data*; US Department of Agriculture, Forest Service, Intermountain Research Station: Ogden, UT, USA, 1993; Volume 297. [CrossRef]
32. Chuvieco, E. *Remote Sensing of Large Wildfires in the European Mediterranean Basin*; Global Forest Resources Assessment 2020: Main Report; FAO: Rome, Italy, 2020; ISBN 3540657673.
33. Cardille, J.A.; Ventura, S.J.; Turner, M.G. Environmental and social factors influencing wildfires in the Upper Midwest, United States. *Ecol. Appl.* **2001**, *11*, 111–127. [CrossRef]
34. Chuvieco, E.; Aguado, I.; Dimitrakopoulos, A.P. Conversion of fuel moisture content values to ignition potential for integrated fire danger assessment. *Can. J. For. Res.* **2004**, *34*, 2284–2293. [CrossRef]
35. Cortez, P.; Morais, A.D.J.R. A Data Mining Approach to Predict Forest Fires Using Meteorological Data. 2007. Available online: http://www3.dsi.uminho.pt/pcortez/fires.pdf (accessed on 20 December 2021).
36. Martínez, J.; Vega-Garcia, C.; Chuvieco, E. Human-caused wildfire risk rating for prevention planning in Spain. *J. Environ. Manag.* **2009**, *90*, 1241–1252. [CrossRef]
37. Ponomarev, E.I.; Ivanov, V.; Korshunov, N. System of Wildfires Monitoring in Russia. In *Wildfire Hazards, Risks and Disasters*; Elsevier: Amsterdam, The Netherlands, 2015; pp. 187–205. [CrossRef]
38. Rodrigues, M.; Jiménez-Ruano, A.; Peña-Angulo, D.; de la Riva, J. A comprehensive spatial-temporal analysis of driving factors of human-caused wildfires in Spain using Geographically Weighted Logistic Regression. *J. Environ. Manag.* **2018**, *225*, 177–192. [CrossRef] [PubMed]

39. Vasilakos, C.; Kalabokidis, K.; Hatzopoulos, J.; Kallos, G.; Matsinos, Y. Integrating new methods and tools in fire danger rating. *Int. J. Wildland Fire* **2007**, *16*, 306–316. [CrossRef]
40. Müller, M.M.; Vilà-Vilardell, L.; Vacik, H. Towards an integrated forest fire danger assessment system for the European Alps. *Ecol. Inform.* **2020**, *60*, 101151. [CrossRef]
41. Mazzeo, G.; De Santis, F.; Falconieri, A.; Filizzola, C.; Lacava, T.; Lanorte, A.; Marchese, F.; Nolè, G.; Pergola, N.; Pietrapertosa, C.; et al. Integrated Satellite System for Fire Detection and Prioritization. *Remote Sens.* **2022**, *14*, 335. [CrossRef]
42. Allen, R.G.; Pereira, L.S.; Raes, D.; Smith, M. *Crop Evapotranspiration: Guidelines for Computing Crop Water Requirements*; FAO: Rome, Italy, 1998; Available online: http://www.fao.org/docrep/x0490e/x0490e00.htm (accessed on 23 March 2022).
43. Bachmann, A.; Allgöwer, B. A consistent wildland fire risk terminology is needed. *Fire Manag. Today* **2001**, *61*, 28–33. Available online: https://www.frames.gov/documents/usfs/fmt/fmt_61-4.pdf#page=28 (accessed on 23 July 2022).
44. Page, M.J.; McKenzie, J.E.; Bossuyt, P.M.; Boutron, I.; Hoffmann, T.C.; Mulrow, C.D.; Shamseer, L.; Tetzlaff, J.M.; Akl, E.A.; Brennan, S.E.; et al. The PRISMA 2020 Statement: An Updated Guideline for Reporting Systematic Reviews. *BMJ Br. Med. J.* **2021**, *372*, n71. [CrossRef]
45. Li, X.; Zhao, G.; Yu, X.; Yu, Q. A comparison of forest fire indices for predicting fire risk in contrasting climates in China. *Nat. Hazards* **2013**, *70*, 1339–1356. [CrossRef]
46. Cruz, M.; Gould, J.S.; Alexander, M.E.; Sullivan, A.; McCaw, W.L.; Matthews, S. Empirical-based models for predicting head-fire rate of spread in Australian fuel types. *Aust. For.* **2015**, *78*, 118–158. [CrossRef]
47. Pagnon Eriksson, C.; Johansson, N. Review of Wildfire Indices: Indices Applicable for a Swedish Context. Division of Fire Safety Engineering, Lund, Sweden: TVBB, No. 3233. 2020. Available online: https://lucris.lub.lu.se/ws/portalfiles/portal/87339976/Review_of_wildfire_indices_Indices_applicable_for_a_Swedish_context.pdf (accessed on 15 April 2022).
48. Baijnath-Rodino, J.A.; Foufoula-Georgiou, E.; Banerjee, T. Reviewing the "Hottest" Fire Indices Worldwide. *ESS Open Archive* **2020**. [CrossRef]
49. Chronopoulos, K.; Matsoukis, A. Meteorological Forest Fire Risk: A brief review. *Acad. Lett.* **2021**. Article 364. [CrossRef]
50. Steinfeld, D.; Peter, A.; Martius, O.; Brönnimann, S. Assessing the performance of various fire weather indices for wildfire occurrence in Northern Switzerland. *EGUsphere* **2022**, *in press*. [CrossRef]
51. Stocks, B.J.; Lawson, B.D.; Alexander, M.E.; Wagner, C.V.; McAlpine, R.S.; Lynham, T.J.; Dube, D.E. The Canadian forest fire danger rating system: An overview. *For. Chron.* **1989**, *65*, 450–457. Available online: https://pubs.cif-ifc.org/doi/pdf/10.5558/tfc65450-6?download=true (accessed on 30 April 2022). [CrossRef]
52. Lawson, B.D.; Armitage, O.B. Weather Guide for the Canadian Forest Fire Danger Rating System. 2008. Available online: https://docs.niwa.co.nz/eco/fwsys/ref/2008CFFDRSWeatherGuide.pdf (accessed on 11 June 2022).
53. De Groot, W.J. *Examples of Canadian Forest Fire Behavior Prediction System fuel types in Saskatchewan*; Canadian Forest Service, Forestry Canada, Northern Forestry Centre: Edmonton, AB, Canada, 1993; Available online: https://cfs.nrcan.gc.ca/publications?id=11038 (accessed on 13 March 2022).
54. Dowdy, A.J. Climatological Variability of Fire Weather in Australia. *J. Appl. Meteorol. Clim.* **2018**, *57*, 221–234. [CrossRef]
55. Van Wagner, C.E. Development and structure of the Canadian Forest Fire Weather Index System. *Can. For. Serv. For. Tech. Rep.* **1987**, *35*, 37. Available online: https://d1ied5g1xfgpx8.cloudfront.net/pdfs/19927.pdf (accessed on 25 November 2021).
56. Van Wagner, C.E.; Pickett, T.L. Equations and FORTRAN program for the Canadian Forest Fire Weather Index System. *Can. For. Serv. Ott. ON. For. Tech. Rep.* **1985**, *33*, 18. Available online: https://d1ied5g1xfgpx8.cloudfront.net/pdfs/19973.pdf (accessed on 25 November 2021).
57. Byram, G.M. Combustion of Forest Fuels. Forest Fire: Control and Use. 1959; pp. 61–89. Available online: https://www.frames.gov/documents/behaveplus/publications/Byram_1959_CombustionOfForestFuels.pdf (accessed on 30 May 2022).
58. Lawson, B.D. Fire Spread in Lodgepole Pine Stands. Master's Thesis, University of Montana, Missoula, MT, USA, 1972. Available online: https://scholarworks.umt.edu/cgi/viewcontent.cgi?article=3376&context=etd (accessed on 20 July 2022).
59. Van Wagner, C.E. Three Experimental Fires in Jack Pine Stands. 1966. Available online: https://d1ied5g1xfgpx8.cloudfront.net/pdfs/24667.pdf (accessed on 25 November 2021).
60. Fogarty, L.G.; Pearce, H.G.; Catchpole, W.R.; Alexander, M.E. Adoption vs Adaptation: Lessons from applying the Canadian Forest Fire Danger Rating System in New Zealand. In Proceedings of the III International Conference on Forest Fire Research, 14th Conference on Fire and Forest Meteorology, Luso, Portugal, 16–20 November 1998; ADAI: Coimbra, Portugal, 1998. Available online: https://d1ied5g1xfgpx8.cloudfront.net/pdfs/18753.pdf (accessed on 13 March 2022).
61. López, A.S.; San-Miguel-Ayanz, J.; Burgan, R.E. Integration of satellite sensor data, fuel type maps and meteorological observations for evaluation of forest fire risk at the pan-European scale. *Int. J. Remote Sens.* **2002**, *23*, 2713–2719. [CrossRef]
62. Groot, W.J.D.; Field, R.D.; Brady, M.A.; Roswintiarti, O.; Mohamad, M. Development of the Indonesian and Malaysian fire danger rating systems. *Mitig. Adapt. Strateg. Glob. Chang.* **2006**, *12*, 165–180. Available online: http://www.columbia.edu/~{}rf2426/index_files/deGrootetal2006MITI.pdf (accessed on 13 March 2022). [CrossRef]
63. Dimitrakopoulos, A.P.; Bemmerzouk, A.M.; Mitsopoulos, I.D. Evaluation of the Canadian fire weather index system in an eastern Mediterranean environment. *Meteorol. Appl.* **2011**, *18*, 83–93. [CrossRef]
64. Giannakopoulos, C.; Lesager, P.; Moriondo, M.; Bindi, M.; Karali, A.; Hatzaki, M.; Kostopoulou, E. Comparison of fire danger indices in the Mediterranean for present day conditions. *Iforest Biogeosci. For.* **2012**, *5*, 197–203. [CrossRef]

65. Bradshaw, L.; Deeming, J.E.; Burgan, R.E.; Cohen, J.D. *1978-National Fire-Danger Rating System—Technical Documentation*; Gen. Tech. Rep. INT 169; Intermountain Forest and Range Experiment Station, Forest Service, U.S. Department of Agriculture: Ogden, UT, USA, 1983; 44p, Available online: https://www.fs.fed.us/rm/pubs_int/int_gtr169.pdf (accessed on 13 March 2022).
66. Burgan, R.E. *1988 Revisions to the 1978 National Fire-Danger Rating System*; Research Paper SE-273; USDA Forest Service, Southeastern Forest Experiment Station: Asheville, NC, USA, 1988; p. 39. Available online: https://www.srs.fs.usda.gov/pubs/rp/rp_se273.pdf (accessed on 18 December 2021).
67. Cohen, J.D.; Deeming, J.E. *The National Fire Danger Rating System: Basic Equations (General Technical Report PSW-GTR-82)*; U.S. Department of Agriculture, Forest Service: Berkeley, CA, USA, 1985. [CrossRef]
68. de Groot, W.J.; Wotton, B.M.; Flannigan, M.D. Wildland Fire Danger Rating and Early Warning Systems. In *Wildfire Hazards, Risks, and Disasters*; Shroder, J.F., Paton, D., Eds.; Elsevier: Dordrecht, The Netherlands, 2015; pp. 207–228. [CrossRef]
69. Rothermel, R.C. *Modeling Moisture Content of Fine Dead Wildland Fuels: Input to the BEHAVE Fire Prediction System (No.359)*; US Department of Agriculture, Forest Service, Intermountain Research Station: Ogden, UT, USA, 1986. [CrossRef]
70. Burgan, R.; Klaver, R.W.; Klaver, J.M. Fuel Models and Fire Potential from Satellite and Surface Observations. *Int. J. Wildland Fire* **1998**, *8*, 159–170. [CrossRef]
71. Helfman, R.S.; Straub, R.J.; Deeming, J.E. *User's Guide to AFFIRMS: Time-Share Computerized Processing for Fire Danger Rating*; Gen. Tech. Rep. INT-82; U.S. Department of Agriculture, Forest Service, Intermountain Forest and Range Experiment Station: Ogden, UT, USA, 1980.
72. Bradshaw, L.; McCormick, E. *FireFamily Plus User's Guide, Version 2*; Gen. Tech. Rep. RMRS-GTR-67; US Department of Agriculture, Forest Service, Rocky Mountain Research Station: Ogden, UT, USA, 2000; p. 67. Available online: https://www.fs.fed.us/rm/pubs/rmrs_gtr067.pdf (accessed on 13 March 2022).
73. Andrews, P.L. *BEHAVE: Fire Behaviour Prediction and Fuel Modeling System: BURN Subsystem, Part 2*; US Department of Agriculture, Forest Service, Intermountain Research Station: Ogden, UT, USA, 1989; Volume 194, Available online: https://www.fs.fed.us/rm/pubs_int/int_gtr260.pdf (accessed on 17 May 2022).
74. Fosberg, M.A.; Rothermel, R.C.; Andrews, P.L. Moisture content calculations for 1000-hour timelag fuels. *For. Sci.* **1981**, *27*, 19–26. Available online: https://www.fs.fed.us/psw/publications/fosberg/psw_1981_fosberg001.pdf (accessed on 13 March 2022).
75. Fosberg, M.A. *Weather in Wildland Fire Management: The Fire Weather Index*; US For Serv Reprints of articles by FS employees: Ogden, UT, USA, 1978; Available online: https://agris.fao.org/agris-search/search.do?recordID=US201302440660 (accessed on 13 March 2022).
76. Goodrick, S.L. Modification of the Fosberg fire weather index to include drought. *Int. J. Wildland Fire* **2002**, *11*, 205–211. Available online: https://www.srs.fs.usda.gov/pubs/ja/ja_goodrick001.pdf (accessed on 4 May 2022). [CrossRef]
77. Andrews, P.L. *BEHAVE: Fire Behaviour Prediction and Fuel Modeling System: BURN Subsystem, Part 1*; US Department of Agriculture, Forest Service, Intermountain Research Station: Ogden, UT, USA, 1986; Volume 194. [CrossRef]
78. Rothermel, R.C. *A Mathematical Model for Fire Spread Predictions in Wildland Fuels*; Res. Pap. INT-1 15; Intermountain Forest and Range Experiment Station, Forest Service, U.S. Department of Agriculture: Ogden, UT, USA, 1972; p. 40. Available online: https://www.fs.fed.us/rm/pubs_int/int_rp115.pdf (accessed on 15 December 2021).
79. Chandler, C.; Cheney, P.; Thomas, P.; Trabaud, L.; Williams, D. *Fire in Forestry. Volume 1. Forest Fire Behavior and Effects. Volume 2. Forest Fire Management and Organization*; John Wiley & Sons, Inc.: Hoboken, NJ, USA, 1983.
80. Srock, A.F.; Charney, J.J.; Potter, B.E.; Goodrick, S.L. The Hot-Dry-Windy Index: A New Fire Weather Index. *Atmosphere* **2018**, *9*, 279. [CrossRef]
81. Haines, D.A. A lower atmospheric severity index for wildland fire. *Natl. Weather. Dig.* **1988**, *13*, 23–27. Available online: https://github.com/Unidata/MetPy/files/1519394/Haines.1988.pdf (accessed on 2 September 2022).
82. Noble, I.R.; Bary, G.A.V.; Gill, A.M. McArthur's fire-danger meters expressed as equations. *Aust. J. Ecol.* **1980**, *5*, 201–203. [CrossRef]
83. Griffiths, D. Improved Formula for the Drought Factor in McArthur's Forest Fire Danger Meter. *Aust. For.* **1999**, *62*, 202–206 [CrossRef]
84. Purton, C.M. *Equations for the McArthur Mark 4 Grassland Fire Danger Meter*; Commonwealth of Australia Bureau of Meteorology Meteorological Note No. 147; Bureau of Meteorology: Melbourne, VC, Australia, 1982; p. 12. Available online: https:// naturaldisaster.royalcommission.gov.au/system/files/exhibit/BOM.9004.0001.0405.pdf (accessed on 30 June 2022).
85. Sneeuwjagt, R.J.; Peet, G.B. *Forest Fire Behaviour Tables for Western Australia*, 3rd ed.; Western Australia Department of Conservation and Land Management: Perth, WA, USA; p. 591985.
86. Beck, J.A. Equations for the Forest Fire Behaviour Tables for Western Australia. *CALM Sci.* **1995**, *1*, 325–348. Available online: https:/ www.victoriasforestryheritage.org.au/pinefire/Beck_1995_EquationsForTheForestFireBehaviourTablesForWesternAustrali pdf (accessed on 5 May 2022).
87. Sharples, J.; McRae, R.; Weber, R.; Gill, A. A simple index for assessing fire danger rating. *Environ. Model. Softw.* **2009**, *24*, 764–77 [CrossRef]
88. Sharples, J.; McRae, R.; Weber, R.; Gill, A. A simple index for assessing fuel moisture content. *Environ. Model. Softw.* **2009**, 2 637–646. [CrossRef]
89. Laing, M.V. *Meteorological Note No.60*; Department of Meteorological Services: Salisbury, Australia, 1978.

90. Meikle, S.; Heine, J. A Fire Danger Index System for the Transvaal Lowveld and Adjoining Escarpment Areas. *South Afr. For. J.* **1987**, *143*, 55–56. [CrossRef]
91. Soares, R.V. Determinação de Um índice de Perigo de Incêndio para a Região Centro Paranaense, Brasil. Master's Thesis, CATIE/IICA, Turrialba, Costa Rica, 1972; p. 72.
92. Nunes, J.R.S.; Fier, I.S.N.; Soares, R.V.; Batista, A.C. Desempenho da fórmula de monte alegre (fma) e da fórmula de monte alegre alterada (fma+) no distrito florestal de monte alegre. *Floresta* **2010**, *40*. [CrossRef]
93. Rodríguez, N.; Moretti, A. *Índice de peligro de propagación de Incendios Forestales. VI Congreso Forestal Argentino*; Tomo III.: Santiago del Estero, Argentine, 1988; 5p.
94. Setzer, A.W.; Sismanoglu, R.A. *Risco de Fogo: Metodologia do Cálculo—Descrição Sucinta da Versão 9*; Instituto Nacional de Pesquisas Espaciais: São José dos Campos, Brazil, 2012; Available online: https://queimadas.dgi.inpe.br/~{}rqueimadas/documentos/RiscoFogo_Sucinto_20130911.pdf (accessed on 13 June 2022).
95. Sampaio, O.B. *Estudo Comparativo de Índices, para Previsão de Incêndios Florestais, na Região de Coronel Fabriciano, Minas Gerais [Dissertação]*; Universidade Federal de Viçosa: Viçosa, MG, USA, 1991.
96. Orieux, A. Conditions météorologiques et incendies en région méditerranéenne. *Rev. For. Française* **1974**, *26 (S)*, 122–129. Available online: https://hal.archives-ouvertes.fr/hal-03395671 (accessed on 27 June 2022).
97. Carrega, P. L'évapotranspiration potentielle et réelle dans le Midi méditerranéen. Son originalité par rapport au reste de le France. *Méditerranée* **1988**, *66*, 3–8. Available online: https://www.persee.fr/doc/medit_0025-8296_1988_num_66_4_2575 (accessed on 20 July 2022). [CrossRef]
98. Drouet, J.C.; Sol, B. *Etude de Nouveaux Indices de Risques Météorologiques D'incendies de forêt en Zone Méditerranéenne. Rapport en 3 tomes*; I.U.T., Aix en Provence, C.I.R.C.O.S.C., Valabre, Service Météorologique Interregional Sud-Est: Marseille-Marignane, France, 1988.
99. Lourenço, L. *Risco Meteorológico de Incêndio Florestal. Colectâneas Cindínicas II*; Núcleo de Investigação Científica de Incêndios Florestais, Faculdade de Letras da Universidade de Coimbra: Coimbra, Portugal, 2004; ISBN 978-972-8330-16-3. Available online: https://www.researchgate.net/publication/318792210_Risco_Risco_Meteorologico_de_Incendio_Florestal_Colectaneas_Cindinicas_II/comments (accessed on 16 June 2022).
100. INMG (Instituto Nacional de Meteorologia e Geofisica). *Nota Explicativa Sobre o Índice de Risco Meteorológico de Incêndios Rurais*; Divisão de Meteorologia Agrícola: Lisbon, Portugal, 1988.
101. Gonçalves, Z.J.; Lourenço, L. Meteorological index of forest fire risk in the portuguese mainland territory. In Proceedings of the International Conference on Forest Fire Research, B.07-1/14. Coimbra, Portugal, 21–24 November 1990; Available online: https://www.researchgate.net/profile/Luciano-Lourenco/publication/313476213_Meteorological_index_of_forest_fire_risk_in_the_Portuguese_Mainland_Territory/links/589baf7faca2721ae1b7ab18/Meteorological-index-of-forest-fire-risk-in-the-Portuguese-Mainland-Territory.pdf (accessed on 3 April 2022).
102. ICONA. Manual de Operaciones Contra Incendios Forestales. Madrid. 5.1/65. 1993. Available online: https://documentacion.fundacionmapfre.org/documentacion/publico/es/media/group/1088234.do (accessed on 17 May 2022).
103. Palmieri, S.; Inghilesi, R.; Siani, A.M. Un indice meteorologico di rischio per incendi boschivi. *Semin. Fight. For. Fires* **1993**, *15*, 49–62.
104. Bovio, G.; Quaglino, A.; Nosenzo, A. Individuazione di un indice di previsione per il pericolo di incendi boschivi. *Monti Boschi* **1984**, *35*, 39–44. Available online: https://iris.unito.it/handle/2318/106685 (accessed on 13 March 2022).
105. Sirca, C.; Spano, D.; Duce, P.; Delogu, G.; Cicalò, G.O. Performance of a newly developed integrated fire rating index in Sardinia, Italy. In Proceedings of the "Wildfire 2007—4th International Wildland Fire Conference", Seville, Spain, 13–17 May 2007; Viegas, X.D., Ed.; Ministry of Environment: Madrid, Spain; Junta de Andalucia: Seville, Spain, 2007.
106. Spano, D.; Georgiadis, T.; Duce, P.; Rossi, F. A Fire Index for Mediterranean Vegetation Based on Micrometeorological and Ecophysiological Measurements. 2001. Available online: https://ams.confex.com/ams/pdfpapers/65497.pdf (accessed on 29 May 2022).
107. Arif, A. *Les Applications Météorologiques dans la lute Contre les Incendies de Forêts—Étude de cas pour la Tunisie*; ESCWA: Beirut, Lebanon, 1994; p. 23.
108. Angstrom, A. Riskerna for skogsbrand och deras beroende av vader och klimat (The risks for forest fires and their relation to weather and climate). *Sven. Skogsvirdsforeningens Tidskr.* **1942**, *40*, 323–343.
109. Baumgartner, A.; Klemmer, L.; Raschke, E.; Waldmann, G. Waldbrände in Bayern 1950 bis 1959. *Mitt. Aus Der Staatsforstverwalt. Bayerns* **1967**, *36*, 57–79.
110. Bruschek, G. Forest areas and forest fires in Brandenburg during the dry and hot summer of 1992. In *The Extreme Summer of 1992 in Northern Germany*; PIK Report No. 2/1; Schellnhuber, H., Enke, W., Flechsig, M., Eds.; Potsdam Institute for Climate Impact Research: Potsdam, Germany, 1994; pp. 245–264.
111. Telicyn, G.P. Logarithmic index of fire weather danger for forests. *Lesn. Khozyaistvo* **1970**, *11*, 1–581970.
112. Nesterov, V.G. *Combustibility of the Forest and Methods for Its Determination*; Goslesbumizdat: Moscow, Russia, 1949; 76p. (In Russian)
113. Käse, H. *Ein Vorschlag für eine Methode zur Bestimmung und Vorhersage der Waldbrandgefährdung mit Hilfe Komplexer Kennziffern*; Akademie Verlag: Berlin, Germany, 1969.

114. Zhdanko, V.A. Scientific basis of development of regional scales and their importance for forest fire management. In *Contemporary Problems of Forest Protection from Fire and Firefighting*; Melekhov, I.S., Ed.; Lesnaya Promyshlennost' Publ.: Moscow, Russia, 1965; pp. 53–86.
115. Venäläinen, A.R.I.; Heikinheimo, M. The spatial variation of long-term mean global radiation in Finland. *Int. J. Climatol. A J. R. Meteorol. Soc.* **1997**, *17*, 415–426. [CrossRef]
116. Munger, T.T. Graphic method of representing and comparing drought intensities. *Mon. Weather. Rev.* **1916**, *44*, 642–643. Available online: https://journals.ametsoc.org/downloadpdf/journals/mwre/44/11/1520-0493_1916_44_642_gmorac_2_0_co_2.xml (accessed on 27 June 2022). [CrossRef]
117. Keetch, J.J.; Byram, G.M. *A Drought Index for Forest Fire Control*; Research Paper SE-38; USDA Forest Service, Southeast Forest Experiment Station: Ashville, NC, USA, 1968; p. 32. Available online: https://www.srs.fs.usda.gov/pubs/rp/rp_se038.pdf (accessed on 5 January 2022).
118. Mount, A.B. *The Derivation and Testing of a Soil Dryness Index Using Run-Off Data*; Bull. 4; Forestry Commission: Tasmania, Australia, 1972; 30p, ISBN 072460006X.
119. Palmer, W.C. *Meteorological Drought*; US Department of Commerce, Weather Bureau: Washington, DC, USA, 1965; Volume 30, Available online: https://www.droughtmanagement.info/literature/USWB_Meteorological_Drought_1965.pdf (accessed on 20 July 2022).
120. Tsakiris, G.; Vangelis, H. Establishing a drought index incorporating evapotranspiration. *Eur. Water* **2005**, *9*, 3–11. Available online: https://www.ewra.net/ew/pdf/EW_2005_9-10_01.pdf (accessed on 24 June 2022).
121. Stephenson, N.L. Climatic control of vegetation distribution—The role of the water balance. *Am. Nat.* **1990**, *135*, 649–670. Available online: https://www.jstor.org/stable/2462028 (accessed on 23 June 2022). [CrossRef]
122. Landsberg, J. *Physiological Ecology of Forest Production*; Academic Press: London, UK, 1986; 198p.
123. Darcy, H. *Les Fontaines Publiques de la Ville de Dijon*; Dalmont: Paris, France, 1856; Available online: https://gallica.bnf.fr/ark:/12148/bpt6k624312/f1n657.pdf (accessed on 20 July 2022).
124. McDowell, N.G.; Allen, C.D. Darcy's law predicts widespread forest mortality under climate warming. *Nat. Clim. Chang.* **2015**, *5*, 669–672. [CrossRef]
125. Tucker, C.J. Red and photographic infrared linear combinations for monitoring vegetation. *Remote Sens. Environ.* **1979**, *8*, 127–150. [CrossRef]
126. Gao, B.-C. NDWI—A normalized difference water index for remote sensing of vegetation liquid water from space. *Remote Sens. Environ.* **1996**, *58*, 257–266. [CrossRef]
127. McFeeters, S.K. The use of the Normalized Difference Water Index (NDWI) in the delineation of open water features. *Int. J. Remote Sens.* **1996**, *17*, 1425–1432. [CrossRef]
128. Hardisky, M.A.; Klemas, V.; Smart, R.M. The influence of soil salinity, growth form, and leaf moisture on the spectral radiance of Spartina alterniflora canopies. *Photogramm. Eng. Remote Sens.* **1983**, *49*, 77–83. Available online: https://www.asprs.org/wp-content/uploads/pers/1983journal/jan/1983_jan_77-83.pdf (accessed on 23 July 2022).
129. Wang, L.; Qu, J.J. NMDI: A normalized multi-band drought index for monitoring soil and vegetation moisture with satellite remote sensing. *Geophys. Res. Lett.* **2007**, *34*, L20405-1-5. [CrossRef]
130. Huete, A.R. A soil-adjusted vegetation index (SAVI). *Remote Sens. Environ.* **1988**, *25*, 295–309. [CrossRef]
131. Huete, A.; Didan, K.; Miura, T.; Rodriguez, E.P.; Gao, X.; Ferreira, L.G. Overview of the radiometric and biophysical performance of the MODIS vegetation indices. *Remote Sens. Environ.* **2002**, *83*, 195–213. [CrossRef]
132. Burgan, R.E.; Andrews, P.L.; Bradshaw, L.S.; Chase, C.H.; Hartford, R.A.; Latham, D.J. Current status of the wildland fire assessment system (WFAS). *Fire Manag. Notes* **1997**, *57*, 14–17. Available online: https://data.fs.usda.gov/wwwbeta/sites/default/files/legacy_files/fire-management-today/057-2_0.pdf#page=14 (accessed on 20 March 2022).
133. Sudiana, D.; Kuze, H.; Takeuchi, N.; Burgan, R.E. Indonesian Forest Fire—A Quantitave Assessment. In Proceedings of the 12th Indonesian Scientific Meeting, Osaka, Japan, 6–September 2003; 2003. Available online: https://staff.ui.ac.id/system/files/users/dodi.sudiana/publication/ti-12-indonesianforestfireaquantitaveassessment.pdf (accessed on 28 June 2022).
134. Huesca, M.; Litago, J.; Palacios-Orueta, A.; Montes, F.; Sebastián-López, A.; Escribano, P. Assessment of forest fire seasonality using MODIS fire potential: A time series approach. *Agric. For. Meteorol.* **2009**, *149*, 1946–1955. Available online: https://www.pc.noaa gov/exper/firecomp/INFO/fosbinfo.html (accessed on 28 June 2022). [CrossRef]
135. Fosberg, M.A. Theory of precipitation effects on dead cylindrical fuels. *For. Sci.* **1972**, *18*, 98–108. [CrossRef]
136. Andrews, P.L. *BehavePlus Fire Modeling System, Version 5: Variables. Gen. Tech. Rep. RMRS-GTR-213 Revised*; Department of Agriculture, Forest Service, Rocky Mountain Research Station: Fort Collins, CO, USA, 2009; p. 111. Available online https://www.fs.fed.us/rm/pubs/rmrs_gtr213.pdf (accessed on 26 July 2022).
137. Andrews, P.L.; Rothermel, R.C. *Charts for Interpreting Wildland Fire Behaviour Characteristics*; US Department of Agriculture Forest Service, Intermountain Forest and Range Experiment Station: Ogden, UT, USA, 1982; Volume 131, Available online https://www.fs.fed.us/rm/pubs_int/int_gtr131.pdf (accessed on 17 May 2022).
138. McDonald, J.M.; Srock, A.F.; Charney, J.J. Development and Application of a Hot-Dry-Windy Index (HDW) Climatolog *Atmosphere* **2018**, *9*, 285. [CrossRef]

139. Kulseth, M.G. *An Evaluation of the Hot-Dry-Windy Fire-Weather Index Using Historical Fire Events and Meteorological Analysis Datasets*; Michigan State University: East Lansing, MI, USA, 2019; Available online: https://d.lib.msu.edu/etd/47779/datastream/OBJ/view/ (accessed on 28 August 2022).
140. Watts, A.; Potter, B.; Charney, J.; Srock, A. *The Hot-Dry-Windy Index: A New Tool for Forecasting Fire Weather. Science Findings 227*; US Department of Agriculture, Forest Service, Pacific Northwest Research Station: Portland, OR, USA, 2020; Available online: https://www.fs.fed.us/pnw/sciencef/scifi227.pdf (accessed on 18 July 2022).
141. Choi, G.; Kim, J.; Won, M.S. Spatial patterns and temporal variability of the Haines Index related to the wildland fire growth potential over the Korean Peninsula. *J. Korean Geogr. Soc.* **2006**, *41*, 168–187. Available online: https://citeseerx.ist.psu.edu/viewdoc/download?doi=10.1.1.486.5239&rep=rep1&type=pdf (accessed on 18 May 2022).
142. Potter, B.E.; Winkler, J.A.; Wilhelm, D.F.; Shadbolt, R.P.; Bian, X. Computing the Low-Elevation Variant of the Haines Index for Fire Weather Forecasts. *Weather Forecast* **2008**, *23*, 159–167. [CrossRef]
143. Barberà, M.J.; Niclòs, R.; Estrela, M.J.; Valiente, J.A. Climatology of the stability and humidity terms in the Haines Index to improve the estimate of forest fire risk in the Western Mediterranean Basin (Valencia region, Spain). *Int. J. Clim.* **2014**, *35*, 1212–1223. [CrossRef]
144. Reges, H.W.; Alden, S. Use of the Haines Index to evaluate large fire potential for interior Alaska. In Proceedings of the Second Conference on Fire and Forest Meteorology, American Meteorological Society, Phoenix, AZ, USA, 4–8 January 1998.
145. Croft, P.J.; Potter, B.E.; Reed, A. The analysis of the Haines Index climatology for the eastern Unites States, Alaska, Hawaii, and Puerto Rico. In *Atmospheric stability environments and fire weather in Australia—Extending the Haines Index, Proceedings of the 4th Symposium on Fire and Forest Meteorology, American Meteorological Society, Reno, NV, USA, 13–15 November 2001*; Technical Report No. 20; Mills, G., McCaw, L., Eds.; The Centre for Australian Weather and Climate Research: Melbourne, Australia, 2002; pp. 242–246. ISBN 978-1-921605-56-7. Available online: https://nla.gov.au/nla.obj-2968722768/view (accessed on 4 March 2022).
146. Potter, B.E.; Goodrick, S. Performance of the Haines Index during August 2000 for Montana. In Proceedings of the 4th Symposium on Fire and Forest Meteorology, Reno, NV, USA, 13–15 November 2001; American Meteorological Society: Boston, MA, USA, 2003; pp. 233–236. Available online: https://www.nrs.fs.fed.us/pubs/jrnl/2003/nc_2003_potter_005.pdf (accessed on 30 May 2022).
147. Mills, G.A.; McCaw, W.L. *Atmospheric Stability Environments and Fire Weather in Australia: Extending the Haines Index*; Centre for Australian Weather and Climate Research: Aspendale, VIC, Australia, 2010; ISBN 978-1-921605-56-7. Available online: https://www.cawcr.gov.au/technical-reports/CTR_020.pdf (accessed on 5 April 2022).
148. Tatli, H.; Türkeş, M. Climatological evaluation of Haines forest fire weather index over the Mediterranean Basin. *Meteorol. Appl.* **2013**, *21*, 545–552. [CrossRef]
149. Stuart, M. A comparison of fire danger rating systems for use in forests. *Aust. Meteorol. Oceanogr. J.* **2009**, *58*, 41–48. [CrossRef]
150. Lucas, C. Fire climates of Australia: Past, present and future. In Proceedings of the 6th Symposium on Fire and Forest Meteorology, Canmore, AB, Canada, 25–27 October 2005; pp. 25–27. Available online: https://www.bushfirecrc.com/sites/default/files/managed/resource/97592.pdf (accessed on 8 July 2022).
151. Gouvas, M. The South African Lowveld Fire Danger Index (LFDI) and its application advantages in Greece (in Greek). In Proceedings of the 19th Hellenic Forestry Conference, Litochoro, Greece, 29 September–2 October 2019; pp. 268–278. Available online: https://www.academia.edu/40551183/The_South_African_Lowveld_Fire_Danger_Index_LFDI_and_its_application_advantages_in_Greece_in_Greek_ (accessed on 5 July 2022).
152. White, L.A.S.; White, B.L.A.; Ribeiro, G.T. Evaluation of Forest Fire Danger Indexes for Eucalypt Plantations in Bahia, Brazil. *Int. J. For. Res.* **2015**, *2015*, 613736. [CrossRef]
153. Cavalcante, R.B.; Souza, B.M.; Ramos, S.J.; Gastauer, M.; Junior, W.R.N.; Caldeira, C.F.; Souza-Filho, P.W. Assessment of fire hazard weather indices in the eastern Amazon: A case study for different land uses. *Acta Amaz.* **2021**, *51*, 352–362. [CrossRef]
154. Torres, F.T.P.; Lima, G.S.; Alvares, B.F. Fire behavior variables and hazard indices of forest fires. *Rev. Árvore* **2018**, *42*. [CrossRef]
155. Torres, F.T.P.; Martins, S.V.; Lima, G.S.; Valverde, S.R. Analysis of efficiency of fire danger indices in forest fire prediction. *Rev. Árvore* **2017**, *41*. [CrossRef]
156. Dentoni, M.C. *Sistema de Evaluación de Peligro de Incendios Forestales, una Experiencia Argentina [en línea]*; Boletín INIA—Instituto de Investigaciones Agropecuarias: Chillán, Chile, 2006; no. 145; Available online: https://hdl.handle.net/20.500.14001/7139 (accessed on 19 July 2022).
157. Martínez Techera, C. *Análisis de Índices de Riesgo de Incendio Forestal y su Aplicabilidad en el Uruguay, Tesis de Grado*; Universidad de la República (Uruguay), Facultad de Ciencias: Montevideo, Uruguay, 2021; Available online: https://hdl.handle.net/20.500.12008/31085 (accessed on 19 July 2022).
158. Setzer, A.; Pereira, M.C.; Pereira, J.R. O uso de satélites NOAA na detecção de queimadas no Brasil. *Climanálise* **1992**, *7*, 40–531992.
159. Setzer, A.W.; Sismanoglu, R.A.; dos Santos, J.G.M. Método do Cálculo do Risco de Fogo do Programa do INPE-Versão 11, junho/2019. CEP, 12, 010. Sao Jose dos Campos, Brasil. 2019. Available online: http://mtc-m21c.sid.inpe.br/col/sid.inpe.br/mtc-21c/2019/11.21.11.03/doc/publicacao.pdf (accessed on 13 June 2022).
160. Torres, F.T.P.; Lima, G.S. Forest Fire Hazard in the Serra do Brigadeiro State Park (MG). *Floresta Ambient.* **2019**, *26*. [CrossRef]
161. Sampaio, O.B. Análise da Eficiência de Quatro Índices na Previsão de Incêndios Florestais Para a Região de Agudos—SP. [tese] Curitiba: Universidade Federal do Paraná. 1999. Available online: https://acervodigital.ufpr.br/bitstream/handle/1884/25618/T%20-%20OTAVIO%20BEZERRA%20SAMPAIO.pdf?sequence=1&isAllowed=y (accessed on 10 June 2022).

162. Pausas, J.G.; Vallejo, R.V. The role of fire in the European Mediterranean ecosystems. In *Remote Sensing of Large Wildfires in the European Mediterranean Basin*; Chuvieco, E., Ed.; Springer-Verlag: Berlin/Heidelberg, Germany, 1999; pp. 3–16. ISBN 3540657673.
163. Westerling, A.L.; Bryant, B.P. Climate change and wildfire in California. *Clim. Chang.* **2007**, *87*, 231–249. [CrossRef]
164. Barriopedro, D.; Fischer, E.M.; Luterbacher, J.; Trigo, R.M.; García-Herrera, R. The Hot Summer of 2010: Redrawing the Temperature Record Map of Europe. *Science* **2011**, *332*, 220–224. [CrossRef]
165. FAO. *Global Forest Resources Assessment 2020: Main Report*; Food and Agriculture Organization: Rome, Italy, 2020.
166. San-Miguel-Ayanz, J.; Schulte, E.; Schmuck, G.; Camia, A.; Strobl, P.; Liberta, G.; Amatulli, G. Comprehensive monitoring of wildfires in Europe: The European forest fire information system (EFFIS). In *Approaches to Managing Disaster-Assessing Hazards, Emergencies and Disaster Impacts*; Tiefenbacher, J.P., Ed.; IntechOpen: London, UK, 2012; Chapter 5; Available online: https://ec.europa.eu/environment/forests/pdf/InTech.pdf (accessed on 12 June 2022).
167. Thornthwaite, C.W.; Mather, J.R. *Instructions and Tables for Computing Potential Evapotranspiration and the Water Balance*; Laboratory of Climatology: Centerton, NJ, USA, 1957; Available online: https://www.wrc.udel.edu/wp-content/publications/ThornthwaiteandMather1957Instructions_Tables_ComputingPotentialEvapotranspiration_Water%20Balance.pdf (accessed on 23 March 2022).
168. Thornthwaite, C.W. An Approach toward a Rational Classification of Climate. *Geogr. Rev.* **1948**, *38*, 55–94. [CrossRef]
169. Ponce, V.M. *Engineering Hydrology: Principles and Practices*; Prentice Hall: Englewood Cliffs, NJ, USA, 1989; Volume 640, Available online: http://ponce.sdsu.edu/330textbook_hydrology_chapters.html (accessed on 29 June 2022).
170. Carrega, P. A Meteorological Index of Forest Fire Hazard in Mediterranean France. *Int. J. Wildland Fire* **1991**, *1*, 79–86. [CrossRef]
171. Sol, B. Estimation du risque météorologique d'incendies de forêts dans le sud-est de la France. *Rev. For. Française Spécial* **1990**, 263–271. Available online: https://hal.archives-ouvertes.fr/hal-03425206/document (accessed on 24 April 2022). [CrossRef]
172. Drouet, J.C.; Sol, B. Mise au point d'un indice numérique de risque météorologique d'incendies de forêts. Dossier: Forêts et incendies 1991–1992. *Mediterranéenne* **1993**, *14*, 155–162. Available online: https://hal.archives-ouvertes.fr/hal-03557194/document (accessed on 20 January 2022).
173. Allen, F.R. *The Management of Risk to Society from Potential Accidents*; Elsevier Applied Science: London, UK, 1998; ISBN 97813158980701992.
174. Camia, A.; Bovio, G. *Description of the Indices Implemented in EUDIC Software for the European Meteorological Forest fire Risk Mapping*; European Commission Join Research Center: Tourin, Italy, 2000; Technical Report.
175. Almeida, R. Noções Básica de Meteorologia e Incêndios Rurais, ISEC. *Pós-Grad. Em Riscos E Proteção Civ.* 2020. Available online: https://fogos.icnf.pt/download/ (accessed on 13 March 2022).
176. WSL Swiss Federal Institute for Forest, Snow and Landscape Research. Available online: https://wikifire.sl.ch/tiki-index515f.html?page=Introduction&structure=Fire (accessed on 30 June 2022).
177. Mestre, A.; Allue, M.; Peral, C.; Santamaria, R.; Lazcano, M. Operational Fire Danger Rating System in Spain. In Proceedings of the International Workshop on Operational Weather Systems for Fire Danger Rating, Edmonton, AT, Canada, 14–16 July 2008.
178. Ventura, F.; Marletto, V.; Zinoni, F. Un metodo per il calcolo dell'indice meteorologico del rischio di incendio forestale. *Sherwood* **2001**, *68*, 13–18.
179. Marletto, V.; Ventura, F. Indici Meteorologici per il Calcolo del Rischio D'incendio Torestale. 2002. Available online: http://www.agrometeorologia.it/documenti/Aiam2002/17_Marletto.pdf (accessed on 25 May 2022).
180. Ricotta, C.; Guglietta, D. Indici di Rischio di INCENDIO Boschivo.IN Il Sistema di Videosorveglianzae Telerilevamento Degli Incendi Boschivi. 2010. Available online: https://www.vigilfuoco.it/aspx/isaViewDoc.aspx?id=21&t=2 (accessed on 24 May 2022).
181. Xiang, K.; Li, Y.; Horton, R.; Feng, H. Similarity and difference of potential evapotranspiration and reference crop evapotranspiration—A review. *Agric. Water Manag.* **2020**, *232*, 106043. [CrossRef]
182. McMahon, T.A.; Peel, M.C.; Lowe, L.; Srikanthan, R.; McVicar, T.R. Estimating actual, potential, reference crop and pan evaporation using standard meteorological data: A pragmatic synthesis. *Hydrol. Earth Syst. Sci.* **2013**, *17*, 1331–1363. Available online: https://hess.copernicus.org/articles/17/1331/2013/hess-17-1331-2013.pdf (accessed on 20 July 2022). [CrossRef]
183. Sirca, C.; Salis, M.; Arca, B.; Duce, P.; Spano, D. Assessing the performance of fire danger indexes in a Mediterranean area. *Iforest Biogeosci. For.* **2018**, *11*, 563–571. [CrossRef]
184. Clément, J. *International Handbook on Forest Fire Protection*; FAO Forestry Department Press: Aix en Provence, France, 2001; 163p. Available online: https://www.fao.org/forestry/27221-06293a5348df37bc8b14e24472df64810.pdf (accessed on 13 June 2022).
185. Sebei, S. Le degré météorologique du risque incendie de forêt (DMRIF): Méthode d'évaluation du risque d'incendie en Tunisie. In Proceedings of the XXVIII Colloque de L'association Internationale de Climatologie, Liège, Belgium, 4 July 2015; Available online: http://climato.be/aic/colloques/actes/ACTES_AIC2015/2%20Climatologie%20appliquee/034-SEBEI-212-217.pdf (accessed on 3 July 2022).
186. Groisman, P.Y.; Sherstyukov, B.G.; Razuvaev, V.N.; Knight, R.W.; Enloe, J.G.; Stroumentova, N.S.; Whitfield, P.H.; Førland, E.; Hannsen-Bauer, I.; Tuomenvirta, H.; et al. Potential forest fire danger over Northern Eurasia: Changes during the 20th century. *Glob. Planet. Chang.* **2007**, *56*, 371–386. [CrossRef]
187. Ou, T. *Droughts and Wildfires in Sweden—Past Variation and Future Projection*; Myndigheten för Samhällsskydd och Beredskap (MSB1112): Karlstad, Sweden, 2017. Available online: https://www.msb.se/siteassets/dokument/publikationer/english-publications/droughts-and-wildfires-in-sweden-past-variation-and-future-projection.pdf (accessed on 12 May 2022).

188. De Rigo, D.; Libertà, G.; Durrant, T.H.; Vivancos, T.A.; San-Miguel-Ayanz, J. Forest Fire Danger Extremes in Europe under Climate Change: Variability and Uncertainty. Ph.D. Thesis, Publications Office of the European Union, Luxembourg, 2017. Available online: https://hal.archives-ouvertes.fr/hal-02906196/document (accessed on 29 September 2022).
189. Knorr, W.; Jiang, L.; Arneth, A. Climate, CO_2 and human population impacts on global wildfire emissions. *Biogeosciences* **2016**, *13*, 267–282. [CrossRef]
190. Langholz, H.; Schmidtmayer, E. Meteorologische Verfaren zur Abschatzung des Waldbrandrisikos. *Allg. Forst* **1993**, *48*, 394–396.
191. Penman, H.L. Physics in Agriculture. *J. Sci. Instrum.* **1948**, *25*, 425–432. [CrossRef]
192. Vadrevu, K.P.; Ohara, T.; Justice, C. *Biomass Burning in South and Southeast Asia: Impacts on the Biosphere*; CRC Press: Boca Raton, FL, USA, 2021; Volume 2, ISBN 9780429022036. [CrossRef]
193. Torres, F.; Ribeiro, G. Índices de risco de incêndios florestais em Juiz de Fora/MG. *Floram* **2008**, *15*, 24–34. Available online: https://www.floram.org/article/588e221ae710ab87018b465c/pdf/floram-15-2-24.pdf (accessed on 30 May 2022).
194. Shetinsky, E.A. Protection of forests and forest pyrology. *Ecology* **1994**, 209. (In Russian)
195. Skvarenina, J.; Mindas, J.; Holecy, J.; Tucek, J. Analysis of the Natural and Meteorological Conditions during Two Largest Forest Fire Events in the Slovak Paradise National Park. Forest Fire in the Wildland-Urban Interface and Rural Areas in Europe: An integral Planning and Management Challenge. Athens. 2003. Available online: http://www.fria.gr/WARM/chapters/warmCh04Skvarenina.pdf (accessed on 20 June 2022).
196. Dominic, A. Evaluation of Meteorological Forest Fire Risk Indices and Projection of Fire Risk for German Federal States. 2011. Available online: https://www.researchgate.net/publication/260487902_Evaluation_of_meteorological_forest_fire_risk_indices_and_projection_of_fire_risk_for_German_federal_states (accessed on 23 June 2022).
197. Monteith, J.L. Evaporation and surface temperature. *Q. J. R. Meteorol. Soc.* **1981**, *107*, 1–27. [CrossRef]
198. Venäläinen, A.R.I.; Heikinheimo, M. The Finnish forest fire index calculation system. In *Early Warning Systems for Natural Disaster Reduction*; Springer: Berlin/Heidelberg, Germany, 2003; pp. 645–647. [CrossRef]
199. Vajda, A.; Venäläinen, A.; Suomi, I.; Junila, P.; Mäkelä, H.M. Assessment of forest fire danger in a boreal forest environment: Description and evaluation of the operational system applied in Finland. *Meteorol. Appl.* **2013**, *21*, 879–887. [CrossRef]
200. Heim, R.R., Jr. A Review of Twentieth-Century Drought Indices Used in the United States. *Bull. Am. Meteorol. Soc.* **2002**, *83*, 1149–1166. [CrossRef]
201. Dowdy, A.J.; Mills, G.A.; Finkele, K.; De Groot, W. *Australian Fire Weather as Represented by the McArthur Forest Fire Danger Index and the Canadian Forest Fire Weather Index*; Centre for Australian Weather and Climate Research: Melbourne, Australia, 2009; p. 91. Available online: https://www.cawcr.gov.au/technical-reports/CTR_010.pdf (accessed on 8 June 2022).
202. Crane, W.J.B. Computing grassland and forest fire behaviour, relative humidity and drought index by pocket calculator. *Aust. For.* **1982**, *45*, 89–97. [CrossRef]
203. Alexander, M.E. Computer calculation of the Keetch-Byram Drought Index-programmers beware. *Fire Manag. Notes* **1990**, *51*, 23–25. Available online: http://www.climateanalyzer.science/python/dashboards/keetch_byram_corrected.pdf (accessed on 20 July 2022).
204. USFS-WFAS. Available online: https://www.fas.net/ (accessed on 24 May 2022).
205. Burrows, N.D. *The Soil Dryness Index for Use in Fire Control in the South-West of Western Australia*; Technical Report; Western Australian Department of Conservation and Land Management: Joondalup, Australia, 1987. Available online: https://library.dbca.wa.gov.au/static/Journals/080075/080075-17.pdf (accessed on 13 March 2022).
206. Kumar, V.; Dharssi, I. Sources of Soil Dryness Measures and Forecasts for Fire Danger Rating. 1987, 15, pp. 2729–2746. Available online: https://www.bnhcrc.com.au/sites/default/files/managed/downloads/brr-009.pdf (accessed on 17 September 2022).
207. Palmer, W.C.; Havens, A.V. A graphical technique for determining evapotranspiration by the thornthwaite method. *Mon. Weather Rev.* **1958**, *86*, 123–128. [CrossRef]
208. Alley, W.M. The Palmer drought severity index: Limitations and assumptions. *J. Appl. Meteorol. Climatol.* **1984**, *23*, 1100–1109. [CrossRef]
209. Heddinghaus, T.R.; Sabol, P. A Review of the Palmer Drought Severity Index and where Do We Go from Here. In *Proceedings of the 7th Conference on Applied Climatology, Salt Lake City, UT, USA, 10–13 September 1991*; American Meteorological Society: Boston, MA, USA, 1991; Available online: https://citeseerx.ist.psu.edu/document?repid=rep1&type=pdf&doi=69512362983873d5a9ffbf816cd1d0ac014aa814 (accessed on 13 March 2022).
210. Bergström, S. The HBV Model—Its Structure and Applications. SMHI Reports RH, No. 4, Norrköping. 1992. Available online: https://www.smhi.se/polopoly_fs/1.83592!/Menu/general/extGroup/attachmentColHold/mainCol1/file/RH_4.pdf (accessed on 8 June 2022).
211. Healy, R.W.; Cook, P. Using groundwater levels to estimate recharge. *Hydrogeol. J.* **2002**, *10*, 91–109. [CrossRef]
212. Nimmo, J.R.; Healy, R.W.; Stonestorm, D.A. *Aquifer Recharge, in Encyclopedia of Hydrological Sciences*; Anderson, M.G., Ed.; John Wiley & Sons Ltd.: Hoboken, NJ, USA, 2005; ISBN 0-471-49103-92005.
213. Felch, R.E. Drought: Characteristics and assessment. In *North American Droughts*; Rosenberg, N.J., Ed.; Amer. Assoc. Adv. Sci. Selected Symp; Routledge: New York, NY, USA, 2019; Volume 15, pp. 25–42. [CrossRef]

214. Morton, F. Operational estimates of areal evapotranspiration and their significance to the science and practice of hydrology. *J. Hydrol.* **1983**, *66*, 1–76. [CrossRef]
215. Tigkas, D.; Vangelis, H.; Tsakiris, G. The RDI as a composite climatic index. *Eur. Water* **2013**, *41*, 17–22. Available online: https://www.ewra.net/ew/pdf/EW_2013_41_02.pdf (accessed on 21 June 2022).
216. Crittenden, J.C.; Trussell, R.R.; Hand, D.W.; Howe, K.J.; Tchobanoglous, G. Appendix C: Physical Properties of Water. In *MWH's Water Treatment: Principles and Design*; John Wiley & Sons: Hoboken, NJ, USA, 2012; pp. 1861–1862. ISBN 9781118131473. [CrossRef]
217. Edwards, W.R.N.; Jarvis, P.G.; Landsberg, J.J.; Talbot, H. A dynamic model for studying flow of water in single trees. *Tree Physiol.* **1986**, *1*, 309–324. [CrossRef]
218. Arndt, N.; Vacik, H.; Koch, V.; Arpaci, A.; Gossow, H. Modeling human-caused forest fire ignition for assessing forest fire danger in Austria. *Iforest Biogeosci. For.* **2013**, *6*, 315–325. [CrossRef]
219. Camia, A.; Bovio, G.; Aguado, I.; Stach, N. Meteorological fire danger indices and remote sensing. In *Remote Sensing of Large Wildfires*; Springer: Berlin/Heidelberg, Germany, 1999; pp. 39–59. Available online: https://www.researchgate.net/profile/Andrea-Camia/publication/268745891_Meteorological_fire_danger_indices_and_remote_sensing/links/57109a6608ae68dc790978f1/Meteorological-fire-danger-indices-and-remote-sensing.pdf (accessed on 14 June 2022).
220. Leblon, B.; Chen, J.; Alexander, M.E.; White, S. Fire danger monitoring using NOAA-AVHRR NDVI images in the case of northern boreal forests. *Int. J. Remote Sens.* **2001**, *22*, 2839–2846. Available online: http://faculty.geog.utoronto.ca/Chen/Chen\T1\textquoterights%20homepage/PDFfiles/InterJournal%20of%20Remote%20SensingVol.22pp2839-2846.pdf (accessed on 30 June 2022). [CrossRef]
221. Paltridge, G.; Barber, J. Monitoring grassland dryness and fire potential in australia with NOAA/AVHRR data. *Remote Sens. Environ.* **1988**, *25*, 381–394. [CrossRef]
222. Gonzalez-Alonso, F.; Cuevas, J.M.; Casanova, J.L.; Calle, A.; Illera, P. A forest fire risk assessment using NOAA AVHRR images in the Valencia area, eastern Spain. *Int. J. Remote Sens.* **1997**, *18*, 2201–2207. [CrossRef]
223. Chuvieco, E.; Allgöwer, B.; Salas, J. Integration of physical and human factors in fire danger assessment. In *Wildland Fire Danger Estimation and Mapping: The Role of Remote Sensing Data*; Chuvieco, E., Ed.; World Scientific Publishing Co., Inc.: Hackensack, NJ, USA, 2003; pp. 197–218. [CrossRef]
224. Dennison, P.E.; Roberts, D.A.; Peterson, S.H.; Rechel, J. Use of Normalized Difference Water Index for monitoring live fuel moisture. *Int. J. Remote Sens.* **2005**, *26*, 1035–1042. [CrossRef]
225. Stow, D.; Niphadkar, M.; Kaiser, J. Time series of chaparral live fuel moisture maps derived from MODIS satellite data. *Int. J. Wildland Fire* **2006**, *15*, 347–360. [CrossRef]
226. Roberts, D.A.; Dennison, P.E.; Peterson, S.; Sweeney, S.; Rechel, J. Evaluation of Airborne Visible/Infrared Imaging Spectrometer (AVIRIS) and Moderate Resolution Imaging Spectrometer (MODIS) measures of live fuel moisture and fuel condition in a shrubland ecosystem in southern California. *J. Geophys. Res. Earth Surf.* **2006**, *111*, GO4S02. [CrossRef]
227. Peterson, S.H.; Roberts, D.A.; Dennison, P. Mapping live fuel moisture with MODIS data: A multiple regression approach. *Remote Sens. Environ.* **2008**, *112*, 4272–4284. [CrossRef]
228. Yebra, M.; Chuvieco, E.; Riaño, D. Estimation of live fuel moisture content from MODIS images for fire risk assessment. *Agric. For. Meteorol.* **2008**, *148*, 523–536. [CrossRef]
229. García, M.; Chuvieco, E.; Nieto, H.; Aguado, I. Combining AVHRR and meteorological data for estimating live fuel moisture content. *Remote Sens. Environ.* **2008**, *112*, 3618–3627. [CrossRef]
230. Yebra, M.; Dennison, P.E.; Chuvieco, E.; Riaño, D.; Zylstra, P.M.; Hunt, E.R., Jr.; Danson, F.M.; Qi, Y.; Jurdao, S. A global review of remote sensing of live fuel moisture content for fire danger assessment: Moving towards operational products. *Remote Sens. Environ.* **2013**, *136*, 455–468. [CrossRef]
231. Burgan, R.E.; Cohen, J.D.; Deeming, J.E. *Manually Calculating Fire-Danger Rating—1978 National Fire Danger Rating System*; Gen. Tech. Rep. INT-40; Intermountain Forest and Range Experiment Station, Forest Service, U.S. Department of Agriculture: Ogden, UT, USA, 1978; 51p, Available online: https://digitalcommons.usu.edu/cgi/viewcontent.cgi?article=1080&context=govdocs_forest (accessed on 18 May 2022).
232. Newnham, G.J.; Verbesselt, J.; Grant, I.F.; Anderson, S.A. Relative Greenness Index for assessing curing of grassland fuel. *Remote Sens. Environ.* **2011**, *115*, 1456–1463. [CrossRef]
233. Gitelson, A.A.; Kaufman, Y.J.; Stark, R.; Rundquist, D. Novel algorithms for remote estimation of vegetation fraction. *Remote Sens. Environ.* **2002**, *80*, 76–87. Available online: https://digitalcommons.unl.edu/cgi/viewcontent.cgi?article=1151&context=natrespapers (accessed on 2 July 2022). [CrossRef]
234. NASA-FIRMS. Available online: https://firms.modaps.eosdis.nasa.gov/ (accessed on 20 July 2022).
235. Matsoukis, A.; Kamoutsis, A.; Chronopoulos, K. Estimation of the meteorological forest fire risk in a mountainous region by using remote air temperature and relative humidity data. *Int. Lett. Nat. Sci.* **2018**, *67*, 1–8. [CrossRef]

236. Gouma, B. Methodology of Spatiotemporal Estimation of Meteorological Danger of Forest Fir—Application in the Mountainous Area of Parnitha. Ph.D. Thesis, Agricultural University of Athens, Athens, Greece, 2001. Available online: https://www.didaktorika.gr/eadd/handle/10442/12994 (accessed on 29 December 2022).
237. Chronopoulos, K. Development and Application of Modern Methods of Analysis and Elaboration for Estimating Environmental Parameters in Topography with Idiosyncratic Geomorphological Characteristics. Ph.D. Thesis, Agricultural University of Athens, Athens, Greece, 2009. Available online: https://www.didaktorika.gr/eadd/handle/10442/28098 (accessed on 29 December 2022).

MDPI
St. Alban-Anlage 66
4052 Basel
Switzerland
Tel. +41 61 683 77 34
Fax +41 61 302 89 18
www.mdpi.com

Land Editorial Office
E-mail: land@mdpi.com
www.mdpi.com/journal/land

www.ingramcontent.com/pod-product-compliance
Lightning Source LLC
Chambersburg PA
CBHW061835210926
51850CB00006B/892
* 9 7 8 3 0 3 6 5 6 7 3 8 9 *